LONDON MATHEMATICAL SOCIETY LECTURE NOTE SERIES

Managing Editor: Professor N.J. Hitchin, Mathematical Institute,
University of Oxford, 24–29 St Giles, Oxford OX1 3LB, United Kingdom

The titles below are available from booksellers, or, in case of difficulty, from Cambridge University Press.

59   Applicable differential geometry, M. CRAMPIN & F.A.E. PIRANI
66   Several complex variables and complex manifolds II, M.J. FIELD
86   Topological topics, I.M. JAMES (ed)
88   FPF ring theory, C. FAITH & S. PAGE
90   Polytopes and symmetry, S.A. ROBERTSON
96   Diophantine equations over function fields, R.C. MASON
97   Varieties of constructive mathematics, D.S. BRIDGES & F. RICHMAN
99   Methods of differential geometry in algebraic topology, M. KAROUBI & C. LERUSTE
100 Stopping time techniques for analysts and probabilists, L. EGGHE
104 Elliptic structures on 3-manifolds, C.B. THOMAS
105 A local spectral theory for closed operators, I. ERDELYI & WANG SHENGWANG
107 Compactification of Siegel moduli schemes, C.-L. CHAI
109 Diophantine analysis, J. LOXTON & A. VAN DER POORTEN (eds)
113 Lectures on the asymptotic theory of ideals, D. REES
116 Representations of algebras, P.J. WEBB (ed)
119 Triangulated categories in the representation theory of finite-dimensional algebras, D. HAPPEL
121 Proceedings of *Groups – St Andrews 1985*, E. ROBERTSON & C. CAMPBELL (eds)
128 Descriptive set theory and the structure of sets of uniqueness, A.S. KECHRIS & A. LOUVEAU
130 Model theory and modules, M. PREST
131 Algebraic, extremal & metric combinatorics, M.-M. DEZA, P. FRANKL & I.G. ROSENBERG (eds)
138 Analysis at Urbana, II, E. BERKSON, T. PECK, & J. UHL (eds)
139 Advances in homotopy theory, S. SALAMON, B. STEER & W. SUTHERLAND (eds)
140 Geometric aspects of Banach spaces, E.M. PEINADOR & A. RODES (eds)
141 Surveys in combinatorics 1989, J. SIEMONS (ed)
144 Introduction to uniform spaces, I.M. JAMES
146 Cohen–Macaulay modules over Cohen–Macaulay rings, Y. YOSHINO
148 Helices and vector bundles, A.N. RUDAKOV *et al*
149 Solitons, nonlinear evolution equations and inverse scattering, M. ABLOWITZ & P. CLARKSON
150 Geometry of low-dimensional manifolds 1, S. DONALDSON & C.B. THOMAS (eds)
151 Geometry of low-dimensional manifolds 2, S. DONALDSON & C.B. THOMAS (eds)
152 Oligomorphic permutation groups, P. CAMERON
153 L-functions and arithmetic, J. COATES & M.J. TAYLOR (eds)
155 Classification theories of polarized varieties, TAKAO FUJITA
158 Geometry of Banach spaces, P.F.X. MÜLLER & W. SCHACHERMAYER (eds)
159 Groups St Andrews 1989 volume 1, C.M. CAMPBELL & E.F. ROBERTSON (eds)
160 Groups St Andrews 1989 volume 2, C.M. CAMPBELL & E.F. ROBERTSON (eds)
161 Lectures on block theory, BURKHARD KÜLSHAMMER
163 Topics in varieties of group representations, S.M. VOVSI
164 Quasi-symmetric designs, M.S. SHRIKANDE & S.S. SANE
166 Surveys in combinatorics, 1991, A.D. KEEDWELL (ed)
168 Representations of algebras, H. TACHIKAWA & S. BRENNER (eds)
169 Boolean function complexity, M.S. PATERSON (ed)
170 Manifolds with singularities and the Adams–Novikov spectral sequence, B. BOTVINNIK
171 Squares, A.R. RAJWADE
172 Algebraic varieties, GEORGE R. KEMPF
173 Discrete groups and geometry, W.J. HARVEY & C. MACLACHLAN (eds)
174 Lectures on mechanics, J.E. MARSDEN
175 Adams memorial symposium on algebraic topology 1, N. RAY & G. WALKER (eds)
176 Adams memorial symposium on algebraic topology 2, N. RAY & G. WALKER (eds)
177 Applications of categories in computer science, M. FOURMAN, P. JOHNSTONE & A. PITTS (eds)
178 Lower K- and L-theory, A. RANICKI
179 Complex projective geometry, G. ELLINGSRUD *et al*
180 Lectures on ergodic theory and Pesin theory on compact manifolds, M. POLLICOTT
181 Geometric group theory I, G.A. NIBLO & M.A. ROLLER (eds)
182 Geometric group theory II, G.A. NIBLO & M.A. ROLLER (eds)
183 Shintani zeta functions, A. YUKIE
184 Arithmetical functions, W. SCHWARZ & J. SPILKER
185 Representations of solvable groups, O. MANZ & T.R. WOLF
186 Complexity: knots, colourings and counting, D.J.A. WELSH
187 Surveys in combinatorics, 1993, K. WALKER (ed)
188 Local analysis for the odd order theorem, H. BENDER & G. GLAUBERMAN
189 Locally presentable and accessible categories, J. ADAMEK & J. ROSICKY
190 Polynomial invariants of finite groups, D.J. BENSON
191 Finite geometry and combinatorics, F. DE CLERCK *et al*
192 Symplectic geometry, D. SALAMON (ed)
194 Independent random variables and rearrangement invariant spaces, M. BRAVERMAN
195 Arithmetic of blowup algebras, WOLMER VASCONCELOS
196 Microlocal analysis for differential operators, A. GRIGIS & J. SJÖSTRAND
197 Two-dimensional homotopy and combinatorial group theory, C. HOG-ANGELONI *et al*
198 The algebraic characterization of geometric 4-manifolds, J.A. HILLMAN
199 Invariant potential theory in the unit ball of $\mathbf{C}^n$, MANFRED STOLL
200 The Grothendieck theory of dessins d'enfant, L. SCHNEPS (ed)
201 Singularities, JEAN-PAUL BRASSELET (ed)

London Mathematical Society Lecture Note Series. 290

Quantum Groups and Lie Theory

Edited by

Andrew Pressley
King's College, University of London

CAMBRIDGE
UNIVERSITY PRESS

CAMBRIDGE
UNIVERSITY PRESS

University Printing House, Cambridge CB2 8BS, United Kingdom

One Liberty Plaza, 20th Floor, New York, NY 10006, USA

477 Williamstown Road, Port Melbourne, VIC 3207, Australia

314-321, 3rd Floor, Plot 3, Splendor Forum, Jasola District Centre, New Delhi - 110025, India

103 Penang Road, #05-06/07, Visioncrest Commercial, Singapore 238467

Cambridge University Press is part of the University of Cambridge.

It furthers the University's mission by disseminating knowledge in the pursuit of education, learning and research at the highest international levels of excellence.

www.cambridge.org
Information on this title: www.cambridge.org/9780521010405

© Cambridge University Press 2001

First published 2001

A catalogue record for this publication is available from the British Library

Library of Congress Cataloging in Publication data
Quantum groups and Lie theory / edited by AndrewPressley.
    p. cm.--(London Mathematical Society lecture note series; 290)
  Includes bibliographical references and index.
  ISBN 0-521-010403
  1. Quantum groups-- Congresses. 2. Lie groups-- Congresses. 3. Mathematical
physics -- Congresses. I. Pressley, Andrew. II. Series.

QC20.7.G76 Q82 2001
530.15′255--dc21

                                                              2001043214

ISBN 978-0-521-01040-5 Paperback

Contents

Contents

Introduction

Since its genesis in the early 1980s, the subject of quantum groups has grown very rapidly. Although much of the groundwork was laid by V. G. Drinfeld in his remarkable talk at the 1986 International Congress of Mathematicians in Berkeley, a number of basic issues in the theory were not resolved until later. In addition, important new developments occurred in the late 1980s and early 1990s, such as the crystal and canonical bases, and the applications of quantum groups in low-dimensional topology. By the late 1990s, however, the theory had reached a stage in which most of the foundational issues had been resolved, and many of the outstanding problems clearly formulated. It was felt that this was an opportune moment to hold a meeting of experts representing all the main strands of the theory, to take stock of what had been achieved so far and to discuss the most fruitful directions for future research. The result was the LMS Durham Symposium on Quantum Groups, which was held at Grey College in the University of Durham from 19 July to 29 July, 1999, and organised by S. Donkin, A. Pressley and A. Sudbery. The present volume is a record of some of the lectures given at the Symposium.

Two lecture series are represented here which form excellent surveys of two important areas. Ariki's 'Lectures on Cyclotomic Hecke Algebras' describe his remarkable realisation of the canonical basis in terms of a Fock space that arises in the study of solvable lattice models in statistical mechanics. Also closely connected with Physics are Etinghof's 'Lectures on the dynamical Yang-Baxter equation'. The Yang-Baxter equation has played a central role in the theory of quantum groups from the beginning, and its dynamical version is a generalization which, like the ordinary Yang-Baxter equation, first appeared in the Physics literature, but has since found many applications, particularly to integrable systems and representation theory. The remaining articles relate to single lectures given at the Symposium, and they cover a wide variety of topics within quantum groups. Several treat the problem of constructing and classifying quantum groups or the associated solutions of the quantum Yang-Baxter equation, including those by Ding & Hodges, Musson and Parashar & McDermott. The papers of Drabant and

Wenzl deal with the tensor categories of representations of quantum groups. Those of Chari & Pressly and Gordon treat the representation theory directly, the first for infinite-dimensional quantum groups and generic q, the latter for finite-dimensional quantum groups and but q a root of unity. The paper of Carter & Marsh gives a new, and partly conjectural, parametrisation of the canonical basis. The papers of Goodearl and Majid take the function algebra approach to quantum groups, from the point of view of algebraic geometry and differential geometry, respectively. And Beggs' paper describes some new ideas which relate to the origins of quantum groups in the theory of integrable systems.

We hope that this volume provides a picture of the state of the theory of quantum groups towards the end of the second millenium and that it also indicates some directions in which the theory can be expected to develop in the next.

Lectures on Cyclotomic Hecke Algebras

Susumu Ariki

1 Introduction

The purpose of these lectures is to introduce the audience to the theory of cyclotomic Hecke algebras of type $G(m, 1, n)$. These algebras were introduced by the author and Koike, Broué and Malle independently. As is well known, group rings of Weyl groups allow certain deformation. It is true for Coxeter groups, which are generalization of Weyl groups. These algebras are now known as (Iwahori) Hecke algebras.

Less studied is its generalization to complex reflection groups. As I will explain later, this generalization is not artificial. The deformation of the group ring of the complex reflection group of type $G(m, 1, n)$ is particularly successful. The theory uses many aspects of very modern development of mathematics: Lusztig and Ginzburg's geometric treatment of affine Hecke algebras, Lusztig's theory of canonical bases, Kashiwara's theory of global and crystal bases, and the theory of Fock spaces which arises from the study of solvable lattice models in Kyoto school.

This language of Fock spaces is crucial in the theory of cyclotomic Hecke algebras. I would like to mention a little bit of history about Fock spaces in the context of representation theoretic study of solvable lattice models. For level one Fock spaces, it has origin in Hayashi's work. The Fock space we use is due to Misra and Miwa. For higher level Fock spaces, they appeared in work of Jimbo, Misra, Miwa and Okado, and Takemura and Uglov. We also note that Varagnolo and Vasserot's version of level one Fock spaces have straight generalization to higher levels and coincide with the Takemura and Uglov's one. The Fock spaces we use are different from them. But they are essential in the proofs.

Since the cyclotomic Hecke algebras contain the Hecke algebras of type A and type B as special cases, the theory of cyclotomic Hecke algebras is also useful to study the modular representation theory of finite classical groups of Lie type.

I shall explain theory of Dipper and James, and its relation to our theory. The relevant Hecke algebras are Hecke algebras of type A. In this case, we have an alternative approach depending on the Lusztig's conjecture on quantum groups, by virtue of Du's refinement of Jimbo's Schur-Weyl reciprocity. Even for this rather well studied case, our viewpoint gives a new insight. This viewpoint first appeared in work of Lascoux, Leclerc and Thibon. This Fock space description looks quite different from the Kazhdan-Lusztig combinatorics, since it hides affine Kazhdan-Lusztig polynomials behind the scene. Inspired by this description, Goodman and Wenzl have found a faster algorithm to compute these polynomials. Leclerc and Thibon are key players in the study of this type A case. I also would like to mention Schiffman and Vasserot's work here, since it makes the relation of canonical bases between modified quantum algebras and quantized Schur algebras very clear.

I will refer to work of Geck, Hiss, and Malle a little if time allows, since we can expect future development in this direction. It is relevant to Hecke algebras of type B. Finally, I will end the lectures with Broué's famous dream.

Detailed references can be found at the end of these lectures. The first three are for overview, and the rest are selected references for the lectures. [i-] implies a reference for the i th lecture.

2 Lecture One

2.1 Definitions

Let k be a field (or an integral domain in general). We define cyclotomic Hecke algebras of type $G(m, 1, n)$ as follows.

Definition 2.1 *Let $v_1, \ldots, v_m, q$ be elements in k, and assume that q is invertible. The Hecke algebra $\mathcal{H}_n(v_1, \ldots, v_m; q)$ of type $G(m, 1, n)$ is the k-algebra defined by the following relations for generators a_i $(1 \leq i \leq n)$. We often write $\mathcal{H}_n$ instead of $\mathcal{H}_n(v_1, \ldots, v_m; q)$. If we want to make the base ring explicit, we write $\mathcal{H}_n/k$.*

$$(a_1 - v_1) \cdots (a_1 - v_m) = 0, \qquad (a_i - q)(a_i + 1) = 0 \quad (i \geq 2)$$

$$a_1 a_2 a_1 a_2 = a_2 a_1 a_2 a_1, \qquad a_i a_j = a_j a_i \quad (j \geq i+2)$$

$$a_i a_{i-1} a_i = a_{i-1} a_i a_{i-1} \ (3 \leq i \leq n)$$

The elements $L_i = q^{1-i} a_i a_{i-1} \cdots a_2 a_1 a_2 \cdots a_i$ $(1 \leq i \leq n)$ are called (Jucy-) Murphy elements or Hoefsmit elements.

---

[0]I would like to thank all the researchers involved in the development. Good interaction with German modular representation group (Geck, Hiss, Malle; Dipper), British combinatorial modular representation group (James, Mathas, Murphy), French combinatorics group (Lascoux, Leclerc, Thibon), modular representation group (Broué, Rouquier; Vigneras), geometric representation group (Varagnolo, Vasserot, Schiffman) and Kyoto solvable lattice model group (Okado, Takemura, Uglov) has nourished the rapid development. We still have some problems to solve, and welcome young people who look for problems.

I also thank Kashiwara, Lusztig, Ginzburg for their theories which we use.

Remark 2.2 *Let $\hat{H}_n$ be the (extended) affine Hecke algebra associated with the general linear group over a non-archimedian field. For each choice of positive root system, we have Bernstein presentation of this algebra. Let $P = \mathbb{Z}\epsilon_1 + \cdots + \mathbb{Z}\epsilon_n$ be the weight lattice as usual. We adopt "geometric choice" for the positive root system. Namely $\{\alpha_i := \epsilon_{i+1} - \epsilon_i\}$ are simple roots. Let S be the associated set of Coxeter generators (simple reflections). Then $\hat{H}_n$ has description via generators X_ϵ ($\epsilon \in P$) and T_s ($s \in S$). We omit the description since it is well known. The following mapping gives rise a surjective algebra homomorphism from $\hat{H}_n$ to $\mathcal{H}_n$.*

$$X_{\epsilon_i} \mapsto L_i, \qquad T_{s_{\alpha_i}} \mapsto a_{i+1}$$

This fact is the reason why we can apply Lusztig's theory to the study of cyclotomic Hecke algebras. Since the module theory for $\mathcal{H}_n$ has been developed by different methods, it has also enriched the theory of affine Hecke algebras.

Remark 2.3 *Let ζ_m be a primitive m th root of unity. If we specialize $q = 1, v_i = \zeta_m^{i-1}$, we have the group ring of $G(m, 1, n)$. $G(m, 1, n)$ is the group of $n \times n$ permutation matrices whose non zero entries are allowed to be m th roots of unity. Under this specialization, L_i corresponds to the diagonal matrix whose i th diagonal entry is ζ_m and whose remaining diagonal entries are 1. We would like to stress two major differences between the group algebra and the deformed algebra $\mathcal{H}_n$.*

(1) $(L_i - v_1) \cdots (L_i - v_m)$ is not necessarily zero for $i > 1$.

(2) If we consider the subalgebra generated by Murphy elements, its dimension is not m^n in general. Further, the dimension depends on parameters $v_1, \ldots, v_m, q$.

Nevertheless, we have the following Lemma. a_w is defined by $a_{i_1} \cdots a_{i_l}$ for a reduced word $s_{i_1} \cdots s_{i_l}$ of w. It is known that a_w does not depend on the choice of the reduced word.

Lemma 2.4 $\{L_1^{e_1} \cdots L_n^{e_n} a_w | 0 \leq e_i < m, w \in \mathfrak{S}_n\}$ *form a basis of $\mathcal{H}_n$.*

(How to prove) We consider $\mathcal{H}_n$ over an integral domain R, and show that $\sum R L_1^{e_1} \cdots L_n^{e_n} a_w$ is a two sided ideal. Then we have that these elements generate $\mathcal{H}_n$ as an R-module. To show that they are linearly independent, it is enough to take $R = \mathbb{Z}[\mathbf{q}, \mathbf{q}^{-1}, \mathbf{v}_1, \ldots, \mathbf{v}_m]$. In this generic parameter case, we embed the algebra into $\mathcal{H}_n/\mathbb{Q}(\mathbf{q}, \mathbf{v}_1, \ldots, \mathbf{v}_m)$. Then we can construct enough simple modules to evaluate the dimension. ∎

An important property of $\mathcal{H}_n$ is the following.

Theorem 2.5 (Malle-Mathas) *Assume that v_i are all invertible. Then $\mathcal{H}_n$ is a symmetric algebra.*

(How to prove) Since $\mathcal{H}_n$ is deformation of the group algebra of $G(m, 1, n)$, we can define a length function $l(w)$ and a_w for a reduced word of w. Unlike the Coxeter group case, a_w does depend on the choice of the reduced word. Nevertheless, the trace function

$$tr(a_w) = \left\{ \begin{array}{ll} 0 & (w \neq 1) \\ 1 & (w = 1) \end{array} \right.$$

is well defined. $(u, v) := tr(uv)$ $(u, v \in \mathcal{H}_n)$ gives the bilinear form with the desired properties. ∎

Remark 2.6 *We have defined deformation algebras for (not all but most of) other types of irreducible complex reflection groups by generators and relations. $(G(m, p, n)$: the author, other exceptional groups: Broué and Malle.)*

The most natural definition of cyclotomic Hecke algebras is given by Broué, Malle and Rouquir. It coincides with the previous definition in most cases.

Let $\mathcal{A}$ be the hyperplane arrangement defined by complex reflections of W. For each $\mathcal{C} \in \mathcal{A}/W$, we can associate the order $e_{\mathcal{C}}$ of the cyclic group which fix a hyperplane in $\mathcal{C}$. Primitive idempotents of this cyclic group are denoted by $\epsilon_j(H)$ $(0 \leq j < e_{\mathcal{C}})$. We set $\mathcal{M} = \mathbb{C}^n \setminus \cup_{H \in \mathcal{A}} H$.

Definition 2.7 *For each hyperplane H, let α_H be the linear form whose kernel is H. It is defined up to scalar multiple. We fix a set of complex numbers $t_{\mathcal{C},j}$. Then the following partial differential equation for $\mathbb{C}W$-valued functions F on $\mathcal{M}$ is called the (generalized) KZ equation.*

$$\frac{\partial F}{\partial x_i} = \frac{1}{2\pi\sqrt{-1}} \sum_{\mathcal{C} \in \mathcal{A}/W} \sum_{j=0}^{e_{\mathcal{C}}-1} \sum_{H \in \mathcal{C}} \frac{\partial(\log \alpha_H)}{\partial x_i} t_{\mathcal{C},j} \epsilon_j(H) F$$

Theorem 2.8 (Broué-Malle-Rouquier) *Assume that parameters are sufficiently generic. Let B be the braid group attached to $\mathcal{A}$. Then the monodromy representation of B with respect to the above KZ equation factors through a deformation ring of $\mathbb{C}W$. If $W = G(m, 1, n)$ for example, it coincides with the cyclotomic Hecke algebra with specialized parameters.*

2.2   Representations

If all modules are projective modules, we say that $\mathcal{H}_n$ is a semi-simple algebra, and call these representations **ordinary representations**. We have

Proposition 2.9 (Ariki(-Koike)) *$\mathcal{H}_n$ is semi-simple if and only if $q^i v_j - v_k$ $(|i| < n, j \neq k)$ and $1 + q + \cdots + q^i$ $(1 \leq i < n)$ are all non zero. In this case, simple modules are parametrized by m-tuples of Young diagrams of*

total size n. For each $\lambda = (\lambda^{(m)}, \ldots, \lambda^{(1)})$, *the corresponding simple module can be realized on the space whose basis elements are indexed by standard tableaux of shape* λ. *The basis elements are simultaneous eigenvectors of Murphy elements, and we have explicit matrix representation for generators* a_i $(1 \le i \le n)$.

These represenations are called **semi-normal form representations**. Hence we have complete understanding of ordinary represcntations. If $\mathcal{H}_n$ is not semi-simple, representations are called **modular representations**. A basic tool to get information for modular representations from ordinary ones is "reduction" procedure.

Definition 2.10 *Let* (K, R, k) *be a modular system. Namely, R is a discrete valuation ring, K is the field of fractions, and k is the residue field. For an* $\mathcal{H}_n/K$-module V, *we take an* $\mathcal{H}_n/R$-lattice V_R *and set* $\overline{V} = V_R \otimes k$. *It is known that* $\overline{V}$ *does depend on the choice of* V_R, *but the composition factors do not depend on the choice of* V_R. *The map between Grothendieck groups of finite dimentional modules given by*

$$dec_{K,k} : K_0(mod\text{--}\mathcal{H}_n/K) \longrightarrow K_0(mod\text{--}\mathcal{H}_n/k)$$

which sends $[V]$ *to* $[\overline{V}]$ *is called a* **decomposition map**. *Since Grothendieck groups have natural basis given by simple modules, we have the matrix representation of the decomposition map with respect to these bases. It is called the* **decomposition matrix**. *The entries are called* **decomposition numbers**.

In the second lecture, we also consider the decomposition map between Grothendieck groups of $KGL(n,q)$–*mod* and $kGL(n,q)$–*mod*.

Remark 2.11 *Decomposition maps are not necessarily surjective even after coefficients are extended to complex numbers. If we take* $m = 1, 2$ *and* $q \in k$ *to be zero, we have counter examples. These are called zero Hecke algebras, and studied by Carter.* **Note that we exclude the case** $q = 0$ **in the definition**. *In the case of group algebras, the theory of Brauer characters ensures that decomposition maps are surjective.*

In the case of cyclotomic Hecke algebras, we have the following result.

Theorem 2.12 (Graham-Lehrer) $\mathcal{H}_n$ *is a cellular algebra. In particular, the decomposition maps are surjective.*

The notion of cellularity is introduced by Graham and Lehrer. It has some resemblance to the definition of quasi hereditary algebras. This is further pursued by König and Changchang Xi.

In this lecture, we follow Dipper, James and Mathas' construction of Specht modules. We first fix notation.

Let $\lambda = (\lambda^{(m)}, \ldots, \lambda^{(1)})$, $\mu = (\mu^{(m)}, \ldots, \mu^{(1)})$ be two m-tuples of Young diagrams. We say λ dominates μ and write $\lambda \trianglerighteq \mu$ if

$$\sum_{j>k} |\lambda^{(j)}| + \sum_{j=1}^{l} \lambda_j^{(k)} \geq \sum_{j>k} |\mu^{(j)}| + \sum_{j=1}^{l} \mu_j^{(k)}$$

for all k, l. This partial order is called **dominance order**.

For each $\lambda = (\lambda^{(m)}, \ldots, \lambda^{(1)})$, we set $a_k = n - |\lambda^{(1)}| - \cdots - |\lambda^{(k)}|$. We have $n \geq a_1 \geq \cdots \geq a_l > 0$ and $a_k = 0$ for $k > l$ for some l. we denote l by $l(a)$. For $a = (a_k)$, we denote by $\mathfrak{S}_a$ the set of permutations which preserve $\{1, \ldots, a_l\}, \ldots, \{a_k + 1, \ldots, a_{k-1}\}, \ldots \{a_1 + 1, \ldots, n\}$. We also set

$$u_a = (L_1 - v_1) \cdots (L_{a_1} - v_1) \times (L_1 - v_2) \cdots (L_{a_2} - v_2) \times \cdots$$
$$\cdots \times (L_1 - v_{l(a)}) \cdots (L_{l(a)} - v_{l(a)})$$

Let t^λ be the canonical tableau. It is the standard tableau on which $1, \ldots, n$ are filled in by the following rule;
$1, \ldots, \lambda_1^{(m)}$ are written in the first row of $\lambda^{(m)}$; $\lambda_1^{(m)} + 1, \ldots, \lambda_1^{(m)} + \lambda_2^{(m)}$ are written in the second row of $\lambda^{(m)}$; $\ldots$; $|\lambda^{(m)}| + 1, \ldots, |\lambda^{(m)}| + \lambda_1^{(m-1)}$ are written in the first row of $\lambda^{(m-1)}$; and so on.

The row stabilizer of t^λ is denoted by $\mathfrak{S}_\lambda$. We set

$$x_\lambda = \sum_{w \in \mathfrak{S}_\lambda} a_w, \quad m_\lambda = x_\lambda u_a = u_a x_\lambda.$$

Let t be a standard tableau of shape λ. If the location of $i_k \in \{1, \ldots, n\}$ in t is the same as the location of k in t^λ, We define $d(t) \in \mathfrak{S}_n$ by $k \mapsto i_k$ $(1 \leq k \leq n)$.

Definition 2.13 *Let* $* : \mathcal{H}_n \to \mathcal{H}_n$ *be the anti-involution induced by* $a_i^* = a_i$. *For each pair* (s, t) *of standard tableaux of shape* λ, *we set* $m_{st} = a_{d(s)}^* m_\lambda a_{d(t)}$.

Remark 2.14 $\{m_{st}\}$ *form a cellular basis of* $\mathcal{H}_n$.

Proposition 2.15 (Dipper-James-Mathas) *Let* (K, R, k) *be a modular system. We set* $\mathcal{I}_\lambda = \sum R m_{st}$ *where sum is over pairs of standard tableaux of shape strictly greater than* λ *(with respect to the dominance order). Then* $\mathcal{I}_\lambda$ *is a two sided ideal of* $\mathcal{H}_n / R$.

(How to prove) It is enough to consider straightening laws for elements $a_i m_{st}$ and $m_{st} a_i$. We can then show that m_{uv} appearing in the expression have greater shapes with respect to the dominance order. ∎

Definition 2.16 *Set* $z_\lambda = m_\lambda \mod \mathcal{I}_\lambda$. *Then the submodule* $S^\lambda = z_\lambda \mathcal{H}_n$ *of* $\mathcal{H}_n / \mathcal{I}_\lambda$ *is called a* **Specht module**.

Theorem 2.17 (Dipper-James-Mathas)
$\{z_\lambda a_{d(t)} | \ t : standard \ of \ shape \ \lambda\}$ *form a basis of* S^λ.

(How to prove) We can show by induction on the dominance order that these generate S^λ. Hence the collection of all these generate $\mathcal{H}_n$. Thus counting argument completes the proof. ■

Definition 2.18 S^λ *is equipped with a bilinear form defined by*

$$\langle z_\lambda a_{d(t)}, z_\lambda a_{d(s)} \rangle m_\lambda = m_\lambda a_{d(s)} a_{d(t)}^* m_\lambda \ \text{mod} \ \mathcal{I}_\lambda$$

Theorem 2.19 (General theory of Specht modules)
(1) $D^\lambda = S^\lambda / \text{rad} \langle \ , \ \rangle$ *is absolutely irreducible or zero module.* $\{D^\lambda \neq 0\}$ *form a complete set of simple* $\mathcal{H}_n$- *modules.*
(2) Assume $D^\mu \neq 0$ *and* $[S^\lambda : D^\mu] \neq 0$. *Then we have* $\mu \trianglelefteq \lambda$.

Remark 2.20 *In the third lecture, we give a criterion for non vanishing of* D^λ.

Theorem 2.21 (Dipper-Mathas) *Let* $\{v_1, \ldots, v_m\} = \sqcup_{i=1}^a S_i$ *be the decomposition such that* v_j, v_k *are in a same* S_i *if and only if* $v_j = v_k q^b$ *for some* $b \in \mathbb{Z}$. *Then we have*

$$mod\text{–}\mathcal{H}_n \simeq \bigoplus_{n_1, \ldots, n_a} mod\text{–}\mathcal{H}_{n_1} \boxtimes \cdots \boxtimes mod\text{–}\mathcal{H}_{n_a}$$

where $\mathcal{H}_n = \mathcal{H}_n(v_1, \ldots, v_m; q)$, $\mathcal{H}_{n_i} = \mathcal{H}_{n_i}(S_i; q)$, *and the sum runs through* $n_1 + \cdots + n_a = n$.

Hence, it is enough to consider the case that v_i are powers of q.

Remark 2.22 *For the classification of simple modules, we can use arguments of Rogawski and Vigneras for the reduction to the case that* v_i *are powers of* q. *Hence we do not need the above theorem for this purpose.*

2.3   First application

Let $k_q^\times = k^\times / \langle q \rangle$. We assume that $q \neq 1$, and denote the multiplicative order of q by r. A **segment** is a finite sequence of consecutive residue numbers which take values in $\mathbb{Z}/r\mathbb{Z}$. A **multisegment** is a collection segments. Assume that a multisegment is given. Take a segment in the multisegment. By adding i ($i \in \mathbb{Z}/r\mathbb{Z}$) to the entries of the segment simultaneously, we have a segment of shifted entries. If all of these r segments appear in the given multisegment, we say that the given multisegment is **periodic**. If it never happens for all segments in the multisegment, we say that the given multisegment is **aperiodic**. We denote by $\mathcal{M}_r^{ap}$ the set of aperiodic multisegments.

Theorem 2.23 (Ariki-Mathas) *Simple modules over $\hat{H}_n/k$ are parametrized by*

$$\mathcal{M}_r^{ap}(k) = \{\lambda : k_q^\times \to \mathcal{M}_r^{ap} | \sum_{x \in k_q^\times} |\lambda(x)| = n\}$$

(How to prove) We consider a setting for reduction procedure, and show that a lower bound and an upper bound for the number of simple modules coincide. To achieve the lower bound, we use the integral module structure of the direct sum of Grothendieck groups of *proj-$\mathcal{H}_n$* with respect to a Kac-Moody algebra action, which will be explained in the second lecture. The upper bound is achieved by cellularity. ∎

Remark 2.24 *The lower bound can be achieved by a different method. This is due to Vigneras.*

Let F be a nonarchimedian local field and assume that the residue field has characteristic different from the characteristic of k. We assume that k is algebraically closed. We consider admissible k-representations of $GL(n, F)$. We take modular system (K, R, k) and consider reduction procedure.

Theorem 2.25 (Vigneras) *All cuspidal representations are obtained by reduction procedure. The admissible dual of k-representations is obtained from the classification of simple $\hat{H}_n/k$-modules.*

Hence we have contribution to the last step of the classification.

Remark 2.26 *Her method is induction from open compact groups and theory of minimal K-types. In the characteristic zero case, it is done by Bushnell and Kutzko. Considering $M := ind_{G,K}(\sigma)$ where (K, σ) is irreducible cuspidal distinguished K-type, she shows that $\mathrm{End}_{kG}(M)$ is isomorphic to product of affine Hecke algebras, and M satisfies the following hypothesis.*

"There exists a finitely generated projective module P and a surjective homomorphism $\beta : P \to M$ such that $\mathrm{Ker}(\beta)$ is $\mathrm{End}_{kG}(P)$-stable."

Then the classification of simple kG-modules reduces to that of simple $\mathrm{End}_{kG}(M)$-modules. This simple fact is known as Dipper's lemma.

3 Lecture Two

3.1 Geometric theory

Let $\mathcal{N}$ be the set of $n \times n$ nilpotent matrices, $\mathcal{F}$ be the set of n-step complete flags in $\mathbb{C}^n$. We define the **Steinberg variety** as follows.

$$Z = \{(N, F_1, F_2) \in \mathcal{N} \times \mathcal{F} \times \mathcal{F} | F_1, F_2 \text{ are } N\text{-stable}\}$$

$G := GL(n, \mathbb{C}) \times \mathbb{C}^\times$ naturally acts on Z via

$$(g, q)(N, F_1, F_2) = (q^{-1} Ad(g)N, gF_1, gF_2).$$

Let $K^G(Z)$ be the Grothendieck group of G-equivariant coherent sheaves on Z. It is an $\mathbb{Z}[\mathbf{q}, \mathbf{q}^{-1}]$- algebra via convolution product.

Theorem 3.1 (Ginzburg)
(1) We have an algebra isomorphism $K^G(Z) \simeq \hat{H}_n$.
(2) Let us consider a central character of the center $\mathbb{Z}[X_{\epsilon_1}^\pm, \ldots, X_{\epsilon_n}^\pm]^{\mathfrak{S}_n}[\mathbf{q}^\pm]$ induced by $\hat{s} : X_{\epsilon_i} \mapsto \lambda_i$. By specializing the center via this linear character, we obtain a specialized affine Hecke algebra. Let s be $\mathrm{diag}(\lambda_1, \ldots, \lambda_n)$. Then $H_*(Z^{(s,q)}, \mathbb{C})$ equipped with convolution product is isomorphic to the specialized affine Hecke algebra. Here the homology groups are Borel-Moore homology groups, and $Z^{(s,q)}$ are fixed points of $(s, q) \in G$.

Remark 3.2 *All simple modules are obtained as simple modules of various specialized affine Hecke algebras.*

Theorem 3.3 (Sheaf theoretic interpretation)
Let $\tilde{\mathcal{N}}$ be $\{(N, F) \in \mathcal{N} \times \mathcal{F} | F \text{ is } N\text{-stable}\}$, $\mu : \tilde{\mathcal{N}} \to \mathcal{N}$ be the first projection. Then
(1) $H_*(Z^{(s,q)}, \mathbb{C}) \simeq Ext^*(\mu_* \mathbb{C}_{\tilde{\mathcal{N}}^{(s,q)}}, \mu_* \mathbb{C}_{\tilde{\mathcal{N}}^{(s,q)}})$.
(2) Let $\mu_* \mathbb{C}_{\tilde{\mathcal{N}}^{(s,q)}} = \oplus_{\mathcal{O}} \oplus_{k \in \mathbb{Z}} L_{\mathcal{O}}(k) \otimes IC(\mathcal{O}, \mathbb{C})[k]$. Then $L_{\mathcal{O}} := \oplus_{k \in \mathbb{Z}} L_{\mathcal{O}}(k)$ is a simple $H_*(Z^{(s,q)}, \mathbb{C})$-module or zero module. Further, non-zero ones form a complete set of simple $H_*(Z^{(s,q)}, \mathbb{C})$-modules. If q is not a root of unity, all $L_{\mathcal{O}}$ are non-zero. If q is a primitive r th root of unity, $L_{\mathcal{O}} \neq 0$ if and only if $\mathcal{O}$ corresponds to a (tuple of) aperiodic multisegments taking residues in $\mathbb{Z}/r\mathbb{Z}$.

In the above theorem, the orbits run through orbits consisting of isomorphic representations of a quiver, which is disjoint union of infinite line quivers or cyclic quivers of length r. The reason is that $\mathcal{N}^{(s,q)}$ is the set of nilpotent matrices N satisfying $sNs^{-1} = qN$, which can be identified with representations of a quiver via considering eigenspaces of s as vector spaces on nodes and N as linear maps on arrows. This is the key fact which relates the affine quantum algebra of type A_∞, $A_{r-1}^{(1)}$ and representations of cyclotomic Hecke algebras.

Definition 3.4 *Let C_n be the full subcategory of mod–$\hat{H}_n$ whose objects are modules which have central character $\hat{s}$ with all eigenvalues of s being powers of q. Let z be an indeterminate and set $c_n(z) = (z - X_{\epsilon_1}) \cdots (z - X_{\epsilon_n})$. We denote by $P_{c_n(z),(z-q^{i_1})\cdots(z-q^{i_n})}(-)$ the exact functor taking generalized eigenspaces of eigenvalue $(z - q^{i_1}) \cdots (z - q^{i_n})$ with respect to $c_n(z)$. We then set*

$$i - Res(M) = \bigoplus_{f(z) \in k[z]} P_{c_{n-1}(z), f(z)/(z-q^i)} \left(Res_{\hat{H}_{n-1}}^{\hat{H}_n} \left(P_{c_n(z), f(z)}(M) \right) \right)$$

This is an exact functor from $\mathcal{C}_n$ *to* $\mathcal{C}_{n-1}$. *We set* $U_n = \mathrm{Hom}_{\mathbb{C}}(K_0(\mathcal{C}_n), \mathbb{C})$, $f_i = (i - Res)^T : U_{n-1} \to U_n$.

I shall give some historical comments here. The motivation to introduce these definitions was Lascoux-Leclerc-Thibon's observation that Kashiwara's global basis on level one modules computes the decomposition numbers of Hecke algebras of type A over the field of complex numbers. The above notions for affine Hecke algebras and cyclotomic Hecke algebras were first introduced by the author in his interpretation of Fock spaces and action of Chevalley generators in LLT observation into (graded dual of) Grothendieck groups of these Hecke algebras and i-restriction and i-induction operations. This is the starting point of a new point of view on the representation theory of affine Hecke algebras and cyclotomic Hecke algebras. As I will explain below, it allows us to give a new application of Lusztig's canonical basis. It triggered intensive studies of canonical bases on Fock spaces. These are carried out mostly in Paris and Kyoto. On the other hand, the research on cyclotomic Hecke algebras are mostly lead by Dipper, James, Mathas, Malle and the author. In the third lecture, these two will be combined to prove theorems on Specht module theory of cyclotomic Hecke algebras.

We now state a key proposition necessary for the proof of the next theorem. In the top row of the diagram, we allow certain infinite sum in $U(\mathfrak{g}(A_\infty))$ in accordance with infinite sum in U_n. Note that we do not have infinite sum in the bottom row.

Proposition 3.5 (Ariki) *There exists a commutative diagram*

$$\begin{array}{ccc} U^-(\mathfrak{g}(A_\infty)) & \simeq & \bigoplus_{n \geq 0} U_n/\mathbf{q} \\ \uparrow & & \uparrow \\ U^-(\mathfrak{g}(A^{(1)}_{r-1})) & \simeq & \bigoplus_{n \geq 0} U_n/q = \sqrt[r]{1} \end{array}$$

such that the left vertical arrow is inclusion, the right vertical arrow is induced by specialization $\mathbf{q} \to q$, *and the bottom horizontal arrow is an* $U^-(\mathfrak{g}(A^{(1)}_{r-1}))$-*module isomorphism. Under this isomorphism, canonical basis elements of* $U^-(\mathfrak{g}(A^{(1)}_{r-1}))$ *map to dual basis elements of* $\{[simple\ module]\}$.

(How to prove) We firstly construct the upper horizontal arrow by using PBW-type basis and dual basis of $\{[$standard module$]\}$ of affine Hecke algebras. Here we use Kazhdan-Lusztig induction theorem. We also use restriction rule for Specht modules. We then appeal to folding argument. On the left hand side, we consider this folding in geometric terms. Since only short explanation was supplied in my original paper, I also refer to Varagnolo-Vasserot's argument for this part. Note that the Hall algebra of the cyclic quiver is realized as the vector space whose basis is given by infinite sums of dual basis elements of $\{[$standard module$]\}$. We then use

[standard module:simple module]=[canonical basis:PBW-type basis] which is a consequence of the Ginzburg's theorem stated above. ∎

We now turn to the cyclotomic case. In this case, we can consider not only negative part of Kac-Moody algebra, but the action of the whole Kac-Moody algebra.

Definition 3.6 *Assume that* $v_i = q^{\gamma_i}$ $(1 \leq i \leq m)$ *and* $q = \sqrt[r]{1}$. *We set*

$$V_n = \mathrm{Hom}_{\mathbb{C}}(K_0(\mathrm{mod}\text{-}\mathcal{H}_n), \mathbb{C}), \quad V = \bigoplus_{n \geq 0} V_n.$$

We define $c_n(z) = (z - L_1) \cdots (z - L_n)$. *Then we can define*

$$i - Res(M) = \bigoplus_{f(z) \in k[z]} P_{c_{n-1}(z), f(z)/(z-q^i)} \left(Res^{\mathcal{H}_n}_{\mathcal{H}_{n-1}} \left(P_{c_n(z), f(z)}(M) \right) \right),$$

$$i - Ind(M) = \bigoplus_{f(z) \in k[z]} P_{c_{n+1}(z), f(z)(z-q^i)} \left(Ind^{\mathcal{H}_{n+1}}_{\mathcal{H}_n} \left(P_{c_n(z), f(z)}(M) \right) \right).$$

These are exact functors and we can define

$$e_i = (i - Ind)^T \quad : V_{n+1} \to V_n$$
$$f_i = (i - Res)^T \quad : V_{n-1} \to V_n$$

Definition 3.7 *Let* $\mathcal{F} = \oplus \mathbb{C}\lambda$ *be a based vector space whose basis elements are* m-*tuples of Young diagrams* $\lambda = (\lambda^{(m)}, \dots, \lambda^{(1)})$.

Assume that $\gamma_i \in \mathbb{Z}/r\mathbb{Z}$ $(1 \leq i \leq m)$ *are given. We introduce the notion of residues of cells as follows: Take a cell in* λ. *If the cell is located on the* (i, j)th *entry of* $\lambda^{(k)}$, *we say that the cell has residue* $-i+j+\gamma_k \in \mathbb{Z}/r\mathbb{Z}$. *Once residues are defined, we can speak of removable* i-*nodes and addable* i-*nodes on* λ: *Convex corners of* λ *with residue* i *are called* **removable** i-**nodes**. *Concave corners of* λ *with residue* i *are called* **addable** i-**nodes**.

We define operators e_i *and* f_i *by* $e_i\lambda$ *(resp.* $f_i\lambda$*) being the sum of all* μ's *obtained from* λ *by removing (resp. adding) a removable (resp. addable)* i-*node. We can extend this action to make* $\mathcal{F}$ *an integrable* $\mathfrak{g}(A^{(1)}_{r-1})$- *module. (If* $r = \infty$, *we consider* $A^{(1)}_{r-1}$ *as* A_∞.)

We call $\mathcal{F}$ *the* **combinatorial Fock space**. *Note that the action of the Kac-Moody algebra depends on* $(\gamma_1, \dots, \gamma_m; r)$.

Theorem 3.8 (Ariki) *We assume* $v_i = q^{\gamma_i}$ $(1 \leq i \leq m)$, $q = \sqrt[r]{1} \in \mathbb{C}$. *We set* $\Lambda = \sum_{i=1}^m \Lambda_{\gamma_i}$. *Then we have the following.*

(1) $L(\Lambda) \simeq V = U(\mathfrak{g}(A^{(1)}_{r-1}))\emptyset \subset \mathcal{F}$.

(2) Through this isomorphism, canonical basis elements of $L(\Lambda)$ *are identified with dual basis elements of simple modules, and the embedding to* $\mathcal{F}$ *is identified with the transpose of the decomposition map.*

(How to prove) We first consider reduction procedure from semi-simple $\mathcal{H}_n/K$ to $\mathcal{H}_n/k$. Note that this is not achieved by $\mathbf{v}_i = \mathbf{q}^{\gamma_i}$ and $\mathbf{q}$ to q. Then V/K can be identified with $\mathcal{F}$. We then consider

$$
\begin{array}{ccc}
U^-(\mathfrak{g}(A^{(1)}_{r-1}))\emptyset & \simeq & V \qquad \subset \mathcal{F} \\
\uparrow & & \uparrow \\
U^-(\mathfrak{g}(A^{(1)}_{r-1})) & \simeq & \oplus\, U_n/q{=}\sqrt[e]{1}
\end{array}
$$

Then the previous proposition and integrality of $\mathcal{F}$ prove the theorem. ∎

Remark 3.9 *The theorem says that we have a new application of Lusztig's canonical bases, which is similar to the application of Kazhdan-Lusztig bases of Hecke algebras to Lie algebras (Kazhdan-Lusztig conjecture) and quantum algebras (Lusztig conjecture). It is interesting to observe that the roles of quantum algebras and Hecke algebras are interchanged: in Lusztig's conjecture, Kazhdan-Lusztig bases of Hecke algebras describe decomposition numbers of quantum algebras at roots of unity; in our case, canonical bases of quantum affine algebras on integrable modules describe decomposition numbers of cyclotomic Hecke algebras at roots of unity. Previously, a positivity result was the only application of canonical bases.*

The fact that affine Kazhdan-Lusztig polynomials appear in geometric construction of quantum algebras and affine Hecke algebras was known to specialists. What was new for affine Hecke algebras is the above proposition, particularly its formulation in terms of Grothendieck groups of affine Hecke algebras.

For canonical bases on integrable modules, the theorem was entirely new, since no one knew the "correct" way of taking quotients of affine Hecke algebras to get the similar Grothendieck group description of canonical bases on integrable modules. It was just after cyclotomic Hecke algebras were introduced.

Remark 3.10 *Let (K, R, k) be a modular system. If we take semi-perfect R, we can identify V with $\oplus_{n\geq 0}K_0(proj{-}\mathcal{H}_n)$, the transpose of the decomposition map with the map induced by lifting idempotents, the dual basis elements of simple modules with principal indecomposable modules, respectively. Here $proj{-}\mathcal{H}_n$ denotes the category of finite dimentional projective $\mathcal{H}_n$-modules. We often use this description since it is more appealing.*

Remark 3.11 *If $m = 1$, namely the Hecke algebra has type A, we have another way to compute decomposition numbers. Let us consider Jimbo's Schur-Weyl reciprocity. It has refinement by Du, and can be considered with specialized parameters. Let us denote the dimension of the natural representation by d, the endomorphism ring $\mathrm{End}_{\mathcal{H}_n}(V^{\otimes n})$ by $\mathcal{S}_{d,n}$. This endomorphism ring is called* **the q-Schur algebra**. *Note that Schur functors embed the decomposition numbers of Hecke algebras into those of q-Schur algebras. Then Du's* *result implies that the decomposition numbers of Hecke algebras are derived from those of quantum algebras $U_q(\mathfrak{gl}_d)$ with $q = \sqrt[m]{1}$. There is a closed formula for decomposition numbers [Weyl module:simple module] of quantum algebras at a root of unity: these are values at 1 of parabolic Kazhdan-Lusztig polynomials for (extended) affine Weyl groups of type A. This formula is known as the Lusztig conjecture for quantum algebras. (This is a theorem of Kazhdan-Lusztig+Kashiwara-Tanisaki. There is another approach for this $m = 1$ case. This is due to Varagnolo-Vasserot and Schiffman.)*

Remark 3.12 *The introduction of combinatorial Fock spaces is due to Misra, Miwa and Hayashi, as I stated in the introduction. We will return to their work on $\mathbf{v}$-deformed Fock spaces in the third lecture.*

3.2 Algorithms

For the case $m = 1$, we have four algorithms to compute decomposition numbers. These are LLT algorithm, LT algorithm, Soergel algorithm, and modified LLT algorithm. For general m, we have Uglov algorithm.

(1) LLT algorithm

This is due to Lascoux, Leclerc and Thibon. It is based on theorem 3.8. Basic idea is to construct "ladder decompostion" of restricted Young diagrams. Then it produces basis $\{A(\lambda)\}$ of the level one module $L(\Lambda_0)$. (I will show an example in the lecture. This is a very simple procedure.)

Once $\{A(\lambda)\}$ is given, we can determine canonical basis elements $G(\lambda)$ recursively. We set

$$G(\lambda) = A(\lambda) - \sum_{\mu \rhd \lambda} c_{\lambda\mu}(v)G(\mu),$$

and find $c_{\lambda\mu}(v)$ by the following condition.

$$c_{\lambda\mu}(v^{-1}) = c_{\lambda\mu}(v), \quad G(\lambda) \in \lambda + \sum_{\mu \rhd \lambda} v\mathbb{Z}[v]\mu$$

Note that we follow the convention that restricted partitions form a basis of $L(\Lambda_0)$.

Remark 3.13 *By a theorem of Leclerc, we can also compute decomposition numbers of q-Schur algebras by using those of Hecke algebras.*

(2) LT algorithm

This is based on Leclerc-Thibon's involution and Varagnolo- Vasserot's reformulation of Lusztig conjecture. It has an advantage that we directly compute all decompoition numbers of q-Schur algebras.

We use fermionic description of the Fock space. Then a simple procedure on basis elements and straightening laws define bar operation on the Fock space. We then compute canonical basis elements by the characterization

$$\overline{G(\lambda)} = G(\lambda), \quad G(\lambda) \in \lambda + \sum_{\mu \rhd \lambda} v\mathbb{Z}[v]\mu$$

(3) Soergel algorithm

It is reformulation of Kazhdan-Lusztig algorithm for parabolic Kazhdan-Lusztig polynomials. Let $\mathcal{A}^+$ be the set of alcoves in the positive Weyl chamber. We consider vector space with basis $\{(A)\}_{A \in \mathcal{A}^+}$. For each simple reflection s, we denote by As the adjacent alcove obtained by the reflection. The Bruhat order determines partial order on $\mathcal{A}^+$. Let C_s be the Kazhdan-Lusztig element corresponding to s (we use $(T_s - v)(T_s + v^{-1}) = 0$ as a defining relation here). Then the action of C_s on this space is given by

$$(A)C_s = \begin{cases} (As) + v(A) & (As \in \mathcal{A}^+, As > A) \\ (As) + v^{-1}(A) & (As \in \mathcal{A}^+, As < A) \\ 0 & (else) \end{cases}$$

We determine Kazhdan-Lusztig basis elements $G(A)$ recursively. For $A \in \mathcal{A}^+$, we take s such that $As < A$. Then we find

$$G(A) = G(As)C_s - \sum_{B < A} c_{A,B}(v)G(B)$$

by the condition

$$c_{A,B}(v^{-1}) = c_{A,B}(v), \quad G(A) \in (A) + \sum_{B < A} v\mathbb{Z}[v](B)$$

(4) modified LLT algorithm

This is an algorithm which improves LLT algorithm. The idea is not to start from the empty Young diagram. This is due to Goodman and Wenzl. Their experiment shows that Soergel's is better than LLT, and modified LLT is much faster than both.

(5) Uglov algorithm

This is generalization of LT algorithm, and it uses the higher level Fock space introduced by Takemura and Uglov.

3.3   Second application

Let us return to the q-Schur algebra. We summarize the previous explanation as follows.

Theorem 3.14 *If $q \neq 1$ is a root of unity in a field of characteristic zero, the decomposition numbers of the q-Schur algebra are computable.*

Corollary 3.15 (Geck) *Let k be a field. We consider the q-Schur algebra over k. If the characteristic of k is sufficiently large, the decomposition numbers of the q-Schur algebra over k are computable. Note that we do not exclude $q = 1$ here.*

It has application to the modular representation theory of $GL(n, q)$. Let q be a power of a prime p, the characteristic of k be $l \neq p$. We assume that k is algebraically closed. This case is called non-describing characteristic case. We want to study $K_0(kGL(n, q)\text{–mod})$.

Theorem 3.16 (Dipper-James) *Assume that the decomposition numbers of q^a-Schur algebras over k for various $a \in \mathbb{Z}$ are known. Then the decomposition numbers of $GL(n, q)$ in non-describing characteristic case are computable.*

We explain how to compute the decomposition numbers of $G := GL(n, q)$. Let (K, R, k) be an l-modular system. James has constructed Specht modules for RG. We denote them by $\{S_R(s, \lambda)\}$. s is a semi-simple element of G. If the degree of s over $\mathbb{F}_q$ is d, λ run through partitions of size n/d.

(1) A complete set of simple KG-modules is given by

$$\left\{ R^G \left(\underset{1 \leq i \leq N}{\boxtimes} S_K(s_i, \lambda^{(i)}) \right) \mid \sum d_i |\lambda^{(i)}| = n \right\}$$

where $R^G(-)$ stands for Harish-Chandra induction, d_i is the degree of s_i, and $\{s_1, \ldots, s_N\}$ run through sets of distinct semi-simple elements. We use Dipper-James' formula

$$[S_k(s, \lambda) : D_k(s, \mu)] = d_{\lambda'\mu'}$$

where $d_{\lambda'\mu'}$ is a decomposition number of the q^d-Schur algebra. Then we rewrite $R^G \left(\underset{1 \leq i \leq N}{\boxtimes} S_k(s_i, \lambda^{(i)}) \right)$ into sum of $R^G \left(\underset{1 \leq i \leq N}{\boxtimes} D_k(s_i, \mu^{(i)}) \right)$.

(2) Let t_i be the l-regular part of s_i, a_i be the degree of t_i, $\nu^{(i)}$ be the Young diagram obtained from $\mu^{(i)}$ by multiplying all columns by d_i/a_i. Then we have $D_k(s_i, \mu^{(i)}) \simeq D_k(t_i, \nu^{(i)})$. This is also due to Dipper and James. Thus we can rewrite $R^G \left(\underset{1 \leq i \leq N}{\boxtimes} D_k(s_i, \mu^{(i)}) \right)$ into $R^G \left(\underset{1 \leq i \leq N}{\boxtimes} D_k(t_i, \nu^{(i)}) \right)$. Assume that $t_i = t_j$. Then we use the inverse of the decomposition matrices of q^{a_i}-Schur algebras of rank $d_i k_i$ and $d_j k_j$ to describe $D_k(t_i, \nu^{(i)}) \boxtimes D_k(t_j, \nu^{(j)})$ as an alternating sum of $S_k(t_i, \eta^{(i)}) \boxtimes S_k(t_j, \eta^{(j)})$. Then the Harish-Chandra induction of this module is explicitly computable by using Littlewood-Richardson rule.

We use the decomposition matrix of the q^{a_i}-Schur algebra of rank $d_i k_i + d_j k_j$ to rewrite it again into the sum of $R^G \left(\underset{1 \leq i \leq N'}{\boxtimes} D_k(t_i, \kappa^{(i)}) \right)$. Continuing this procedure, we reach the case that all t_i are mutually distinct.

(3) The final result of the previous step already gives the answer since the following set is a complete set of simple kG-modules.

$$\left\{ R^G \left(\underset{1 \leq i \leq N'}{\boxtimes} D_k(t_i, \kappa^{(i)}) \right) \mid \sum a_i |\kappa^{(i)}| = n \right\}$$

where $\{t_1, \ldots, t_{N'}\}$ run through sets of distinct l-regular semi-simple elements.

4   Lecture Three

4.1   Specht modules and v-deformed Fock spaces

We now **v**-deform the setting we have explained in the second lecture. The view point which has emerged is that behind the representation theory of cyclotomic Hecke algebras, there is the same crystal structure as integrable modules over quantum algebras of type $A_{r-1}^{(1)}$, and this crystal structure is induced by canonical bases of integrable modules. As a corollary to this viewpoint, Mathas and the author have parametrized simple $\mathcal{H}_n$-modules over an arbitrary field using crystal graphs. Since the canonical basis is defined in the **v**-deformed setting, It further lead to intensive study of canonical bases on various **v**-deformed Fock spaces.

The purpose of the third lecture is to show the compatibility of this crystal structure with Specht module theory. The above mentioned studies on canonical bases on **v**-deformed Fock spaces are essential in the proof.

Before going to this main topic, I shall mention related work recently done in Vazirani's thesis. This can be understood in the above context. As I have explained in the second lecture, this viewpoint has origin in Lascoux, Leclerc and Thibon's work, which I would like to stress here again.

Theorem 4.1 (Vazirani-Grojnowski) *Let $\tilde{e}_i(M) = soc(i - Res(M))$. If M is irreducible, then $\tilde{e}_i(M)$ is irreducible or zero module.*

The case $m = 1$ is included in Kleshchev and Brundan's modular branching rule. It is natural to think that the socle series would explain the canonical basis in the crystal structure. This observation was first noticed by Rouquier as was explained in [2b], and adopted in this Vazirani's thesis.

We now start to explain how Specht module theory fits in the description of higher level Fock spaces.

Let $\mathcal{F}_{\mathbf{v}} = \oplus \mathbb{C}(\mathbf{v})\lambda$ be the **v**-deformed Fock space. It has $U_{\mathbf{v}}(\mathfrak{g}(A_{r-1}^{(1)}))$-module structure which is deformation of $U(\mathfrak{g}(A_{r-1}^{(1)}))$-module structure on $\mathcal{F}$. To explain it, we introduce notation.

Let x be a cell on $\lambda = (\lambda^{(m)}, \ldots, \lambda^{(1)})$. Assume that it is the (a, b)th cell of $\lambda^{(c)}$. We say that a cell is **above** x if it is on $\lambda^{(k)}$ for some $k > c$, or if it is on $\lambda^{(c)}$ and the row number is strictly smaller than a. We denote the set of addable (resp. removable) i-nodes of λ which are above x by $A_i^a(x)$ (resp. $R_i^a(x)$). In a similar way, we say that a cell is **below** x if it is on $\lambda^{(k)}$ for some $k < c$, or if it is on $\lambda^{(c)}$ and the row number is strictly greater than a. We denote the set of addable (resp. removable) i-nodes of λ which are below x by $A_i^b(x)$ (resp. $R_i^b(x)$). The set of all addable (resp. removable) i-nodes of λ is denoted by $A_i(\lambda)$ (resp. $R_i(\lambda)$). We then set

$$N_i^a(x) = |A_i^a(x)| - |R_i^a(x)|, \quad N_i^b(x) = |A_i^b(x)| - |R_i^b(x)|$$

$$N_i(\lambda) = |A_i(\lambda)| - |R_i(\lambda)|$$

We denote the number of all 0-nodes in λ by $N_d(\lambda)$. Then the $U_{\mathbf{v}}(\mathfrak{g}(A_{r-1}^{(1)}))$-module structure given to $\mathcal{F}_{\mathbf{v}}$ is as follows.

$$e_i\lambda = \sum_{\lambda/\mu=\boxed{i}} \mathbf{v}^{-N_i^a(\lambda/\mu)}\mu, \quad f_i\lambda = \sum_{\mu/\lambda=\boxed{i}} \mathbf{v}^{N_i^b(\mu/\lambda)}\mu$$

$$\mathbf{v}^{h_i}\lambda = \mathbf{v}^{N_i(\lambda)}\lambda, \quad \mathbf{v}^d\lambda = \mathbf{v}^{-N_d(\lambda)}\lambda$$

This action is essentially due to Hayashi.

Set $V_{\mathbf{v}} = U_{\mathbf{v}}(\mathfrak{g}(A_{r-1}^{(1)}))\emptyset$. It is considered as the $\mathbf{v}$-deformed space of $V = \oplus_{n\geq 0} K_0(proj-\mathcal{H}_n)$.

Remark 4.2 *If we apply a linear map $(\lambda^{(m)}, \ldots, \lambda^{(1)}) \mapsto (\lambda^{(1)'}, \ldots, \lambda^{(m)'})$, we have Kashiwara's lower crystal base which is compatible with his coproduct Δ_-.*

On the other hand, if an anti-linear map $(\lambda^{(m)}, \ldots, \lambda^{(1)}) \mapsto (\lambda^{(m)'}, \ldots, \lambda^{(1)'})$ is applied, we have Lusztig's basis at ∞ which is compatible with his coproduct. We denote it by $\mathcal{F}_{\mathbf{v}^{-1}}^{-\gamma}$.

Set $L = \oplus \mathbb{Q}[\mathbf{v}]_{(\mathbf{v})}\lambda$ and $B = \{\lambda \bmod \mathbf{v}\}$. Then it is known that (L, B) is a crystal base of $\mathcal{F}_{\mathbf{v}}$. We nextly set $L_0 = V_{\mathbf{v}} \cap L$, and $B_0 = (L_0/\mathbf{v}L_0) \cap B$. Then general theory concludes that (L_0, B_0) is a crystal base of $V_{\mathbf{v}}$.

Definition 4.3 *We say that λ is $(\gamma_1, \ldots, \gamma_m; r)$- Kleshchev if $\lambda \bmod \mathbf{v} \in B_0$. We often drop parameters and simply says λ is Kleshchev.*

It has the following combinatorial definition. We say that a node on λ is **good** if there is $i \in \mathbb{Z}/r\mathbb{Z}$ such that if we read addable i-nodes (write A in short) and removable i-nodes (write R in short) from the top row of $\lambda^{(m)}$ to the bottom row of $\lambda^{(1)}$ and do RA deletion as many as possible, then the node sits in the left end of the remaining R's. (I will give an example in the lecture.)

Definition 4.4 λ *is called* $(\gamma_1, \ldots, \gamma_m; r)$- **Kleshchev** *if there is a standard tableau T of shape λ such that for all k,* $\boxed{k}$ *is a good node of the subtableau $T_{\leq k}$ which consists of nodes* $\boxed{1}, \ldots, \boxed{k}$ *by definition.*

Theorem 4.5 (Ariki) *We assume that $v_i = q^{\gamma_i}, q = \sqrt[r]{1}$. Then $D^\lambda \neq 0$ if and only if λ is $(\gamma_1, \ldots, \gamma_m; r)$-Kleshchev.*

(How to prove) We show that canonical basis elements $G(\lambda)$ (λ=Kleshchev) have the form

$$G(\lambda) = \lambda + \sum_{\mu \triangleright \lambda} c_{\lambda\mu}(\mathbf{v})\mu$$

On the other hand, the Specht module theory constructed by Dipper-James-Mathas shows that the principal indecomposable module P^λ for $D^\lambda \neq 0$ has the form

$$P^\lambda = S^\lambda + \sum_{\mu \triangleright \lambda} m_{\lambda\mu}S^\mu$$

Comparing these, and recalling that $\lambda \in \mathcal{F}$ is identified with S^λ, we have the result. ■

To know the form of $G(\lambda)$, we have to understand higher level $\mathbf{v}$-deformed Fock spaces.

Definition 4.6 *Take $\gamma = (\gamma_1, \ldots, \gamma_m) \in (\mathbb{Z}/r\mathbb{Z})^m$. If $\tilde{\gamma} = (\tilde{\gamma}_1, \ldots, \tilde{\gamma}_m) \in \mathbb{Z}^m$ satisfies $\tilde{\gamma}_k \bmod r = \gamma_k$ for all k, we say that $\tilde{\gamma}$ is a **lift** of γ.*

Theorem 4.7 (Takemura-Uglov) *For each $\tilde{\gamma} \in \mathbb{Z}^m$, we can construct higher level $\mathbf{v}$-deformed Fock space, whose underlying space is the same as $\mathcal{F}_{\mathbf{v}}$.*

It has geometric realization due to Varagnolo and Vasserot. For reader's convenience, I also add it here. Let V be a $\mathbb{Z}$-graded $\mathbb{C}$-vector space whose dimension type is $(d_i)_{i \in \mathbb{Z}}$. We denote by $\overline{V}$ the $\mathbb{Z}/r\mathbb{Z}$-graded vector space defined by $\overline{V}_{\bar{i}} = \oplus_{j \in \bar{i}}V_j$. We set $\overline{V}_{j \geq i} = \oplus_{j \geq i}V_j$. Let

$$E_V = \bigoplus_{i \in \mathbb{Z}} \mathrm{Hom}_{\mathbb{C}}(V_i, V_{i+1}), \quad E_{\overline{V}} = \bigoplus_{\bar{i} \in \mathbb{Z}/r\mathbb{Z}} \mathrm{Hom}_{\mathbb{C}}(V_{\bar{i}}, V_{\overline{i+1}}).$$

and define $E_{\overline{V},V} = \{x \in \overline{V} \mid x(\overline{V}_{j \geq i}) \subset \overline{V}_{j \geq i}\}$. Then we have a natural diagram

$$E_V \xleftarrow{\kappa} E_{\overline{V},V} \xrightarrow{\iota} E_V$$

We consider $\gamma_d := \kappa_! \iota^*[\text{shift}]$. Then it defines a map from the derived category which is used to construct $U_{\mathbf{v}}^-(\mathfrak{g}(A_{r-1}^{(1)}))$ to the derived category which is used to construct $U_{\mathbf{v}}^-(\mathfrak{g}(A_\infty))$. Let η be anti-involutions on both quantum algebras which sends f_i to f_i respectively.

Recall that $\mathcal{F}_{\mathbf{v}^{-1}}^{-\tilde{\gamma}}$ is a $U_{\mathbf{v}}(\mathfrak{g}(A_\infty))$-module. We then have the following.

Theorem 4.8 (Varagnolo-Vasserot) *For each* $x \in U_{\mathbf{v}}^-(\mathfrak{g}(A_{r-1}^{(1)}))$, *we set*

$$x\lambda = \sum_d \eta(\gamma_d(\eta(x)))\mathbf{v}^{-\sum_{i<j, i \equiv j} d_i h_j} \lambda$$

Then $\mathcal{F}_{\mathbf{v}^{-1}}^{-\tilde{\gamma}}$ *becomes an* $U_{\mathbf{v}}^-(\mathfrak{g}(A_{r-1}^{(1)}))$-*module.*

Remark 4.9 *If we take* $\tilde{\gamma}_i \gg \tilde{\gamma}_{i+1}$, *the canonical basis elements on these three Fock spaces coincide as long as the size of the Young diagrams indexing these canonical basis elements is not too large.*

Remark 4.10 *If we take* $0 \leq -\tilde{\gamma}_1 \leq \cdots \leq -\tilde{\gamma}_m < r$ *in the above Fock space, we have Jimbo-Misra-Miwa-Okado higher level Fock space. This Fock space is the first example of higher level Fock spaces.*

By the above remark, we can use these Fock spaces to compute canonical basis elements on $\mathcal{F}_{\mathbf{v}}$ if we suitably care about the choice of $\tilde{\gamma}$.

Theorem 4.11 (Uglov) *The Takemura-Uglov Fock space has a bar operation such that* $\overline{\emptyset} = \emptyset$, $\overline{f_i \lambda} = f_i \overline{\lambda}$ *and* $\overline{\lambda}$ *has the form* $\lambda + (higher\ terms)$ *with respect to a dominance order.*

The relation between the dominance order in the above theorem and the dominance order we use is well understood by using "abacus". As a conclusion, we can prove that $G(\lambda) = \lambda + \sum_{\mu \rhd \lambda} c_{\lambda \mu}(\mathbf{v})\mu$ as desired.

We have explained that how crystal base theory on higher level Fock spaces fits in the modular representation theory of cyclotomic Hecke algebras. In particular, Kac q-dimension formula gives the generating function of the number of simple $\mathcal{H}_n$-modules. Even for type B Hecke algebras, it was new.

4.2   Future direction and Broué's dream

The original motivation of Broué and Malle to introduce cyclotomic Hecke algebras is the study of modular representation theory of finite classical groups of Lie type over fields of non-describing characteristics. For example, Geck, Hiss and Malle's result towards classification of simple modules inspires many future problems. I may mention more in the lecture on demand.

I would like to end these lectures with Broué's famous dream. Let B be a block of a group ring of a finite group G, and assume that it has an abelian defect group D. Let b be the Brauer correspondent in the group ring of $DC_G(D) = C_G(D) \subset N_G(D)$. $((D, b)$ is called a maximal subpair or Brauer pair.) Then he conjectures that $D^b(B\text{-}mod) \simeq D^b(N_G(D, b)b\text{-}mod)$,i.e. B and $N_G(D, b)b$ are derived equivalent (Rickard equivalent). To be more precise on its base ring, let (K, R, k) be a modular system. He conjectures the derived equivalence over R.

Let q be a power of a prime p, $G = G(q)$ be the general linear group $GL(n, q)$, and k be an algebraically closed field of characteristic $l \neq p$, (K, R, k) be a l-modular system. Assume that $l > n$, and take d such that $d | \Phi_d(q)$, $\Phi_d(q) | q^{n(n-1)/2}(q^n - 1) \cdots (q - 1) = |G(q)|$, where $\Phi_d(\mathbf{q})$ is a cyclotomic polynomial. We take B to be a unipotent block. In this case, unipotent blocks are paramerized by d-cuspidal pairs $(L(q), \lambda)$ up to conjugacy. Here $L(q)$ is a Levi subgroup, λ is an irreducible cuspidal $KL(q)$-module. Further, D is the l-part of the center of $L(q)$. ($L(q)$ is the centralizer of a "Φ_d-torus" $S(q)$.) If we set $W(D, \lambda) := N_G(D, \lambda)/C_G(D)$, it is isomorphic to $G(d, 1, a)$ for some a. $W(D, \lambda)$ is called **cyclotomic Weyl group**. These are due to Broué, Malle and Michel.

In this setting, Broué, Malle and Michel give an explicit conjecture on the bimodule complex which induces the Rickard equivalence between B and $N_G(D, b)b$. It is given in terms of a variety which appeared in Deligne-Lusztig theory to trivialize a $L(q)$-bundle on a Deligne-Lusztig variety. Going down to the Deligne-Lusztig variety itself, it naturally conjectures the existence of a bimodule complex which induces derived equivalence between B and a deformation ring of the group ring of the semi-direct of $S(q)_l$ with $W(D, \lambda) \simeq G(d, 1, a)$. This conjecture is supported by the fact that they are isotypic in the sense of Broué.

It is expected that the deformation of $W(D, \lambda)$ is the cyclotomic Hecke algebra we have studied in these lectures. Hence, we expect that cyclotomic Hecke algebras with m not restricted to 1 or 2 will have applications in this field. We remark that the Broué conjecture is not restricted to $GL(n, q)$ only.

References

[Gen1] S.Ariki, Representations over Quantum Algebras of type $A_{r-1}^{(1)}$ and Combinatorics of Young Tableaux, Sophia University Lecture Notes Series, to appear. (in Japanese)

[Gen2] R.Dipper, M.Geck, G.Hiss and G.Malle, Representations of Hecke algebras and finite groups of Lie type, Algorithmic Algebra and Number Theory, B.H.Matzat, G.M.Greuel, and G.Hiss eds. (1999), Springer-Verlag.

[Gen3] M.Geck, Representations of Hecke algebras at roots of unity, Séminaire Bourbaki n°**836** (1997-98).

[1a] K.Bremke and G.Malle, Reduced words and a length function for $G(e, 1, n)$, Indag.Math. **8** (1997), 453-469.

[1b] G.Malle and A.Mathas, Symmetric cyclotomic Hecke algebras, J.Algebra **205** (1998), 275-293.

[1c]   M.Broué, G.Malle and R.Rouquier, On Complex reflection groups and their associated braid groups, Representations of Groups, B.N.Allison and G.H.Cliff eds. C.M.S. Conference Proceedings **16** (1995), 1-13.

[1d]   J.Graham and G.Lehrer, Cellular algebras, Invent.Math. **123** (1996), 1-34.

[1e]   R.Dipper, G.James and A.Mathas, Cyclotomic q-Schur algebras, Math.Zeit. to appear.

[1f]   S.Ariki and A.Mathas, The number of simple modules of the Hecke algebras of type $G(r, 1, n)$, Math.Zeit. to appear.

[1g]   R.Dipper and A.Mathas, Morita equivalences of Ariki-Koike algebras, in preparation.

[1h] * M-F.Vigneras, Induced R-representations of p-adic reductive groups, Selecta Mathematica, **New Series 4** (1998), 549-623.

[2a]   S.Ariki, On the decomposition numbers of the Hecke algebra of type $G(m, 1, n)$, J.Math.Kyoto Univ. **36** (1996), 789-808.

[2b]   A.Lascoux, B.Leclerc and J-Y.Thibon, Hecke algebras at roots of unity and crystal bases of quantum affine algebras, Comm.Math.Phys. **181** (1996), 205-263.

[2c]   F.M.Goodman and H.Wenzl, Crystal bases of quantum affine algebras and affine Kazhdan-Lusztig polynomials, Int.Math.Res.Notices **5** (1999), 251-275.

[2d]   B.Leclerc and J-Y.Thibon, Littlewood-Richardson coefficients and Kazhdan-Lusztig polynomials, **math.QA/9809122**.

[2e]   R.Dipper and G.James, The q-Schur algebra, Proc.London Math.Soc.(3) **59** (1989), 23-50.

[2f]   M.Vazirani, Irreducible modules over the affine Hecke algebras: a strong multiplicity one result, Thesis, U.C.Berkeley (1999).

[3a]   M.Varagnolo and E.Vasserot, On the decomposition matrices of the quantized Schur algebra, Duke Math.J., to appear, **math.QA/9803023**.

[3b]   K.Takemura and D.Uglov, Representations of the quantum toroidal algebra on highest weight modules of the quantum affine algebra of type $\mathfrak{gl}_N$, **math.QA/9806134**.

[3c]   D.Uglov, Canonical bases of higher-level q-deformed Fock spaces, short version in **math.QA/9901032**; full version in **math.QA/9905196**.

[3d]	S.Ariki, On the classification of simple modules for cyclotomic Hecke algebras of type $G(m, 1, n)$ and Kleshchev multipartitions, preprint.

[3e]	M.Geck, G.Hiss and G.Malle, Towards a classification of the irreducible representations in non-defining characteristic of a finite group of Lie type, Math.Zeit. **221** (1996), 353-386.

AN INTRODUCTION TO GROUP DOUBLECROSS PRODUCTS AND SOME USES

Edwin Beggs

Department of Mathematics
University of Wales, Swansea
Singleton Park, Swansea SA2 8PP, UK

1 INTRODUCTION

Factorisations of groups have been sudied for a long time, and it is well known that Hopf algebras can be constructed from them [16, 11]. In this article I shall review this material in the finite group case, and then comment on some more recent developments on quantum doubles and duality [2, 5]. Then I shall discuss the relation between group factorisations and integrable models, including the Hamiltonian structure and some speculations on the quantum theory. This is based on the inverse scattering process [8, 14, 15], using a formalism emphasising the algebraic structure [3, 4]. Finally I shall mention some recent work connecting group doublecrossproducts and T-duality in sigma models in classical field theory [9, 10, 6].

I have worked in these areas jointly with S. Majid (on Hopf algebras and T-duality) and with P.R. Johnson (integrable models). I would like to thank the organisers of the symposium for inviting me to speak.

2 Group doublecross product

A group doublecross product is a group X which has two subgroups G and M so that every element $x \in X$ can be factored uniquely as $x = su$ for $s \in M$ and $u \in G$, and also as $x = vt$ for $t \in M$ and $v \in G$. We use the notation $X = G \bowtie M$ to denote a doublecross product. For a finite group X we need only find two subgroups G and M so that $G \cap M$ is just the identity, and where the product of the orders of G and M is the order of X.

Example 2.1 *Consider the group $X = S_3 \times S_3$ as the permutations of 6 objects labelled 1 to 6, where the first factor $S_3 \times \{e\}$ leaves the last 3 objects fixed, and the second factor $\{e\} \times S_3$ leaves the first 3 objects fixed. We take G to be the cyclic group of order 6 generated by the permutation $1_G = (123)(45)$,*

*and M to be the cyclic group of order 6 generated by the permutation $1_M =$
(12)(456). Our convention is that permutations act on objects on their right,
and 1_G applied to 1 gives 2. The intersection of G and M is just the identity
permutation, and counting elements shows that $GM = MG = S_3 \times S_3$. We
write each cyclic group additively, for example $G = \{0_G, 1_G, 2_G, 3_G, 4_G, 5_G\}$.*

Example 2.2 *To show that even a simple group can be a doublecross product,
consider $X = A_n$ for n odd, which is simple for $n \geq 5$. This is the doublecross
product of the subgroup G, a cyclic group generated by the n-cycle $(12 \ldots n)$,
and the subgroup M, consisting of the permutations in A_n leaving the first
object fixed.*

We call the group doublecross product $X = G \bowtie M$ *self dual* if there
is a group automorphism $\theta : X \to X$ which has the property that $\theta G =
M$ and $\theta M = G$. Such an automorphism θ we will call a *factor reversing*
automorphism. The first example above has a factor reversing automorphism
given by conjugation with the permutation $(14)(25)(36)$. The second example
does not have one, as the sizes of the subgroups G and M are different.

The factorisation gives rise to group actions as follows. If we take $su \in X$,
where $u \in G$ and $s \in M$, then we can uniquely write this as $su = vt$, where
$t \in M$ and $v \in G$. If we choose to write these unique elements as $v = s{\triangleright}u$
and $t = s{\triangleleft}u$, then we have $su = (s{\triangleright}u)(s{\triangleleft}u)$. The usual group laws for
X give the following rules for the binary operations ${\triangleright} : M \times G \to G$ and
${\triangleleft} : M \times G \to M$. Firstly ${\triangleright} : M \times G \to G$ is a left action of the group M on
G, and ${\triangleleft} : M \times G \to M$ is a right action of the group G on M. The identity
element is left fixed by these actions, and finally $s{\triangleright}uw = (s{\triangleright}u)((s{\triangleleft}u){\triangleright}w)$ and
$ps{\triangleleft}u = (p{\triangleleft}(s{\triangleright}u))(s{\triangleleft}u)$.

In our first example the action of the element 1_M on G is seen to be given
by the permutation $(1_G, 5_G)(2_G, 4_G)$, and that of 1_G on M is given by the
permutation $(1_M, 5_M)(2_M, 4_M)$.

These actions give an alternative way to describe group doublecross prod-
ucts by specifying the groups G and M, and the actions ${\triangleright}$ and ${\triangleleft}$. The group
X which is factored by G and M can be realised as the set $G \times M$ with group
operations $(u, s)(v, t) = (u(s{\triangleright}v), (s{\triangleleft}v)t)$ and $(u, s)^{-1} = (s^{-1}{\triangleright}u^{-1}, s^{-1}{\triangleleft}u^{-1})$.

3   Hopf algebra bicrossproducts

There are two obvious constructions of a Hopf algebra from a group. One
is the group algebra, and the other is the commutative function algebra. If
we have a group which is factored into two subgroups, we can try to use the
group algebra construction on one subgroup and the commutative function
algebra on the other subgroup. These interact via the actions discussed in the
last section, and the result is an example of a Hopf algebra bicrossproduct. If

we combine the group algebra kM (with basis $s \in M$) and the commutative function algebra $k(G)$ (with basis δ_u, the function taking the value 1 at $u \in G$ and zero elsewhere), we get the Hopf algebra bicrossproduct $H = kM{\blacktriangleright\!\!\triangleleft}k(G)$, with basis $s \otimes \delta_u$ and operations

$$(s \otimes \delta_u)(t \otimes \delta_v) = \delta_{u, t \triangleright v}(st \otimes \delta_v), \quad \Delta(s \otimes \delta_u) = \sum_{xy=u} s \otimes \delta_x \otimes s \triangleleft x \otimes \delta_y$$

$$1 = \sum_u e \otimes \delta_u, \quad \epsilon(s \otimes \delta_u) = \delta_{u,e},$$

$$S(s \otimes \delta_u) = (s \triangleleft u)^{-1} \otimes \delta_{(s \triangleright u)^{-1}}, \quad (s \otimes \delta_u)^* = s^{-1} \otimes \delta_{s \triangleright u}.$$

Here Δ denotes the coproduct, ϵ the counit and S the antipode of the Hopf algebra. Additionally we have given a formula for the star operation. This Hopf algebra is, in general, neither commutative nor cocommutative. The dual $H^* = k(M){\blacktriangleright\!\!\triangleleft}kG$ can also be made into a Hopf algebra as follows:

$$(\delta_s \otimes u)(\delta_t \otimes v) = \delta_{s \triangleleft u, t}(\delta_s \otimes uv), \quad \Delta(\delta_s \otimes u) = \sum_{ab=s} \delta_a \otimes b \triangleright u \otimes \delta_b \otimes u$$

$$1 = \sum_s \delta_s \otimes e, \quad \epsilon(\delta_s \otimes u) = \delta_{s,e},$$

$$S(\delta_s \otimes u) = \delta_{(s \triangleleft u)^{-1}} \otimes (s \triangleright u)^{-1}, \quad (\delta_s \otimes u)^* = \delta_{s \triangleleft u} \otimes u^{-1}.$$

If the group doublecross product has a factor reversing automorphism $\theta : X = G \bowtie M \to X$, there is a Hopf algebra isomorphism $\tilde{\theta} : kM{\blacktriangleright\!\!\triangleleft}k(G) \to k(M){\blacktriangleright\!\!\triangleleft}kG$ given by $\tilde{\theta}(s \otimes \delta_u) = \delta_{\theta(s \triangleright u)} \otimes \theta(s \triangleleft u)$, so we can say that the bicrossproducts are self-dual.

From the group doublecross product $X = G \bowtie M$ we can construct another group doublecross product, $Y{\bowtie}X$. Here $Y = G \times M^{op}$ with group law $(us).(vt) = uvts$, and the actions (for $us \in X$ and $vt \in Y$) are

$$us \tilde{\triangleleft} vt = ((s \triangleleft v)ts^{-1} \triangleright u^{-1})^{-1}(s \triangleleft v)$$

$$us \tilde{\triangleright} vt = us(vt)(us)^{-1} = u(s \triangleright v)((s \triangleleft v)ts^{-1} \triangleright u^{-1})((s \triangleleft v)ts^{-1} \triangleleft u^{-1}).$$

Now we can construct the bicrossproduct $kX{\blacktriangleright\!\!\triangleleft}k(Y)$ by the previous formulae. In fact $kX{\blacktriangleright\!\!\triangleleft}k(Y)$ is isomorphic to the Drinfeld double $D(H) = H^{*op} \bowtie H$ (with actions the mutual coadjoint actions), by the map

$$\psi : D(H) \to kX{\blacktriangleright\!\!\triangleleft}k(Y), \quad \psi(\delta_s \otimes u \otimes t \otimes \delta_v) = (s \triangleright u)^{-1}t \otimes \delta_{v(t \triangleleft v)^{-1}s^{-1}t}.$$

As $D(H)$ is quasitriangular, $kX{\blacktriangleright\!\!\triangleleft}k(Y)$ is also quasitriangular, with

$$\mathcal{R} = \sum_{u,v \in G,\ s,t \in M} v^{-1} \otimes \delta_{us} \otimes s^{-1} \otimes \delta_{(s \triangleright v)t}.$$

The Hopf algebra $kX{\blacktriangleright\!\!\triangleleft}k(Y)$ is typically rather large (of dimension the order of X squared). In the self dual case we can consider a much smaller

Hopf algebra (of dimension the order of X), constructed as follows. There is a subgroup X^θ of X consisting of those elements x for which $\theta x = x$, and a subgroup Y^θ of Y consisting of those elements y for which $\theta y = y^{-1}$ (inverse in X). The actions $\tilde{\triangleright}, \tilde{\triangleleft}$ restrict to X^θ, Y^θ, forming a double cross product group $Y^\theta \bowtie X^\theta$ factorising into Y^θ, X^θ. The corresponding bicrossproduct Hopf algebra $kX^\theta \blacktriangleright\!\!\triangleleft k(Y^\theta)$ has an isomorphic coalgebra to that of $kM \blacktriangleright\!\!\triangleleft k(G)$. There remains the problem of whether $kX^\theta \blacktriangleright\!\!\triangleleft k(Y^\theta)$ is always quasitriangular, and I would be very interested in a proof or counterexample.

4   Integrable models in field theory

Here is a (not entirely standard) definition of an integrable classical field theory. I do not claim that it is equivalent to any other definition of integrability, but it has its uses, for example in constructing some of the 'missing charges' in Affine Toda theory [4]. We begin with:

    1)   A group doublecross product $X = GM = MG$. Here G is called the *classical phase space*.

    2)   A set S, called the *classical space-time*.

    3)   A function $a : S \to M$, which we shall call the *classical vacuum map*.

Each $\phi_0 \in G$ contains the information necessary to describe a solution of the classical field theory at every point in space-time. To recover the field at a point $x \in S$, we perform a factorisation in X:

$$a(x)\,\phi_0 \;=\; \phi(x)\,b(x)\,, \quad \phi(x) \in G\,, \; b(x) \in M\,.$$

The field at the point $x \in S$ is encoded into $b(x)$ or $\phi(x)$, in a manner which is dependent on the example of the field theory being considered.

To see that this definition of integrability is related to other definitions, we show how it gives rise to a 'linear system' for the theory. If we assume that S is a manifold, and differentiate the equation $a(x)\,\phi_0 = \phi(x)\,b(x)$ along the vector $(x; g)$ in S, we get (using subscripts for directional derivatives)

$$\phi_g \;=\; a_g a^{-1}\phi \;-\; \phi b_g b^{-1}\,.$$

If the group X was a loop group, with a complex parameter λ, then this equation would look like the usual sort of linear system for a matrix valued function $\phi(x, \lambda)$.

I shall now consider a particular example, the sine-Gordon equation,

$$\frac{\partial^2 u}{\partial t^2} - \frac{\partial^2 u}{\partial x^2} \;=\; -\frac{4m^2}{\beta}\,\sin(\beta u)\,,$$

for $u(t, x) \in \mathbb{R}$, and m and β constants. This is a nonlinear wave equation in one space and one time variable, which has soliton solutions. Though it does not exhibit the interesting phenomena of soliton decay and coalsecence seen in the principal chiral model, it is still an interesting case for study, and the quantum theory is still very difficult.

We shall give a method of solution for the sine-Gordon model which gives only the soliton solutions. Take a basis of the Lie algebra su_2 to be

$$s_1 = \frac{1}{4}\begin{pmatrix} 0 & i \\ i & 0 \end{pmatrix}, \qquad s_2 = \frac{1}{4}\begin{pmatrix} 0 & 1 \\ -1 & 0 \end{pmatrix}, \qquad s_3 = \frac{1}{4}\begin{pmatrix} i & 0 \\ 0 & -i \end{pmatrix}.$$

We define the group M to consist of analytic functions $c : \mathbb{C}^* \to M_2(\mathbb{C})$, which are unitary on $\mathbb{R}^*$ and satisfy the symmetry condition

$$s_3 \, c(\lambda) \, s_3^{-1} = f(\lambda) \, c(-\lambda) \,,$$

where $f(\lambda)$ is any scalar valued function. The group G consists of meromorphic functions $\phi : \mathbb{C}_\infty \to M_2(\mathbb{C})$, which are unitary on $\mathbb{R}_\infty$ and also satisfy the symmetry condition. The solution $u(t,x)$ to the sine-Gordon equation is encoded into the meromorphic loops $\phi(\lambda)$ by the additional normalisation condition that $\phi(\infty) = \exp(\beta u s_3)$ and $\phi(0) = N \exp(-\beta u s_3)$, for N a constant matrix commuting with s_1.

The classical vacuum map $a : \mathbb{R}^{1,1} \to M$ is defined by

$$a(t,x)(\lambda) = \exp(-2m\lambda s_1 (t+x) - 2m\lambda^{-1}s_1 (t-x)) \,.$$

The elements of the group G can be written as products of a constant matrix and factors of the form

$$P_i^\perp + \frac{\lambda - \bar{\alpha}_i}{\lambda - \alpha_i} P_i \,,$$

where $\{\alpha_1, \ldots, \alpha_n\}$ is the set of poles of the element, and every P_i is a Hermitian projection on $\mathbb{C}^2$. (We skirt over the implications of the symmetry condition and the resulting distinction between solitons and breather solutions at this point.) Note that the projections change if the poles are taken in a different order in the product. In terms of the solitons in the model, the positions of the solitons are encoded into the projections, and their momenta in the pole positions.

We can now consider the algebra of observable functions, which classically is just $\mathbb{C}(G)$, the commutative algebra of functions on the classical phase space. This can be enlarged to the Hopf algebra $\mathbb{C}M \triangleright\!\!\blacktriangleleft \mathbb{C}(G)$ by including the group algebra of M, which contains things like space and time translations.

In the case of the quantum theory, we could conjecture that all we had to do to recover the quantum theory was to q-deform the Hopf algebra to get $\mathbb{C}M_q \triangleright\!\!\blacktriangleleft \mathbb{C}(G)_q$, and take this to contain the *complete* algebra of quantum observables. However in the example of the sine-Gordon model this would require deforming the algebra of functions on a meromorphic loop group, and I believe that this has not been done. On an even more speculative note, as the space-time is imbedded in the group double by the classical vacuum map, if we deform the double we should also deform the vacuum map and possibly also the space-time. In this manner the space-time could become a non-commutative object, with its geometry determined by the field theory [7].

I should elaborate on what I mean by a *complete* quantisation. This is a method of solution to the quantisation of a system which does not destroy information (though the usual rules about similtaneous measurement given by the uncertainty principle must still apply). For example the quantum inverse scattering method for $1 + 1$ dimensional integrable field theories proceeds by quantising the scattering of the system. The scattering contains all the information present in the dispersive component of the solution, that is the part of the solution which will eventually decay to the vacuum when viewed by an observer moving at any velocity. However the scattering does not contain all the information on the solitons in the system, it forgets their positions (and in the case of breathers, their phases). For a more physically important example, consider the standard model in particle physics. Here the scattering can be calculated by Feynman diagrams and renormalisation. However there are still outstanding problems in the theory, outside the scope of existing methods of solution. These include quark confinement, and calculating the binding energy of the deuterium nucleus. Someone who doubts this should consider why a lot of effort is put into lattice models and other approximate methods to consider these questions. The problem is again that the method of quantisation does not consider the positions of particles on an equal footing with the momenta, and it is simply not possible to add the positions later as an afterthought. Calculating with positions included brings us into the realm of *non-perturbative* methods, as where distances are small the fields tend to be high.

The sine-Gordon model is a Hamiltonian system, and we can calculate its symplectic form in terms of the loop groups G and M [3]. Take a point ϕ_0 in the phase space G and two vectors (changes in ϕ_0), $(\phi_0; v_0)$ and $(\phi_0; w_0)$. Then we can differentiate the factorisation $a(t, x)\,\phi_0 = \phi(t, x)\,b(t, x)$ with respect to ϕ_0 to get $av_0 = \phi_v b + \phi b_v$ and $aw_0 = \phi_w b + \phi b_w$. Now the symplectic form is

$$\omega(\phi_0; v_0, w_0) = \frac{1}{4\pi i} \lim_{R \to \infty} \oint_\eta \text{Trace}\Big[b_v b^{-1} \phi^{-1} \phi_w - b_w b^{-1} \phi^{-1} \phi_v\Big]_{x=-R}^{R} \frac{d\lambda}{\lambda} \,,$$

where the contour η is a small clockwise circle and a large anti-clockwise circle around the origin. Here 'small' means that all the poles of the meromorphic function ϕ_0 lie outside the contour, and 'large' means that all the poles lie inside it. The points $x = -R$ and $x = R$ are taken to lie on either side of the 'interesting' region, that is where the fields are substantially different from the vacuum. The value of the formula is independent of t.

Because of the role of the order of the poles in the factorisation, the group G has a rather unusual algebraic structure:

Definition 4.1 *The order actions $\hat{\triangleright}$ and $\hat{\triangleleft}$ of G on G are defined by order reversal in the following manner: If $\phi \in G$ has poles at a set of complex values $\{\alpha_1, \ldots, \alpha_n\}$ and $\psi \in G$ has poles at $\{\beta_1, \ldots, \beta_m\}$, we have $\phi\psi = (\phi\hat{\triangleright}\psi)(\phi\hat{\triangleleft}\psi)$, where $\phi\hat{\triangleright}\psi \in G$ has poles at $\{\beta_1, \ldots, \beta_m\}$ and $\phi\hat{\triangleleft}\psi \in G$ has poles at $\{\alpha_1, \ldots, \alpha_n\}$. This action is only defined when the sets of poles are disjoint, but this is true for a dense subset of $G \times G$.*

From an explicit calculation of the symplectic form for sine-Gordon solitons in terms of the pole positions and the projections [1, 3] we find the following method to calculate the symplectic form of the multi-soliton solutions. If we take $\phi_0 \in G$, for any pole α of ϕ_0 we can write $\phi_0 = \zeta_\alpha \eta_\alpha$, where ζ_α has only a pole at α, and η_α is regular at α. If we take v_0 and w_0 to be changes in ϕ_0, then

$$\omega(\phi_0; v_0, w_0) \;=\; \sum_\alpha \omega(\zeta_\alpha; \zeta_{\alpha v}, \zeta_{\alpha w}) \,, \tag{1}$$

where $(\zeta_\alpha; \zeta_{\alpha w})$ is the vector corresponding to the vector $(\phi_0; w_0)$. Now we would like to study the corresponding Poisson structure on G. This is a section $\gamma = \sum \gamma_1 \otimes \gamma_2$ of $TG \otimes TG$ defined by

$$\omega(\phi_0; v_0, \gamma_1(\phi_0)) \, \gamma_2(\phi_0) \;=\; v_0 \,,$$

for any vector $(\phi_0; v_0)$, omitting the summation. To see what cocycle condition is obeyed by γ, we suppose that ϕ_0 has poles at the points in $\underline{\alpha} \cup \underline{\beta}$, where $\underline{\alpha}$ and $\underline{\beta}$ are disjoint finite subsets of $\mathbb{C}$. Then we can write $\phi_0 = \chi_\alpha \psi_\beta = \chi_\beta \psi_\alpha$, where χ_α and ψ_α have poles in $\underline{\alpha}$ and χ_β and ψ_β have poles in $\underline{\beta}$. Then, using (1),

$$\omega(\phi_0; v_0, w_0) \;=\; \omega(\chi_\alpha; \chi_{\alpha v}, \chi_{\alpha w}) \;+\; \omega(\chi_\beta; \chi_{\beta v}, \chi_{\beta w}) \,.$$

The correspondence between the left ordered coordinates $(\chi_\alpha, \chi_\beta)$ and the original group element is $\phi_0 = \chi_\alpha(\chi_\alpha^{-1}\hat{\triangleright}\chi_\beta) = \chi_\beta(\chi_\beta^{-1}\hat{\triangleright}\chi_\alpha)$. We see that the vector corresponding to $(0, \gamma_1(\chi_\beta))$ in left-ordered coordinates is $\chi_\alpha(\chi_\alpha^{-1}\hat{\triangleright}\gamma_1(\chi_\beta))$, and that corresponding to $(\gamma_1(\chi_\alpha), 0)$ is $\chi_\beta(\chi_\beta^{-1}\hat{\triangleright}\gamma_1(\chi_\alpha))$. Now if we take

$$\begin{aligned}
\gamma(\phi_0) \;=\;& \chi_\alpha(\chi_\alpha^{-1}\hat{\triangleright}\gamma_1(\chi_\beta)) \otimes \chi_\alpha(\chi_\alpha^{-1}\hat{\triangleright}\gamma_2(\chi_\beta)) \\
&+ \chi_\beta(\chi_\beta^{-1}\hat{\triangleright}\gamma_1(\chi_\alpha)) \otimes \chi_\beta(\chi_\beta^{-1}\hat{\triangleright}\gamma_2(\chi_\alpha)) \,,
\end{aligned} \tag{2}$$

we find that

$$\begin{aligned}
\omega(\phi_0; v_0, \gamma_1(\phi_0)) \, \gamma_2(\phi_0) \;=\;& \omega(\chi_\alpha; \chi_{\alpha v}, \gamma_1(\chi_\alpha)) \, \chi_\beta(\chi_\beta^{-1}\hat{\triangleright}\gamma_2(\chi_\alpha)) \\
&+ \omega(\chi_\beta; \chi_{\beta v}, \gamma_1(\chi_\beta)) \, \chi_\alpha(\chi_\alpha^{-1}\hat{\triangleright}\gamma_2(\chi_\beta)) \\
=\;& \chi_\beta(\chi_\beta^{-1}\hat{\triangleright}\chi_{\alpha v}) \;+\; \chi_\alpha(\chi_\alpha^{-1}\hat{\triangleright}\chi_{\beta v}) \;=\; v_0 \,,
\end{aligned}$$

as required. We see that (2) does not look at all like the standard cocycle condition for a Poisson-Lie group.

5   T-duality in sigma models

In this article I will mainly mention material from [9] and [10]. T-duality had its origin in String Theory, but I shall just consider it within classical field theory.

Suppose that we have a doublecross product group $X = G \bowtie M$, where G and M are Lie groups. Then we can split the Lie algebra of X as $\mathfrak{d} = \mathfrak{g} + \mathfrak{m}$, where $\mathfrak{g}$ and $\mathfrak{m}$ are the Lie algebras of G and M respectively. In addition suppose that there is an adjoint-invariant bilinear form $\langle , \rangle$ on $\mathfrak{d}$ which is zero on restriction to $\mathfrak{g}$ and $\mathfrak{m}$. This means that $\mathfrak{m} = \mathfrak{g}^{*op}$, that the factorisation is a coadjoint matched pair and that $\mathfrak{d} = D(\mathfrak{g})$, the Drinfeld double of $\mathfrak{g}$. On $\mathbb{R}^2$ we use light cone coordinates $x_+ = t + x$ and $x_- = t - x$, where t and x are the standard time-space coordinates, and use subscript $\pm$ for the partial derivatives with respect to the light cone coordinates.

To specify the model we suppose that $\mathfrak{d}$ is the direct sum of two perpendicular subspaces $\mathcal{E}_-$ and $\mathcal{E}_+$. The solution to the model is given by a function $k : \mathbb{R}^2 \to G \bowtie M$, with the properties that $k_+ k^{-1}(x_+, x_-) \in \mathcal{E}_-$ and $k_- k^{-1}(x_+, x_-) \in \mathcal{E}_+$ for all $(x_+, x_-) \in \mathbb{R}^2$. Then we see that, if we factor $k = us$ for $u \in G$ and $s \in M$,

$$u^{-1}u_\pm + s_\pm s^{-1} \in u^{-1}\mathcal{E}_\mp u .$$

If the projection $\pi_\mathfrak{g} : \mathfrak{d} \to \mathfrak{g}$ (with kernel $\mathfrak{m}$) is 1-1 and onto when restricted to $u^{-1}\mathcal{E}_- u$ and $u^{-1}\mathcal{E}_+ u$, we can find graph coordinates $E_u : \mathfrak{g} \to \mathfrak{m}$ and $T_u : \mathfrak{g} \to \mathfrak{m}$ so that

$$\{\xi + E_u(\xi) : \xi \in \mathfrak{g}\} = u^{-1}\mathcal{E}_+ u \quad \text{and} \quad \{\xi + T_u(\xi) : \xi \in \mathfrak{g}\} = u^{-1}\mathcal{E}_- u .$$

It follows that $s_- s^{-1} = E_u(u^{-1}u_-)$ and $s_+ s^{-1} = T_u(u^{-1}u_+)$. From the identity

$$(s_+ s^{-1})_- - (s_- s^{-1})_+ = [s_- s^{-1}, s_+ s^{-1}]$$

we deduce that $u(x_+, x_-)$ satisfies the equation

$$(T_u(u^{-1}u_+))_- - (E_u(u^{-1}u_-))_+ = [E_u(u^{-1}u_-), T_u(u^{-1}u_+)] , \qquad (3)$$

which is of the type known as a *sigma model*. Klimčík shows that the Lagrangian density

$$\mathcal{L} = \langle E_u(u^{-1}u_-), u^{-1}u_+ \rangle \qquad (4)$$

gives rise to (3) as its equation of motion.

The dual theory is given by the factorisation $k = tv$, where $t \in M$ and $v \in G$. If we let $\hat{E}_t : \mathfrak{m} \to \mathfrak{g}$ and $\hat{T}_t : \mathfrak{m} \to \mathfrak{g}$ be the graph coordinates of $t^{-1}\mathcal{E}_+ t$ and $t^{-1}\mathcal{E}_- t$ respectively, then $t(x_+, x_-)$ obeys the dual equation

$$(\hat{T}_t(t^{-1}t_+))_- - (\hat{E}_t(t^{-1}t_-))_+ = [\hat{E}_t(t^{-1}t_-), \hat{T}_t(t^{-1}t_+)] . \qquad (5)$$

These are the equations of motion for a sigma model with Lagrangian

$$\hat{\mathcal{L}} = \langle \hat{E}_t(t^{-1}t_-), t^{-1}t_+ \rangle. \tag{6}$$

These two equations are for functions t and u into different groups, but are both descriptions of the model defined on X. The (u, s) and (t, v) variables are related by the actions of the double cross product group structure, $tv = (t \triangleright v)(t \triangleleft v) = us$.

References

[1] O. Babelon and D. Bernard. The sine-Gordon solitons as an N-body problem. *Phys. Lett. B*, 317:363–368, 1993.

[2] E. Beggs, J. Gould and S. Majid. Finite group factorisations and braiding. *J. Algebra*, 181:112–151, 1996.

[3] E. Beggs and P.R. Johnson. Loop groups and the symplectic form for solitons in integrable theories. *Nonlinearity*, vol 12 no 4:1053–1070, 1999.

[4] E. Beggs and P.R. Johnson. Inverse scattering and solitons in A_{n-1} affine Toda field theories II. *Nucl. Phys. B*, 529 no 3:567–587, 1998.

[5] E. Beggs and S. Majid. Quasitriangular and differential structures on bicrossproduct Hopf algebras. *J. Algebra*, vol 219 no 2:682–727, 1999.

[6] E. Beggs and S. Majid. Poisson-Lie T-duality for Quasitriangular Lie bialgebras. *Hep-th*, June 1999.

[7] A. Connes. Non-commutative differential geometry. Technical Report 62, IHES, 1986.

[8] L.D. Faddeev and L.A. Takhtajan. Hamiltonian methods in the theory of solitons. *Springer-Verlag*, 1987.

[9] C. Klimčík. Poisson-Lie T-duality. *Nucl. Phys. B (Proc. Suppl.)*, 46:116–121, 1996.

[10] C. Klimčík and P. Severa. Poisson-Lie T-duality and loop groups of Drinfeld doubles. *Phys. Lett. B*, 372:65–71, 1996.

[11] S. Majid. Physics for algebraists: Non-commutative and non-cocommutative Hopf algebras by a bicrossproduct construction. *J. Algebra*, 130:17–64, 1990. From PhD Thesis, Harvard, 1988.

[12] S. Majid. *Foundations of Quantum Group Theory*. Cambridge Univeristy Press, 1995.

[13] S. Majid. The quantum double as quantum mechanics. *J. Geom. Phys.*, 13:169–202, 1994.

[14] S.P. Novikov, S.V. Manakov, L.P. Pitaevskii and V.E. Zakharov. Theory of solitons. *Contemporary Soviet Mathematics, Consultants Bureau*, 1984.

[15] M.A. Semenov-Tian-Shansky. Dressing transformations and Poisson group actions. *Publ. RIMS (Kyoto)*, 21:1237–1260, 1985.

[16] M. Takeuchi. Matched pairs of groups and bismash products of Hopf algebras. *Commun. Alg.*, 9:841, 1981.

Canonical Bases and Piecewise-linear Combinatorics

Roger Carter

Mathematics Institute, University of Warwick,
Coventry CV4 7AL, England
E-mail: rwc@maths.warwick.ac.uk

Robert Marsh

Department of Mathematics and Computer Science,
University of Leicester, University Road,
Leicester LE1 7RH, England
E-mail: R.Marsh@mcs.le.ac.uk

Abstract

Let U_q be the quantum group associated to a Lie algebra $\mathbf{g}$ of rank n. The negative part U_q^- of U_q has a canonical basis $\mathbf{B}$ with favourable properties (see Kashiwara [3] and Lusztig [6, §14.4.6]). The approaches of Lusztig and Kashiwara lead to a set of alternative parametrizations of the canonical basis, one for each reduced expression for the longest word in the Weyl group of $\mathbf{g}$. We describe the authors' recent work establishing close relationships between the Lusztig cones, canonical basis elements and the regions of linearity of reparametrization functions arising from the above parametrizations in type A_4 and give some speculations for type A_n.

Keywords: Quantum group, Lie algebra, Canonical basis, Tight monomials, Weyl group, Piecewise-linear functions.

1 Introduction

Let $\mathbf{g}$ be a finite-dimensional simple Lie algebra over $\mathbb{C}$ and $U_q(\mathbf{g})$ be the corresponding quantized enveloping algebra over $\mathbb{C}[v, v^{-1}]$. Let

$$U_q(\mathbf{g}) = U_q^- \otimes U_q^0 \otimes U_q^+$$

be the triangular decomposition of $U_q(\mathbf{g})$. Let $\mathbf{B}$ be the canonical basis of U_q^-, introduced independently by Lusztig and Kashiwara. It is natural to ask how the elements of $\mathbf{B}$ are expressed in terms of the generators $F_1, F_2, \ldots, F_n$ of U_q^-. This is a difficult question which is known completely in only a few low rank cases, but the attempt to understand it has led to a remarkable theory of piecewise-linear combinatorics associated with the canonical basis.

In this paper we consider only the case in which **g** has type A_n. We describe the situation for small values of n before giving some speculations for arbitrary n.

First suppose that **g** has type A_1. We write:

$$\mathbb{N} = \{a \in \mathbb{Z} : a \geq 0\},$$
$$[a] = \frac{v^a - v^{-a}}{v - v^{-1}}, \quad \text{for } a \in \mathbb{N},$$
$$[a]! = [1][2] \cdots [a], \quad \text{for } a \in \mathbb{N},$$

and $F_1^{(a)} = F_1^a / [a]!$.

It was shown by Lusztig [4] that $\mathbf{B} = \{F_1^{(a)} : a \in \mathbb{N}\}$. Thus the canonical basis elements are the quantized divided powers of the generator F_1.

Next, suppose that **g** has type A_2. The canonical basis in this case was also determined by Lusztig [4]. We have

$$\mathbf{B} = \{F_1^{(a)} F_2^{(b)} F_1^{(c)} : a, b, c \in \mathbb{N}, \ b \geq a + c\}$$
$$\cup \{F_2^{(a)} F_1^{(b)} F_2^{(c)} : a, b, c \in \mathbb{N}, \ b \geq a + c\}.$$

When $b = a + c$ one has the relation

$$F_1^{(a)} F_2^{(a+c)} F_1^{(c)} = F_2^{(c)} F_1^{(a+c)} F_2^{(a)},$$

but apart from this the above elements of **B** are all distinct. Thus in this case each canonical basis element may be written as a monomial in the generators F_1, F_2. The two types of monomials which arise are related to the two reduced decompositions of the longest element w_0 of the Weyl group W of **g**. We have $W = \langle s_1, s_2 \rangle$ and $w_0 = s_1 s_2 s_1 = s_2 s_1 s_2$. Each of these reduced words for w_0 gives rise to a type of monomial in the canonical basis.

We now turn to the case when **g** has type A_3. Lusztig [5] obtained many, but not all, elements of **B** as monomials in the generators F_1, F_2, F_3 and gave an example of an element of **B** which could not be written as a monomial in F_1, F_2, F_3. The remaining elements of **B** were determined by Xi [10]. In order to describe the monomials in **B** we consider reduced words for w_0. In type A_3 w_0 has 16 reduced expressions. However it is natural to divide them into equivalence classes called commutation classes; two reduced words being equivalent if one can be obtained from the other by a succession of commutations in the Coxeter generators of W. For example, $s_1 s_3 s_2 s_1 s_3 s_2$ is in the same commutation class as $s_3 s_1 s_2 s_3 s_1 s_2$, where the generators are numbered as in the Dynkin diagram (see Figure 1).

Figure 1: Dynkin diagram of type A_3.

There are 8 commutation classes of reduced words for w_0 in type A_3. Each of these gives rise to a family of monomials in F_1, F_2, F_3 which lie in $\mathbf{B}$. For example, the reduced word $w_0 = s_1 s_3 s_2 s_1 s_3 s_2$ gives rise to monomials

$$F_1^{(a)} F_3^{(b)} F_2^{(c)} F_1^{(d)} F_3^{(e)} F_2^{(f)}, \qquad a, b, c, d, e, f \in \mathbb{N},$$

and such a monomial was shown by Lusztig to lie in the canonical basis $\mathbf{B}$ provided

$$c \geq a + d, \quad c \geq b + e, \quad d + e \geq c + f.$$

These inequalities come from considering two consecutive occurrences of a given Coxeter generator s_i in the given reduced word. The sum of the exponents corresponding to these occurrences of s_i is less than or equal to the sum of the exponents corresponding to Coxeter generators s_j between these two occurrences of s_i such that s_j does not commute with s_i. Lusztig obtained 8 families of monomials in $\mathbf{B}$ in this way, and the remaining elements of $\mathbf{B}$ determined by Xi are linear combinations of monomials with coefficients which are quantum binomial coefficients.

We now suppose that $\mathbf{g}$ has type A_4. Here the situation is more involved and has been investigated by the authors. We shall outline the situation in the present paper, and hope to publish the proofs in a subsequent article. For each reduced word $w_0 = s_{i_1} s_{i_2} \cdots s_{i_k}$, where $k = \ell(w_0)$, we write $\mathbf{i} = (i_1, i_2, \ldots, i_k)$. For each such $\mathbf{i}$, we define a subset $C_{\mathbf{i}}$ of $\mathbb{N}^k$ whose definition is motivated by the rule mentioned above for a monomial to lie in $\mathbf{B}$ in type A_3. We define $C_{\mathbf{i}}$ to be the set of those $\mathbf{a} \in \mathbb{N}^k$ such that for each pair $t, t' \in [1, k]$ with $t < t'$, $i_t = i_{t'}$, $i_p \neq i_t$ for $t < p < t'$, we have $\sum_p a_p \geq a_t + a_{t'}$, summed over all p with $t < p < t'$ such that s_p does not commute with s_{i_t}.

The cone $C_{\mathbf{i}}$ will be called the *Lusztig cone* associated with $\mathbf{i}$. The elements of $C_{\mathbf{i}}$ give rise to monomials. Let

$$M_{\mathbf{i}} = \{F_{i_1}^{(a_1)} F_{i_2}^{(a_2)} \cdots F_{i_k}^{(a_k)} : \mathbf{a} = (a_1, a_2, \ldots, a_k) \in C_{\mathbf{i}}\}.$$

It was shown by Marsh [7] that when $\mathbf{g}$ has type A_4 we have $M_{\mathbf{i}} \subseteq \mathbf{B}$ for each $\mathbf{i}$. In type A_4 there are 62 commutation classes of reduced words for w_0, and we obtain in this way 62 families of monomials in the canonical basis. These are far from being the only elements of $\mathbf{B}$, however, and the remaining elements are not known as expressions in terms of F_1, F_2, F_3, F_4.

2   PBW-type Bases and the Canonical Basis

We shall now recall Lusztig's approach to the canonical basis. For each reduced word $\mathbf{i}$ for w_0, Lusztig defined [4] a PBW-type basis $B_{\mathbf{i}}$ of U_q^-. The reduced word $w_0 = s_{i_1} s_{i_2} \cdots s_{i_k}$ gives rise to a total order on the set Φ^+ of positive roots of $\mathbf{g}$. We have

$$\Phi^+ = \{\alpha^1, \alpha^2, \ldots, \alpha^k\},$$

where

$$\alpha^1 = \alpha_{i_1}, \ \alpha^2 = s_{i_1}(\alpha_{i_2}), \ldots \alpha^k = s_{i_1} s_{i_2} \cdots s_{i_{k-1}}(\alpha_{i_k}),$$

and $\alpha_1, \alpha_2, \ldots, \alpha_n$ are the simple roots. By using braid group operations, Lusztig defines for each $\alpha \in \Phi^+$ a root vector $F_\alpha \in U_q^-$, beginning with $F_{\alpha_i} = F_i$. We write $F_{\mathbf{i}}^{\mathbf{c}} = F_{\alpha^1}^{(c_1)} F_{\alpha^2}^{(c_2)} \cdots F_{\alpha^k}^{(c_k)}$ for $\mathbf{c} = (c_1, c_2, \ldots, c_k) \in \mathbb{N}^k$. Then the set $B_{\mathbf{i}} = \{F_{\mathbf{i}}^{\mathbf{c}} : \mathbf{c} \in \mathbb{N}^k\}$ is a basis for U_q^-, called the PBW-type basis associated to $\mathbf{i}$.

The lattice $\mathcal{L} = \mathbb{C}[v]B_{\mathbf{i}}$ was shown by Lusztig to be independent of $\mathbf{i}$ and there is a bijective map

$$
\begin{array}{ccc}
\mathbf{B} & \longrightarrow & B_{\mathbf{i}} \\
b & \mapsto & F_{\mathbf{i}}^{\mathbf{c}}
\end{array}
$$

such that $b \equiv F_{\mathbf{i}}^{\mathbf{c}} \mod v\mathcal{L}$. We write $\varphi_{\mathbf{i}}(b) = \mathbf{c}$. Then the map $\varphi_{\mathbf{i}} : \mathbf{B} \to \mathbb{N}^k$ is bijective. This gives, for each $\mathbf{i}$, a parametrization of $\mathbf{B}$ by elements of $\mathbb{N}^k$.

Lusztig introduced [6] two particular reduced words $\mathbf{j}$, $\mathbf{j}'$ for w_0. In type A_n these are as follows. Suppose the vertices of the Dynkin diagram of A_n are labelled as in Figure 2.

Figure 2: Dynkin diagram of type A_n.

Let $\mathbf{j}$ be $135 \cdots 246 \cdots 135 \cdots 246 \cdots$, where $k = \frac{1}{2}n(n+1)$ factors are taken, and let $\mathbf{j}'$ be $246 \cdots 135 \cdots 246 \cdots 135 \cdots$, where again k factors are taken. Then $\mathbf{j}, \mathbf{j}'$ are reduced words for w_0 and their commutation classes are as far apart as possible, in the sense that they give opposite orderings on Φ^+. Consider the parametrizations $\varphi_{\mathbf{j}} : \mathbf{B} \to \mathbb{N}^k$ and $\varphi_{\mathbf{j}'} : \mathbf{B} \to \mathbb{N}^k$ of $\mathbf{B}$ corresponding to $\mathbf{j}, \mathbf{j}'$ and let $R = \varphi_{\mathbf{j}'} \varphi_{\mathbf{j}}^{-1} : \mathbb{N}^k \to \mathbb{N}^k$ be the bijective map which relates them.

The function R is the restriction to $\mathbb{N}^k$ of a piecewise-linear map $R : \mathbb{R}^k \to \mathbb{R}^k$ and the results in low rank cases suggest that the regions of linearity of R are related to the different types of canonical basis elements.

For example, in type A_2, we have $\mathbf{j} = 121$, $\mathbf{j}' = 212$, and R has two regions of linearity. These correspond to the two types of monomial in the canonical basis.

In type A_3, the function R has 10 regions of linearity, and 8 of these regions correspond to the 8 families of monomials in the canonical basis. The remaining 2 regions correspond to the remaining non-monomial elements of $\mathbf{B}$ obtained by Xi. These two regions may be distinguished from the other 8 regions as follows.

In type A_3, we have $R : \mathbb{R}^6 \to \mathbb{R}^6$. The 8 regions of linearity of R which give rise to monomials in $\mathbf{B}$ are each defined by 3 inequalities whereas the 2 remaining regions of linearity are each defined by 4 inequalities. This suggests that regions of linearity defined by the minimum number of inequalities might give rise to canonical basis elements of a particularly favourable form.

This turns out to be the case in type A_4 also. This time we have a piecewise-linear map $R : \mathbb{R}^{10} \to \mathbb{R}^{10}$, which was shown by Carter to have 144 regions of linearity. Of these regions, 62 are defined by 6 inequalities, 70 by 7 inequalities, 10 by 8 inequalities, and 2 by 11 inequalities. It is striking that the number of regions defined by the minimum number of inequalities is equal to the number of commutation classes of reduced words for w_0.

The authors have shown that in type A_4 there is a natural bijection between commutation classes of reduced words for w_0 and regions of linearity for R defined by the minimum number of inequalities. This comes about as follows. For each reduced word $\mathbf{i}$ for w_0 we have a corresponding family $M_{\mathbf{i}}$ of monomials in $\mathbf{B}$, as described above. We consider the set of points in $\mathbb{N}^{10}$ which parametrize these elements of $\mathbf{B}$ under the map $\varphi_{\mathbf{j}}$: let $X_{\mathbf{i}}^+ = \varphi_{\mathbf{j}}(M_{\mathbf{i}})$. Then one can show:

(a) $X_{\mathbf{i}}^+$ is the set of all points in a region of linearity $X_{\mathbf{i}}$ of R with coordinates in $\mathbb{N}$.

(b) $X_{\mathbf{i}}$ is a region defined by the minimum number of inequalities.

(c) The map $\mathbf{i} \to X_{\mathbf{i}}$ is a bijection between commutation classes of reduced words for w_0 and regions of R defined by the minimum number of inequalities.

In fact, both the set of commutation classes of reduced words for w_0 and the set of regions of R defined by the minimum number of inequalities can be given a natural graph structure, and the map $\mathbf{i} \to X_{\mathbf{i}}$ is then an isomorphism of graphs.

3 Rectangle Calculus

It is natural to ask whether the region $X_{\mathbf{i}}$ can be described in terms of $\mathbf{i}$ without using ideas concerned with the canonical basis, but simply in combinatorial terms. This can be done by a form of combinatorics which we call the rectangle calculus. The basic idea is to associate with the reduced word $\mathbf{i}$ a set of linearly independent vectors such that $X_{\mathbf{i}}$ is the set of all linear combinations of these vectors with non-negative coefficients. These vectors will be called *spanning vectors* of $X_{\mathbf{i}}$.

We first introduce the idea of a partial quiver. This is a Dynkin diagram with arrows on certain edges, such that the set of edges with arrows in non-empty and connected. An example of a partial quiver of type A_7 is given in Figure 3.

Figure 3: A partial quiver of type A_7.

We are concerned with partial quivers of type A_n, and shall use L or R to indicate whether an arrow goes left or right. Thus the above partial quiver is denoted $--RLL-$.

The possible partial quivers of type A_4 are LLL, RLL, LRL, LLR, LRR, RLR, RRL, RRR, $LL-$, $-LL$, $LR-$, $-LR$, $RL-$, $-RL$, $RR-$, $-RR$, $L--$, $-L-$, $--L$, $R--$, $-R-$, $--R$.

We shall now explain a procedure by which each reduced word $\mathbf{i}$ for w_0 in type A_n gives a set of $\frac{1}{2}n(n-1)$ partial quivers. We first write down the braid diagram of $\mathbf{i}$. This determines a set of chambers, and for each bounded chamber we write down the corresponding chamber set, which is the subset of $\{1, 2, \ldots, n+1\}$ corresponding to the strings which pass below the chamber. We illustrate this by the example in which $n = 4$ and $w_0 = s_2 s_3 s_4 s_3 s_1 s_2 s_1 s_3 s_2 s_4$ — see Figure 4.

Each chamber set obtained in this way is a subset of $\{1, 2, \ldots, n+1\}$ which is not an initial or terminal subset, i.e. not of form $\{1, 2, \ldots, i\}$ or $\{i, i+1, \ldots, n+1\}$ for any i.

Let $\mathcal{S}$ be the set of all subsets of $\{1, 2, \ldots, n+1\}$ which are not initial or terminal subsets and $\mathcal{P}$ be the set of all partial quivers of type A_n. We shall

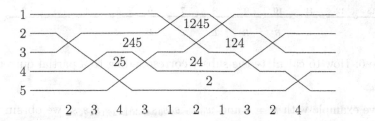

Figure 4: A chamber diagram.

describe a bijection from $\mathcal{S}$ to $\mathcal{P}$. We first number the edges of the Dynkin diagram as shown in Figure 5.

Figure 5: Edge numbering of Dynkin diagram.

Thus the edges are numbered $2, 3, \ldots, n$ from right to left and $1, n+1$ are regarded as virtual edges.

Let $S \in \mathcal{S}$ and P be the corresponding partial quiver. P is obtained from S by the following rules.

(a) If $1, n+1 \notin S$, the entries in S give the edges of type L in P, edges intermediate between those of type L having type R. The leftmost and rightmost labelled edges of P have type L.
(b) Now suppose $1 \in S$. Let i be such that $1, 2, \ldots, i \in S$ but $i+1 \notin S$. Then edge $i+1$ is labelled R and is the rightmost labelled edge in P.
(c) Now suppose $n+1 \in S$. Let i be such that $i, i+1, \ldots, n+1$ lie in S but $i-1 \notin S$. Then edge $i-1$ of P is labelled R and is the leftmost labelled edge in P.
(d) The elements of S not in an initial segment as in (b) or a terminal segment as in (c) give rise to edges L or R in P as in (a).

Example.
Let $n = 13$. If $S = \{1, 2, 3, 4, 7, 8, 11\}$, then $P = -\ -\ LRRLLRR\ -\ --$.
See Figure 6 for an explanatory diagram.

Applying this bijection $\mathcal{S} \to \mathcal{P}$ to the chamber sets obtained from a reduced word $\mathbf{i}$ we obtain a set of $\frac{1}{2}n(n-1)$ partial quivers associated with $\mathbf{i}$. In

14 13 12 11 10 9 8 7 6 5 4 3 2 1
– – L R R L L R R – – –

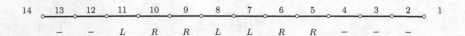

Figure 6: How to calculate the subset corresponding to a partial quiver.

the above example with $n = 4$ and $w_0 = s_2s_3s_4s_3s_1s_2s_1s_3s_2s_4$ we obtain the partial quivers shown in Figure 7.

1
2
3
4
5
–R–
–RL
LR–
RRL
LRL
––L
2 3 4 3 1 2 1 3 2 4

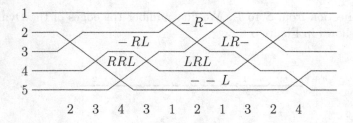

Figure 7: Partial quivers for $s_2s_3s_4s_3s_1s_2s_1s_3s_2s_4$.

We denote by $\mathcal{P}(\mathbf{i})$ the set of partial quivers obtained from $\mathbf{i}$ in this way.

We now introduce the rectangles which we shall be considering. Let $i, j, k, l \in \mathbb{N}$ satisfy:

$$i < j < l, \quad i < k < l, \quad i + l = j + k.$$

An (i, j, k, l)-rectangle is a rectangle with corners on levels i, j, k, l. It is most convenient to illustrate this idea by means of an example. See Figure 8.

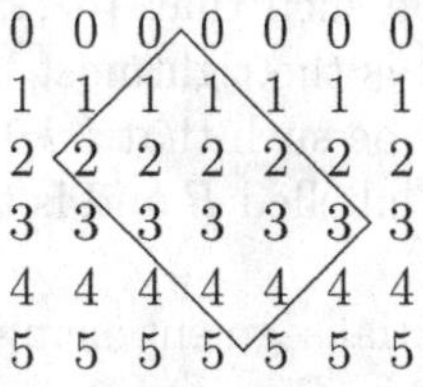

0 0 0 0 0 0 0
1 1 1 1 1 1 1
2 2 2 2 2 2 2
3 3 3 3 3 3 3
4 4 4 4 4 4 4
5 5 5 5 5 5 5

Figure 8: Drawing a $(0, 2, 3, 5)$-rectangle.

The sides of the rectangle have gradient $\pm\pi/4$. The (i, j, k, l)-rectangle contains alternate columns of integers starting with the first column if the entry in it is odd and the second column otherwise. See Figure 9.

Figure 9: A $(0, 2, 3, 5)$-rectangle.

The columns of integers in the rectangle are interpreted as positive roots; for example the $(0, 2, 3, 5)$-rectangle contains roots $\alpha_1 + \alpha_2 + \alpha_3$, $\alpha_2 + \alpha_3 + \alpha_4$.

We next describe how each partial quiver determines a configuration of rectangles. It is again most convenient to explain this by means of an example. Consider the case of type A_{10} in which we take the partial quiver $P = -LLRRRLRR$ — see Figure 10.

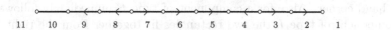

Figure 10: The partial quiver $P = -LLRRRLRR$.

The edges of the partial quiver are numbered as shown. We first divide the partial quiver into its components, i.e. the maximal subquivers containing a set of consecutive L's and R's. The components of our given partial quiver P are:

$$- \quad L \quad L \quad - \quad - \quad - \quad - \quad - \quad -$$

$$- \quad - \quad - \quad R \quad R \quad R \quad - \quad - \quad -$$

$$- \quad - \quad - \quad - \quad - \quad - \quad L \quad - \quad -$$

$$- \quad - \quad - \quad - \quad - \quad - \quad - \quad R \quad R$$

For each component K of P we define positive integers $a(K)$, $b(K)$ with $a(K) < b(K)$. The integer $a(K)$ is the number of the edge following the rightmost arrow of K and $b(K)$ is the number of the edge preceding the leftmost arrow of K. In the above example the numbers $a(K), b(K)$ are as follows:

K									a(K)	b(K)
$-$	L	L	$-$	$-$	$-$	$-$	$-$	$-$	7	10
$-$	$-$	$-$	R	R	R	$-$	$-$	$-$	4	8
$-$	$-$	$-$	$-$	$-$	$-$	L	$-$	$-$	3	5
$-$	$-$	$-$	$-$	$-$	$-$	$-$	R	R	1	4

For each component K of type L we take a $(0, a, n+2-b, n+a-b+2)$-rectangle and for each component K of type R we take a $(b-a-1, b-1, n+1-a, n+1)$-rectangle, where $a = a(K)$, $b = b(K)$. Thus the 4 components of our partial quiver $P = - LLRRRLRR$ give the 4 rectangles shown in Figure 11.

We now observe that for a component of type L immediately followed by a component of type R the two corresponding rectangles fit together from the left hand corner. Also, for a component of type R immediately followed by a component of type L the two rectangles fit together from the right hand corner. This can be observed in the four rectangles above, in which the first and the second fit from the left, the second and the third from the right, and the third and the fourth from the left.

We use these rules to superimpose the rectangles obtained from the components of the given partial quiver. In the case of the partial quiver $P = - LLRRRLRR$ we obtain the configuration of rectangles shown in Figure 12.

We next define the *centre* of such a configuration. First consider the rectangles in the diagonals from north-west to south-east. The number of such rectangles in these diagonals in the above configuration is $1, 3, 4, 2$. It is always the case that one obtains a set of odd numbers followed by a set of even numbers or vice versa. We draw the diagonal line separating the diagonal blocks giving odd and even numbers of rectangles. This is the line ℓ in Figure 13. A similar phenomenon occurs for the diagonals from northeast to southwest. The number of rectangles in such diagonals in the above configuration is $2, 4, 3, 1$ and we draw the diagonal line separating the diagonal blocks giving odd and even numbers of rectangles. This is the line ℓ' in Figure 13. The point O in which ℓ and ℓ' intersect is called the centre of the configuration and we draw in the vertical line through O, called the central line, m. See Figure 13.

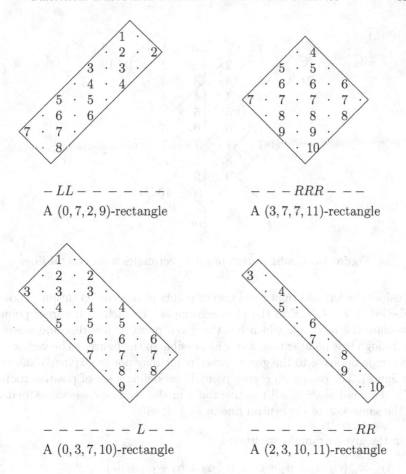

$- LL - - - - - -$

A $(0, 7, 2, 9)$-rectangle

$- - - RRR - - -$

A $(3, 7, 7, 11)$-rectangle

$- - - - - - L - -$

A $(0, 3, 7, 10)$-rectangle

$- - - - - - - RR$

A $(2, 3, 10, 11)$-rectangle

Figure 11: The 4 rectangles for $- LLRRRLRR$.

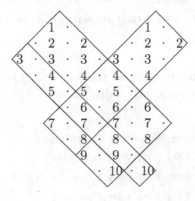

Figure 12: Configuration of all 4 rectangles.

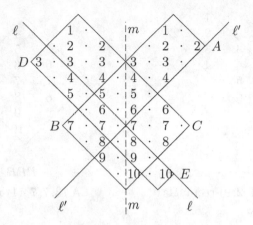

Figure 13: Configuration of all 4 rectangles with central line.

We consider the left and right hand corner points in this configuration. These are labelled A, B, C, D, E in the above example. To each such corner point we associate the rectangle which has the given point as a vertex and whose edges through this point extend as far as possible in the figure. (The vertex of the rectangle opposite to the given corner point may not be explicitly shown in the figure). For each such corner point V we define a set of positive roots $\Phi^+(V)$. This is the set of all positive roots in the rectangle associated with V on the same side of the central line m as V itself.

Thus in the given example, we have:

$$\Phi^+(A) = \{\alpha_2, \alpha_1 + \alpha_2 + \alpha_3 + \alpha_4, \alpha_3 + \alpha_4 + \alpha_5 + \alpha_6\}$$
$$\Phi^+(B) = \{\alpha_7, \alpha_5 + \alpha_6 + \alpha_7 + \alpha_8 + \alpha_9\}$$
$$\Phi^+(C) = \{\alpha_6 + \alpha_7 + \alpha_8, \alpha_4 + \alpha_5 + \alpha_6 + \alpha_7 + \alpha_8 + \alpha_9 + \alpha_{10}\}$$
$$\Phi^+(D) = \{\alpha_3, \alpha_1 + \alpha_2 + \alpha_3 + \alpha_4 + \alpha_5, \alpha_2 + \alpha_3 + \alpha_4 + \alpha_5 + \alpha_6 + \alpha_7\}$$
$$\Phi^+(E) = \{\alpha_{10}, \alpha_8 + \alpha_9\}$$

We then define $\Phi^+(P)$ to be the union of the sets $\Phi^+(V)$ for all corner points V in the configuration. This is always a disjoint union. In the given example we have:

$$\begin{aligned}
\Phi^+(P) = \{ & \alpha_2, \alpha_1 + \alpha_2 + \alpha_3 + \alpha_4, \alpha_3 + \alpha_4 + \alpha_5 + \alpha_6, \alpha_7, \\
& \alpha_5 + \alpha_6 + \alpha_7 + \alpha_8 + \alpha_9, \alpha_6 + \alpha_7 + \alpha_8, \\
& \alpha_4 + \alpha_5 + \alpha_6 + \alpha_7 + \alpha_8 + \alpha_9 + \alpha_{10}, \alpha_3, \alpha_1 + \alpha_2 + \alpha_3 + \alpha_4 + \alpha_5, \\
& \alpha_2 + \alpha_3 + \alpha_4 + \alpha_5 + \alpha_6 + \alpha_7, \alpha_{10}, \alpha_8 + \alpha_9 \}.
\end{aligned}$$

We now define a vector $v_P \in \mathbb{N}^k$ whose coordinates are all 0 or 1. Let $\mathbf{j}$ be

the reduced word
$$135 \cdots 246 \cdots 135 \cdots 246 \cdots ,$$
considered above, and let $\alpha^1, \alpha^2, \ldots, \alpha^k$ be the corresponding order on the set of positive roots. We define v_P as the vector whose ith coordinate is 1 if $\alpha^i \in \Phi^+(P)$ and is 0 otherwise. We also define vectors $v_j \in \mathbb{N}^k$ for $j = 1, 2, \ldots, n$, where the ith component of v_j is 1 if the ith letter in $\mathbf{j}$ is j and is 0 otherwise.

These vectors v_P, $P \in \mathcal{P}(\mathbf{i})$ and v_j, $j \in \{1, 2, \ldots, n\}$ turn out to be our required spanning vectors for the region $X_{\mathbf{i}}$.

Proposition 3.1 *Suppose that* $\mathbf{g}$ *has type* A_4 *and let* $\mathbf{j}$ *be the reduced word* 1324132413 *for* w_0. *Let* $\mathbf{i}$ *be any reduced word for* w_0 *and* $\mathcal{P}(\mathbf{i})$ *be the set of partial quivers associated with* $\mathbf{i}$. *We have* $|\mathcal{P}(\mathbf{i})| = 6$. *Then the region* $X_{\mathbf{i}}^+$ *associated with* $\mathbf{i}$ *is the set of all non-negative integral combinations of the vectors* v_P, $P \in \mathcal{P}(\mathbf{i})$, *and* v_j *for* $1 \leq j \leq 4$.

This Proposition explains how the regions $X_{\mathbf{i}}^+$ can be described by the rectangle combinatorics.

4 Speculations for type A_n

It is natural to ask whether the set $X_{\mathbf{i}}$ defined as the set of non-negative linear combinations of the vectors v_P, $P \in \mathcal{P}(\mathbf{i})$ and v_j, $1 \leq j \leq n$ is a region of linearity for $R : \mathbb{R}^k \to \mathbb{R}^k$ in type A_n. No counter-example is known to the authors. If so, are the $X_{\mathbf{i}}$ the only regions of linearity of R defined by the minimum number of inequalities? This would give a bijection in type A_n between commutation classes of reduced words for w_0 and regions of linearity for R defined by the minimum number of inequalities.

The set $M_{\mathbf{i}}$ of monomials corresponding to points in the Lusztig cone $C_{\mathbf{i}}$ is not in general contained in the canonical basis in type A_n. It is nevertheless possible to consider a subset of $\mathbf{B}$ in bijective correspondence with $C_{\mathbf{i}}$ by means of Kashiwara's approach to the canonical basis. Kashiwara [3] defines certain root operators $\widetilde{F}_i$ which lead to a parametrization of the canonical basis $\mathbf{B}$ for each $\mathbf{i}$ by a certain subset $K_{\mathbf{i}} \subseteq \mathbb{N}^k$ which we call the *string cone*. This gives a bijection
$$\psi_{\mathbf{i}} : \mathbf{B} \to K_{\mathbf{i}}.$$
It has been shown independently by Marsh [8] and by Premat [9] that $C_{\mathbf{i}} \subseteq K_{\mathbf{i}}$ for each $\mathbf{i}$, i.e. that the Lusztig cone lies in the string cone. Thus we obtain

a subset $\psi_{\mathbf{i}}^{-1}(C_{\mathbf{i}}) \subseteq \mathbf{B}$. This subset is equal to the set of monomials $M_{\mathbf{i}}$ when $n \leq 4$, but does not consist of monomials in general. Using Lusztig's parametrization

$$\varphi_{\mathbf{j}} : \mathbf{B} \to \mathbb{N}^k$$

where $\mathbf{j} = 135 \cdots 246 \cdots 135 \cdots 246 \cdots$, this subset $\psi_{\mathbf{i}}^{-1}(C_{\mathbf{i}})$ of $\mathbf{B}$ corresponds to a certain subset of $\mathbb{N}^k$. Let

$$S_{\mathbf{i}}^{\mathbf{j}} : K_{\mathbf{i}} \to \mathbb{N}^k$$

be the transition map given by $S_{\mathbf{i}}^{\mathbf{j}} = \varphi_{\mathbf{j}} \psi_{\mathbf{i}}^{-1}$. It is known that in type A_n, the images of the spanning vectors of the Lusztig cone $C_{\mathbf{i}}$ under $S_{\mathbf{i}}^{\mathbf{j}}$ are the vectors v_P, $P \in \mathcal{P}(\mathbf{i})$ and v_j, $1 \leq j \leq n$ given by the rectangle combinatorics. This can be proved using a transition function introduced by Berenstein, Fomin and Zelevinsky [1].

Finally, what can be said about canonical basis elements corresponding under $\varphi_{\mathbf{j}}$ to regions of linearity of R not defined by the minimum number of inequalities? The results of Xi [10] in type A_3 are interesting in this respect. In type A_3, Lusztig's function $R : \mathbb{R}^6 \to \mathbb{R}^6$ has 10 regions of linearity, 8 of which are the regions $X_{\mathbf{i}}$ for the different commutation classes of reduced words $\mathbf{i}$ for w_0. These are all defined by 3 inequalities. The remaining two are defined by 4 inequalities and we denote these by X_9 and X_{10}. We now define $\widetilde{X_{\mathbf{i}}^+}$ to be the set of points in $X_{\mathbf{i}}$ whose coordinates are all real and nonnegative. Thus we have

$$X_{\mathbf{i}}^+ \subseteq \widetilde{X_{\mathbf{i}}^+} \subseteq X_{\mathbf{i}}$$

and $X_{\mathbf{i}}^+$ is the set of integral points in $\widetilde{X_{\mathbf{i}}^+}$. We define $\widetilde{X_9^+}$ and $\widetilde{X_{10}^+}$ similarly. We have additional inequalities defining $X_{\mathbf{i}}^+$ asserting that all coordinates are non-negative, but some of these inequalities will be redundant. In fact each of the 8 regions $\widetilde{X_{\mathbf{i}}^+}$ can be defined by 6 inequalities and, for suitable numbering, $\widetilde{X_9^+}$ can be defined by 8 inequalities and $\widetilde{X_{10}^+}$ by 9 inequalities. The regions $X_{\mathbf{i}}^+$ are called simplicial regions as the number of defining inequalities is equal to the dimension of the ambient space. $\widetilde{X_9^+}$ can be written as the union of two simplicial regions, and $\widetilde{X_{10}^+}$ as the union of four simplicial regions.

Xi obtains 8 families of monomials in $\mathbf{B}$, which are parametrized by the integral points in the 8 simplicial regions $\widetilde{X_{\mathbf{i}}^+}$. In addition he obtains 6 families of elements in $\mathbf{B}$ which are not monomials. These are parametrized by the integral points in the two simplicial regions whose union is $\widetilde{X_9^+}$ and the four simplicial regions whose union is $\widetilde{X_{10}^+}$. The canonical basis elements corresponding to a given region are all of the same type, i.e. they can all be expressed as a linear combination of monomials corresponding to a fixed reduced expression with quantum binomial coefficients and exponents lying on a line segment in $\mathbb{N}^6$.

The authors plan to give the proofs of the results described in this article in a forthcoming paper.

References

[1] A. Berenstein, S. Fomin, and A. Zelevinsky. Parametrizations of canonical bases and totally positive matrices. *Adv. Math.*, 122(1):49–149, 1996.

[2] R. W. Carter and R. J. Marsh. Regions of linearity, Lusztig cones and canonical basis elements for the quantized enveloping algebra of type A_4. In preparation.

[3] M. Kashiwara. On crystal bases of the q-analogue of universal enveloping algebras. *Duke Math. J.*, 63(2):465–516, 1991.

[4] G. Lusztig. Canonical bases arising from quantized enveloping algebras. *J. Amer. Math. Soc.*, 3:447–498, 1990.

[5] G. Lusztig. Introduction to quantized enveloping algebras. *Progr. Math.*, 105:49–95, 1992.

[6] G. Lusztig. *Introduction to Quantum Groups*. Birkhäuser, Boston, 1993.

[7] R. J. Marsh. More tight monomials in quantized enveloping algebras. *J. Alg.*, 204:711–732, 1998.

[8] R. J. Marsh. Rectangle diagrams for the Lusztig cones of quantized enveloping algebras of type A. Preprint, 1998.

[9] A. Premat. The Lusztig cone and the image of the Kashiwara map. Private communication, 1999.

[10] N. Xi. Canonical basis for type A_3. Preprint, 1998.

INTEGRABLE AND WEYL MODULES
FOR QUANTUM AFFINE sl_2.

VYJAYANTHI CHARI AND ANDREW PRESSLEY

0. INTRODUCTION

Let t be an arbitrary symmetrizable Kac-Moody Lie algebra and $\mathbf{U}_q(t)$ the corresponding quantized enveloping algebra of t defined over $\mathbf{C}(q)$. One can associate to any dominant integral weight μ of t an irreducible integrable $\mathbf{U}_q(t)$-module $L(\mu)$. These modules have many interesting properties and are well understood, [K], [L1].

More generally, given any integral weight λ, Kashiwara [K] defined an integrable $\mathbf{U}_q(t)$-module $V^{max}(\lambda)$ generated by an extremal vector v_λ. If w is any element of the Weyl group W of t, then one has $V^{max}(\lambda) \cong V^{max}(w\lambda)$. Further, if λ is in the Tits cone, then $V^{max}(\lambda) \cong L(w_0\lambda)$, where $w_0 \in W$ is such that $w_0\lambda$ is dominant integral. In the case when λ is not in the Tits cone, the module $V^{max}(\lambda)$ is not irreducible and very little is known about it, although it is known that it admits a crystal basis, [K].

In the case when t is an affine Lie algebra, an integral weight λ is not in the Tits cone if and only if λ has level zero. Choose $w_0 \in W$ so that $w_0\lambda$ is dominant with respect to the underlying finite-dimensional simple Lie algebra of t. In as yet unpublished work, Kashiwara proves that $V^{max}(\lambda) \cong W_q(w_0\lambda)$, where $W_q(w_0\lambda)$ is an integrable $\mathbf{U}_q(t)$-module defined by generators and relations analogous to the definition of $L(\mu)$.

In [CP4], we studied the modules $W_q(\lambda)$ further. In particular, we showed that they have a family $W_q(\boldsymbol{\pi})$ of non-isomorphic finite-dimensional quotients which are maximal, in the sense that any other finite-dimensional quotient is a proper quotient of some $W_q(\boldsymbol{\pi})$. In this paper, we show that, if t is the affine Lie algebra associated to sl_2 and $\lambda = m \in \mathbf{Z}^+$, the modules $W_q(\boldsymbol{\pi})$ all have the same dimension 2^m. This is done by showing that the modules $W_q(\boldsymbol{\pi})$, under suitable conditions, have a $q = 1$ limit, which allows us to reduce to the study of the corresponding problem in the classical case carried out in [CP4]. The modules $W_q(\boldsymbol{\pi})$ have a unique irreducible quotient $V_q(\boldsymbol{\pi})$, and we show that these are all the irreducible finite-dimensional $\mathbf{U}_q(t)$-modules. In [CP1], [CP2], a similar classification was obtained by regarding q as a complex number and $\mathbf{U}_q(t)$ as an algebra over $\mathbf{C}$; in the present situation, we have to allow modules defined over finite extensions of $\mathbf{C}(q)$.

We are then able to realize the modules $W_q(m)$ as the space of invariants of an action of the Hecke algebra $\mathcal{H}_m$ on the tensor product $(V \otimes \mathbf{C}(q)[t, t^{-1}])^{\otimes m}$, where V is a two-dimensional vector space over $\mathbf{C}(q)$. Again, this is done by reducing to the case of $q = 1$.

In the last section, we indicate how to extend some of the results of this paper to the general case. We conjecture that the dimension of the modules $W_q(\pi)$ depends only on λ, and we give a formula for this dimension.

1. PRELIMINARIES AND SOME IDENTITIES

Let sl_2 be the complex Lie algebra with basis $\{x^+, x^-, h\}$ satisfying

$$[x^+, x^-] = h, \quad [h, x^\pm] = \pm 2x^\pm.$$

Let $\mathfrak{h} = \mathbf{C}h$ be the Cartan subalgebra of sl_2, let $\alpha \in \mathfrak{h}^*$ the positive root of sl_2, given by $\alpha(h) = 2$, and set $\omega = \alpha/2$. Let $s : \mathfrak{h}^* \to \mathfrak{h}^*$ be the simple reflection given by $s(\alpha) = -\alpha$.

The extended loop algebra of sl_2 is the Lie algebra

$$L^e(\}) = sl_2 \otimes \mathbf{C}[t, t^{-1}] \oplus \mathbf{C}d,$$

with commutator given by

$$[d, x \otimes t^r] = rx \otimes t^r, \quad [x \otimes t^r, y \otimes t^s] = [x, y] \otimes t^{r+s}$$

for $x, y \in sl_2$, $r, s \in \mathbf{Z}$. The loop algebra $L(\})$ is the subalgebra $sl_2 \otimes \mathbf{C}[t, t^{-1}]$ of $L^e(\})$. Let $\mathfrak{h}^e = \mathfrak{h} \oplus \mathbf{C}d$. Define $\delta \in (\mathfrak{h}^e)^*$ by

$$\delta(\mathfrak{h}) = 0, \quad \delta(d) = 1.$$

Extend $\lambda \in \mathfrak{h}^*$ to an element of $(\mathfrak{h}^e)^*$ by setting $\lambda(d) = 0$. Set $P^e = \mathbf{Z}\omega \oplus \mathbf{Z}\delta$, and define P^e_+ in the obvious way. We regard s as acting on $(\mathfrak{h}^e)^*$ by setting $s(\delta) = \delta$.

For any $x \in sl_2$, $m \in \mathbf{Z}$, we denote by x_m the element $x \otimes t^m \in L^e(\})$. Set

$$e_1^\pm = x^\pm \otimes 1, \quad e_0^\pm = x^\mp \otimes t^{\pm 1}.$$

Then, the elements $e_i^\pm$, $i = 0, 1$, and d generate $L^e(\})$.

For any Lie algebra $\mathfrak{a}$, the universal enveloping algebra of $\mathfrak{a}$ is denoted by $\mathbf{U}(\mathfrak{a})$. We set

$$\mathbf{U}(L^e(\})) = \mathbf{U}^e, \quad \mathbf{U}(L(\})) = \mathbf{U}, \quad \mathbf{U}(\}) = \mathbf{U}^{\mathrm{fin}}.$$

Let $\mathbf{U}(<)$ (resp. $\mathbf{U}(>)$) be the subalgebra of $\mathbf{U}$ generated by the x_m^- (resp. x_m^+) for $m \in \mathbf{Z}$. Set $\mathbf{U}^{\mathrm{fin}}(<) = \mathbf{U}(<) \cap \mathbf{U}^{\mathrm{fin}}$ and define $\mathbf{U}^{\mathrm{fin}}(>)$ similarly. Finally, let $\mathbf{U}(0)$ be the subalgebra of $\mathbf{U}$ generated by the h_m for all $m \neq 0$. We have

$$\mathbf{U}^{\mathrm{fin}} = \mathbf{U}^{\mathrm{fin}}(<)\mathbf{U}(\mathfrak{h})\mathbf{U}^{\mathrm{fin}}(>),$$

$$\mathbf{U}^e = \mathbf{U}(<)\mathbf{U}(0)\mathbf{U}(\mathfrak{h}^e)\mathbf{U}(>).$$

Now let q be an indeterminate, let $\mathbf{K} = \mathbf{C}(q)$ be the field of rational functions in q with complex coefficients, and let $\mathbf{A} = \mathbf{C}[q, q^{-1}]$ be the subring of Laurent polynomials. For $r, m \in \mathbf{N}$, $m \geq r$, define

$$[m] = \frac{q^m - q^{-m}}{q - q^{-1}}, \quad [m]! = [m][m-1]\ldots[2][1], \quad \begin{bmatrix} m \\ r \end{bmatrix} = \frac{[m]!}{[r]![m-r]!}.$$

Then, $\begin{bmatrix} m \\ r \end{bmatrix} \in \mathbf{A}$ for all $m \geq r \geq 0$.

Let $\mathbf{U}_q^e$ be the quantized enveloping algebra over $\mathbf{K}$ associated to $L^e(\})$. Thus, $\mathbf{U}_q^e$ is the quotient of the quantum affine algebra obtained by setting the central generator equal to 1. It follows from [Dr], [B], [J] that $\mathbf{U}_q^e$ is the algebra with generators $\mathbf{x}_r^\pm$ $(r \in \mathbf{Z})$, $K^{\pm 1}$, $\mathbf{h}_r$ $(r \in \mathbf{Z}\backslash\{0\})$, $D^{\pm 1}$, and the following defining relations:

$$KK^{-1} = K^{-1}K = 1, \quad DD^{-1} = D^{-1}D = 1, \quad DK = KD,$$

$$K\mathbf{h}_r = \mathbf{h}_r K, \quad K\mathbf{x}_r^\pm K^{-1} = q^{\pm 2}\mathbf{x}_r^\pm,$$

$$D\mathbf{x}_r^\pm D^{-1} = q^r \mathbf{x}_r^\pm, \quad D\mathbf{h}_r D^{-1} = q^r \mathbf{h}_r,$$

$$[\mathbf{h}_r, \mathbf{h}_s] = 0, \quad [\mathbf{h}_r, \mathbf{x}_s^\pm] = \pm\frac{1}{r}[2r]\mathbf{x}_{r+s}^\pm,$$

$$\mathbf{x}_{r+1}^\pm \mathbf{x}_s^\pm - q^{\pm 2}\mathbf{x}_s^\pm \mathbf{x}_{r+1}^\pm = q^{\pm 2}\mathbf{x}_r^\pm \mathbf{x}_{s+1}^\pm - \mathbf{x}_{s+1}^\pm \mathbf{x}_r^\pm,$$

$$[\mathbf{x}_r^+, \mathbf{x}_s^-] = \frac{\psi_{r+s}^+ - \psi_{r+s}^-}{q - q^{-1}},$$

where the $\psi_r^\pm$ are determined by equating powers of u in the formal power series

$$\sum_{r=0}^\infty \psi_{\pm r}^\pm u^{\pm r} = K^{\pm 1}\exp\left(\pm(q - q^{-1})\sum_{s=1}^\infty \mathbf{h}_{\pm s}u^{\pm s}\right).$$

Define the q-divided powers

$$(\mathbf{x}_k^\pm)^{(r)} = \frac{(\mathbf{x}_k^\pm)^r}{[r]!},$$

for all $k \in \mathbf{Z}$, $r \geq 0$.

Define

$$\Lambda^\pm(u) = \sum_{m=0}^\infty \Lambda_{\pm m}u^m = \exp\left(-\sum_{k=1}^\infty \frac{\mathbf{h}_{\pm k}}{[k]}u^k\right).$$

The subalgebras $\mathbf{U}_q$, $\mathbf{U}_q^{\text{fin}}$, $\mathbf{U}_q(<)$, $\mathbf{U}(0)$ etc., are defined in the obvious way. Let $\mathbf{U}_q^e(0)$ be the subalgebra of $\mathbf{U}_q^e$ generated by $\mathbf{U}(0)$, $K^{\pm 1}$ and $D^{\pm 1}$. The following result is a simple corollary of the PBW theorem for $\mathbf{U}_q^e$, [B].

Lemma 1.1. $\mathbf{U}_q^e = \mathbf{U}_q(<)\mathbf{U}_q^e(0)\mathbf{U}_q(>)$. $\qquad\qquad\qquad\qquad\qquad\qquad\Box$

For any invertible element $x \in \mathbf{U}_q^e$ and any $r \in \mathbf{Z}$, define

$$\begin{bmatrix} x \\ r \end{bmatrix} = \frac{xq^r - x^{-1}q^{-r}}{q - q^{-1}}.$$

Let $\mathbf{U}_\mathbf{A}^e$ be the $\mathbf{A}$-subalgebra of $\mathbf{U}_q^e$ generated by the $K^{\pm 1}$, $(\mathbf{x}_k^\pm)^{(r)}$ $(k \in \mathbf{Z}$, $r \geq 0)$, $D^{\pm 1}$ and $\begin{bmatrix} D \\ r \end{bmatrix}$ $(r \in \mathbf{Z})$. Then, [L1], [BCP],

$$\mathbf{U}_q^e \cong \mathbf{U}_\mathbf{A}^e \otimes_\mathbf{A} \mathbf{K}.$$

Define $\mathbf{U_A}(<)$, $\mathbf{U_A}(0)$ and $\mathbf{U_A}(>)$ in the obvious way. Let $\mathbf{U_A^e}(0)$ be the $\mathbf{A}$-subalgebra of $\mathbf{U_A}$ generated by $\mathbf{U_A}(0)$ and the elements $K^{\pm 1}$, $D^{\pm 1}$, $\begin{bmatrix} K \\ r \end{bmatrix}$ and $\begin{bmatrix} D \\ r \end{bmatrix}$ $(r \in \mathbf{Z})$. The following is proved as in Proposition 2.7 in [BCP].

Proposition 1.1. $\mathbf{U_A^e} = \mathbf{U_A}(<)\mathbf{U_A}(0)\mathbf{U_A^e}(\mathfrak{h})\mathbf{U_A}(>)$. $\square$

The next lemma is easily checked.

Lemma 1.2.

(i) *There is a unique $\mathbf{C}$-linear anti-automorphism Ψ of $\mathbf{U_q^e}$ such that $\Psi(q) = q^{-1}$ and*

$$\Psi(K) = K, \quad \Psi(D) = D,$$
$$\Psi(x_r^{\pm}) = x_r^{\pm}, \quad \Psi(h_r) = -h_r,$$

for all $r \in \mathbf{Z}$.

(ii) *There is a unique $\mathbf{K}$-algebra automorphism Φ of $\mathbf{U_q^e}$ such that*

$$\Phi(\mathbf{x}_r^{\pm}) = \mathbf{x}_{-r}^{\pm}, \quad \Phi(\Lambda^{\pm}(u)) = \Lambda^{\mp}(u).$$

(iii) *For $0 \neq a \in \mathbf{K}$, there exists a $\mathbf{K}$-algebra automorphism τ_a of $\mathbf{U_q^e}$ such that*

$$\tau_a(\mathbf{x}_r^{\pm}) = a^r \mathbf{x}_r^{\pm}, \quad \tau_a(\mathbf{h}_r) = a^r \mathbf{h}_r, \quad \tau_a(K) = K, \quad \tau_a(D) = D,$$

for $r \in \mathbf{Z}$. Moreover,

$$\tau_a(\Lambda_r) = a^r \Lambda_r.$$

2. The Modules $W_q(m)$

In this section, we recall the definition and elementary properties of the modules $W_q(\lambda)$ from [CP4], and state the main theorem of this paper.

Definition 2.1. A $\mathbf{U_q^e}$-module V_q is said to be of *type 1* if

$$V_q = \bigoplus_{\lambda \in P^e} (V_q)_{\lambda},$$

where the weight space

$$(V_q)_{\lambda} = \{ v \in V_q : K.v = q^{\lambda(h)}v, \quad D.v = q^{\lambda(d)}v \}.$$

A $\mathbf{U_q^e}$-module of type 1 is said to be *integrable* if the elements $\mathbf{x}_k^{\pm}$ act locally nilpotently on V_q for all $k \in \mathbf{Z}$. The analogous definitions for $\mathbf{U^e}$, $\mathbf{U^{fin}}$ and $\mathbf{U_q^{fin}}$ are clear.

We shall only be interested in modules of type 1 in this paper. It is well known [L1] that, if $m \geq 0$, there is a unique irreducible $\mathbf{U_q^{fin}}$-module $V_q^{fin}(m)$, of dimension $m + 1$, generated by a vector v such that

$$K.v = q^m v, \quad x_0^+.v = 0, \quad (x_0^-)^{m+1}.v = 0.$$

Recall [L2] that, if V_q is any integrable sl_2-module, then

$$\dim_{\mathbf{K}}(V_q)_n = \dim_{\mathbf{K}}(V_q)_{-n},$$

for all $n \in \mathbf{Z}$. Let $V(m)$ denote the $(m+1)$-dimensional irreducible representation of sl_2.

Define the following generating series in an indeterminate u with coefficients in $\mathbf{U}_q$:

$$\tilde{\mathbf{X}}^-(u) = \sum_{m=-\infty}^{\infty} \mathbf{x}_m^- u^{m+1}, \qquad \mathbf{X}^-(u) = \sum_{m=1}^{\infty} \mathbf{x}_m^- u^m,$$

$$\mathbf{X}^+(u) = \sum_{m=0}^{\infty} \mathbf{x}_m^+ u^m, \qquad \mathbf{X}_0^-(u) = \sum_{m=0}^{\infty} \mathbf{x}_m^- u^{m+1},$$

$$\tilde{\mathbf{H}}(u) = \sum_{m=-\infty}^{\infty} \mathbf{h}_m u^{m+1}, \qquad \mathbf{\Lambda}^{\pm}(u) = \sum_{m=0}^{\infty} \mathbf{\Lambda}_{\pm m} u^m = \exp\left(-\sum_{k=1}^{\infty} \frac{\mathbf{h}_{\pm k}}{[k]} u^k\right).$$

Given a power series f in u, we let f_s denote the coefficient of u^s in f.

For any integer $m \geq 0$, let $I_q^e(m)$ be the left ideal in $\mathbf{U}_q^e$ generated by the elements

$$\mathbf{x}_k^+ \ (k \in \mathbf{Z}), \ K - q^m, \ D - 1,$$

$$\mathbf{\Lambda}_r \ (|r| > m), \ \mathbf{\Lambda}_m \mathbf{\Lambda}_{-r} - \mathbf{\Lambda}_{m-r} \ (1 \leq r \leq m),$$

$$\left(\tilde{\mathbf{X}}_i^-(u)\mathbf{\Lambda}^+(u)\right)_r \mathbf{U}(0) \ (r \in \mathbf{Z}), \ \left(\mathbf{X}_0^-(u)^r \mathbf{\Lambda}^+(u)\right)_s \mathbf{U}(0) \ (r \geq 1, \ |s| > m).$$

The ideal $I_q(m)$ in $\mathbf{U}_q$ is defined in the obvious way (by omitting D from the definition).

Set

$$W_q(m) = \mathbf{U}_q^e / I_q^e(m) \cong \mathbf{U}_q / I_q(m).$$

Clearly, $W_q(m)$ is a left $\mathbf{U}_q^e$-module and a right $\mathbf{U}_q(0)$-module. Further, the left and right actions of $\mathbf{U}_q(0)$ on $W_q(m)$ commute. Let w_m denote the image of 1 in $W_q(m)$. If $I_q(m,0)$ (resp. $I_{\mathbf{A}}(m,0)$) is the left ideal in $\mathbf{U}_q(0)$ (resp. $\mathbf{U}_{\mathbf{A}}(0)$) generated by the elements $\mathbf{\Lambda}_m$ $(|m| > \lambda(h))$ and $\mathbf{\Lambda}_{\lambda(h)}\mathbf{\Lambda}_{-m} - \mathbf{\Lambda}_{\lambda(h)-m}$ $(1 \leq m \leq \lambda(h))$, then

$$\mathbf{U}_q(0).w_m \cong \mathbf{U}_q(0)/I_q(m,0) \quad (\text{resp.} \ \mathbf{U}_{\mathbf{A}}(0).w_m \cong \mathbf{U}_{\mathbf{A}}(0)/I_{\mathbf{A}}(m,0))$$

as $\mathbf{U}_q(0)$-modules (resp. as $\mathbf{U}_{\mathbf{A}}(0)$-modules). The $\mathbf{U}^e$-modules $W(m)$ are defined in the analogous way.

Let $\mathbf{U}_q(+)$ be the subalgebra of $\mathbf{U}_q$ generated by the $\mathbf{x}_k\pm$ for $k \geq 0$. The subalgebras $\mathbf{U}_{\mathbf{A}}(+)$ and $\mathbf{U}(+)$ of $\mathbf{U}_{\mathbf{A}}$ and $\mathbf{U}$, respectively, are defined in the obvious way. The following proposition was proved in [CP4].

Proposition 2.1. *Let $m \geq 1$.*

(i) *We have*

$$\mathbf{U}_q(0)/I_q(m,0) \cong \mathbf{K}[\mathbf{\Lambda}_1, \mathbf{\Lambda}_2, \cdots, \mathbf{\Lambda}_m, \mathbf{\Lambda}_m^{-1}],$$

$$\mathbf{U}_{\mathbf{A}}(m,0)/I_{\mathbf{A}}(m,0) \cong \mathbf{A}[\mathbf{\Lambda}_1, \mathbf{\Lambda}_2, \cdots, \mathbf{\Lambda}_m, \mathbf{\Lambda}_m^{-1}],$$

as algebras over **K** and **A**, respectively.

(ii) *The $\mathbf{U}^e$-module $W_q(m)$ is integrable for all $m \geq 0$.*

(iii) $W_q(m) = \mathbf{U}_q(+).w_m$. *In fact, $W_q(m)$ is spannned over* **K** *by the elements*

$$(x_0^-)^{(r_0)}(x_1^-)^{(r_1)} \cdots (x_{m-1}^-)^{(r_{m-1})}\mathbf{U}_q(0).w_m,$$

where $r_j \geq 0$, $\sum_j r_j \leq m$.

Analogous results hold for the **U**-*modules $W(m)$.* $\qquad\square$

Let $\mathcal{P}_m$ be the Laurent polynomial ring in m variables with complex coefficients. The symmetric group Σ_m acts on it in the obvious way; let $\mathcal{P}_m^{\Sigma_m}$ be the ring of symmetric Laurent polynomials. In view of Proposition 2.1, we see that

$$\mathbf{U}_q(0)/I_q(m,0) \cong \mathbf{K}\mathcal{P}_m^{\Sigma_m}, \quad \mathbf{U_A}/I_\mathbf{A}(m,0) \cong \mathbf{A}\mathcal{P}_m^{\Sigma_m},$$

where $\mathbf{K}\mathcal{P}_m^{\Sigma_m}$ denotes $\mathcal{P}_m^{\Sigma_m} \otimes \mathbf{K}$, etc.

Let V be the two-dimensional irreducible sl_2-module with basis v_0, v_1 such that

$$x^+.v_0 = 0, \quad h.v_0 = v_0, \quad x^-.v_0 = v_1,$$
$$x^+.v_1 = v_0, \quad h.v_1 = -v_1, \quad x^-.v_1 = 0.$$

Let $L(V) = V \otimes \mathbf{C}[t, t^{-1}]$ be the $L(sl_2)$-module defined in the obvious way. Let $T^m(L(V))$ be the m-fold tensor power of $L(V)$ and let $S^m(L(V))$ be its symmetric part. Then, $T^m(L(V))$ is a left **U**-module and a right $\mathcal{P}_m$-module, and $S^m(L(V))$ is a left **U**-module and a right $\mathcal{P}_m^{\Sigma_m}$-module. The following was proved in [CP4].

Theorem 1. *As left* **U**-*modules and right $\mathcal{P}_m^{\Sigma_m}$-modules, we have*

$$W(m) \cong S^m(L(V)).$$

In particular, $W(m)$ is a free $\mathcal{P}_m^{\Sigma_m}$-module of rank 2^m. $\qquad\square$

Our goal in this paper is to prove an analogue of this result for the $W_q(m)$. To do this, we introduce a suitable quantum analogue of $S^m(L(V))$ by using the Hecke algebra and a certain quantum symmetrizer.

Definition 2.2. The *Hecke algebra $\mathcal{H}_m$* is the associative unital algebra over $\mathbf{C}(q)$ generated by elements T_i ($i = 1, 2, \ldots, m-1$) with the following defining relations:

$$(T_i + 1)(T_i - q^2) = 0,$$
$$T_i T_{i+1} T_i = T_{i+1} T_i T_{i+1},$$
$$T_i T_j = T_j T_i \quad \text{if } |i - j| > 1.$$

Set $L_q(V) = L(V) \otimes \mathbf{K}$. It is easily checked that the following formulas define an action of $\mathbf{U}_q^e$ on $L_q(V)$:

$$(2.1) \qquad x_k^{\pm}.(v_{\pm} \otimes t^r) = 0, \quad x_k^{\pm}.(v_{\mp} \otimes t^r) = v_{\pm} \otimes t^{k+r},$$

$$(2.2) \qquad \Psi^{+}(u).(v_{\pm} \otimes t^r) = v_{\pm} \otimes \frac{q^{\pm 1} - q^{\mp 1}tu}{1 - tu}t^r,$$

$$(2.3) \qquad \Psi^{-}(u).(v_{\pm} \otimes t^r) = v_{\pm} \otimes \frac{q^{\mp 1} - q^{\pm 1}t^{-1}u}{1 - t^{-1}u}t^r.$$

The m-fold tensor product $T^m(L_q(V))$ is a left $\mathbf{U}_q^e$-module (the action being given by the comultiplication of $\mathbf{U}_q$) and a right $\mathcal{P}_m$-module (in the obvious way). Now, as a vector space over $\mathbf{K}$,

$$L_q(V)^{\otimes m} \cong V^{\otimes m} \otimes_{\mathbf{K}} \mathbf{K}[t_1^{\pm 1}, \ldots, t_m^{\pm 1}],$$

and Σ_m acts naturally (on the right) on both $V^{\otimes m}$ and $\mathbf{K}[t_1^{\pm 1}, \ldots, t_m^{\pm 1}]$ by permuting the variables. If $\mathbf{v} \in V^{\otimes m}$ and $f \in \mathbf{K}[t_1^{\pm 1}, \ldots, t_m^{\pm 1}]$, denote the action of $\sigma \in \Sigma_m$ by $\mathbf{v}^{\sigma}$ and f^{σ}, respectively. Let σ_i be the transposition $(i, i+1) \in \Sigma_m$.

Proposition 2.2. *([KMS, Section 1.2]) The Hecke algebra $\mathcal{H}_m$ acts on $L_q(V)^{\otimes m}$ on the right, the action of the generators being given as follows :*

$$(v_{t_1} \otimes \cdots \otimes v_{t_m} \otimes f).T_i = \begin{cases} -q(v_{t_1} \otimes \cdots \otimes v_{t_m})^{\sigma_i} \otimes f^{\sigma_i} \\ \quad -(q^2 - 1)(v_{t_1} \otimes \cdots \otimes v_{t_m}) \otimes \frac{t_{i+1}f^{\sigma_i} - t_i f}{t_i - t_{i+1}} \\ \quad \text{if } t_i = +, \ t_{i+1} = -, \\ -v_{t_1} \otimes \cdots \otimes v_{t_m} \otimes f^{\sigma_i} \\ \quad -(q^2 - 1)(v_{t_1} \otimes \cdots \otimes v_{t_m}) \otimes \frac{t_i(f^{\sigma_i} - f)}{t_i - t_{i+1}} \\ \quad \text{if } t_i = t_{i+1}, \\ -q(v_{t_1} \otimes \cdots \otimes v_{t_m})^{\sigma_i} \otimes f^{\sigma_i} \\ \quad -(q^2 - 1)(v_{t_1} \otimes \cdots \otimes v_{t_m}) \otimes \frac{t_i(f^{\sigma_i} - f)}{t_i - t_{i+1}} \\ \quad \text{if } t_i = -, \ t_{i+1} = +. \end{cases}$$

Moreover, this action commutes with the left action of $\mathbf{U}_q^e$ on $L_q(V)$ and the right action of $\mathbf{K}\mathcal{P}_m^{\Sigma_m}$. $\qquad \square$

As is well known, the second and third relations in the definition of $\mathcal{H}_m$ imply that, if $\sigma = \sigma_{i_1} \ldots \sigma_{i_N}$ is a reduced expression for $\sigma \in \Sigma_m$, so that N is the length $\ell(\sigma)$, the element $T_{\sigma} = T_{i_1} \ldots T_{i_N} \in \mathcal{H}_m$ depends only on σ, and is independent of the choice of its reduced expression. We define the symmetrizing operator

$$\mathcal{S}^{(m)} : L_q(V)^{\otimes m} \to L_q(V)^{\otimes m}$$

by

$$\mathcal{S}^{(m)} = \frac{1}{[m]!} \sum_{\sigma \in \Sigma_m} (-q^{-2})^{\ell(\sigma)} T_{\sigma}.$$

Proposition 2.3. *As left $\mathbf{U}_q^e$-modules and right $\mathbf{K}\mathcal{P}_m^{\Sigma_m}$-modules, we have*

$$L_q(V)^{\otimes m} = im(\mathcal{S}^{(m)}) \oplus ker(\mathcal{S}^{(m)}).$$

Proof. It is clear from Proposition 2.2 that $im(\mathcal{S}^{(m)})$ and $ker(\mathcal{S}^{(m)})$ are sub-modules for both the right and left actions.

The following proof is adapted from that of Proposition 1.1 in [KMS]. For each $i = 1, \ldots, m-1$, we have a factorization

$$\mathcal{S}^{(m)} = \left(\sum_{\sigma'} (-q^{-2})^{\ell(\sigma')} T_{\sigma'} \right) (1 - q^{-2} T_i),$$

where σ' ranges over $\Sigma_m / \{1, \sigma_i\}$. From this and the first of the defining relations of $\mathcal{H}_m$, it follows that

$$\mathcal{S}^{(m)}(T_i + 1) = 0.$$

In other words, T_i acts on the right on $im(\mathcal{S}^{(m)})$ as multiplication by -1. It follows that $\mathcal{S}^{(m)}$ acts on $im(\mathcal{S}^{(m)})$ by multiplication by the scalar

$$\frac{1}{[m]!} \sum_{\sigma \in \Sigma_m} (q^{-2})^{\ell(\sigma)} = \frac{1}{[m]!} \prod_{l=1}^m \frac{1 - q^{-2l}}{1 - q^{-2}} = q^{-m(m-1)/2}.$$

Hence,

$$\mathcal{S}^{(m)}(\mathcal{S}^{(m)} - q^{-m(m-1)/2}) = 0,$$

and this implies the proposition. $\qquad\qquad\square$

As in [KMS], define an ordered basis $\{u_m\}_{m \in \mathbf{Z}}$ of $L_q(V)$ by setting

$$u_{-2r} = v_+ \otimes t^r, \qquad u_{1-2r} = v_- \otimes t^r.$$

Let $u_{r_1} \otimes_S \cdots \otimes_S u_{r_m}$ be the image of $u_{r_1} \otimes \cdots \otimes u_{r_m}$ under the projection of $L_q(V)^{\otimes m}$ onto $L_q(V)^{\otimes m}/ker(\mathcal{S}^{(m)})$. By Proposition 2.3, this can be identified with an element, which we also denote by $u_{r_1} \otimes_S \cdots \otimes_S u_{r_m}$, in $im(\mathcal{S}^{(m)})$.

Proposition 2.4. *The set $\{u_{r_1} \otimes_S \cdots \otimes_S u_{r_m} : r_1 \geq \cdots \geq r_m\}$ is a basis of the vector space $im(\mathcal{S}^{(m)})$. Further, $im(\mathcal{S}^{(m)})$ is a free $\mathbf{K}\mathcal{P}_m^{\Sigma_m}$-module on 2^m generators.*

Proof. The first statement in proved as in [KMS], Proposition 1.3. As for the second, for any $0 \leq s \leq m$, let $im(\mathcal{S}^{(m)})_s$ be the subspace spanned by $u_{r_1} \otimes_S \cdots \otimes_S u_{r_m}$, where exactly s of the r_i are even. This space is naturally isomorphic as a right $\mathbf{K}\mathcal{P}_m^{\Sigma_m}$-module to $\mathbf{K}\mathcal{P}_m^{\Sigma_s \times \Sigma_{m-s}}$. But this module is well-known to be free of rank $\binom{m}{s}$. $\qquad\qquad\square$

Let $\mathbf{w} = u_0 \otimes_S \cdots \otimes_S u_0$. Then, $\mathbf{w}$ satisfies the defining relations of $W_q(m)$, so we have a map of left $\mathbf{U}_q^e$-modules and right $\mathbf{K}\mathcal{P}_m^{\Sigma_m}$-modules $\eta_m : W_q(m) \to im(\mathcal{S}^{(m)})$ that takes w_m to $\mathbf{w}$. The main theorem of this paper is

Theorem 2. *The map η_m is an isomorphism. In particular, $W_q(m)$ is a free $\mathbf{K}\mathcal{P}_m^{\Sigma_m}$-module of rank 2^m.*

The theorem is deduced from the following two lemmas.

Lemma 2.1. *Let* $\mathfrak{m}$ *be any maximal ideal in* $\mathbf{K}\mathcal{P}_m^{\Sigma_m}$, *and let* d *be the degree of the field extension* $\mathbf{K}\mathcal{P}_m^{\Sigma_m}/\mathfrak{m}$ *of* $\mathbf{K}$. *Then,*

$$dim_{\mathbf{K}}\frac{W_q(m)}{W_q(m)\mathfrak{m}} = 2^m d.$$

Lemma 2.2. *The map* η_m *is surjective.*

We defer the proofs of these lemmas to the next section. Once we have these two lemmas, the proof of Theorem 2 is completed in exactly the same way as Theorem 1. We include it here for completeness.

Proof of Theorem 2. Let K be the kernel of η_m. Since $\mathrm{im}(\mathcal{S}^{(m)})$ is a free, hence projective, right $\mathbf{K}\mathcal{P}_m^{\Sigma_m}$-module by Proposition 2.4, it follows that

$$W_q(m) = \mathrm{im}(\mathcal{S}^{(m)}) \oplus K,$$

as right $\mathbf{K}\mathcal{P}_m^{\Sigma_m}$-modules. Let $\mathfrak{m}$ be any maximal ideal in $\mathbf{K}\mathcal{P}_m^{\Sigma_m}$. It follows from Lemma 2.1 and Proposition 2.4 that

$$K/K\mathfrak{m} = 0$$

as vector spaces over $\mathbf{K}$. Since this holds for all maximal ideals $\mathfrak{m}$, Nakayama's lemma implies that $K = 0$, proving the theorem. $\square$

3. PROOF OF LEMMAS 2.1 AND 2.2

In preparation for the proof of Lemma 2.1, we first show that the modules in question are finite-dimensional. Recall that a maximal ideal in $\mathbf{K}\mathcal{P}_m^{\Sigma_m}$ is defined by an m-tuple of points $\boldsymbol{\pi} = (\pi_1, \cdots, \pi_m)$, with $\pi_m \neq 0$, in an algebraic closure $\overline{\mathbf{K}}$ of $\mathbf{K}$, i.e., it is the kernel of the homomorphism $ev_{\boldsymbol{\pi}}$: $\mathbf{K}\mathcal{P}_m^{\Sigma_m} \to \overline{\mathbf{K}}$ that sends $\Lambda_i \to \pi_i$. Let $\mathbf{F}_{\boldsymbol{\pi}}$ be the smallest subfield of $\overline{\mathbf{K}}$ containing $\mathbf{K}$ and $\pi_1, \cdots, \pi_m$. Clearly, $\mathbf{F}_{\boldsymbol{\pi}}$ is a finite-rank $\mathbf{U}_q(0)$-module. Set

$$W_q(\boldsymbol{\pi}) = W_q(m) \otimes_{\mathbf{U}_q(0)} \mathbf{F}_{\boldsymbol{\pi}},$$

and let $w_{\boldsymbol{\pi}} = w_m \otimes 1$. The $\mathbf{U}$-modules $W(\boldsymbol{\pi})$ are defined similarly (with $\boldsymbol{\pi} \in \mathbf{C}^m$).

The following lemma is immediate from Proposition 2.1.

Lemma 3.1. *We have*

$$\mathbf{U}_q(0).w_{\boldsymbol{\pi}} = \mathbf{F}_{\boldsymbol{\pi}}w_{\boldsymbol{\pi}}.$$

Further, $W_q(\boldsymbol{\pi})$ *is spanned over* $\mathbf{F}_{\boldsymbol{\pi}}$ *by the elements*

$$(x_0^-)^{(r_0)}(x_1^-)^{(r-1)}\cdots(x_{m-1}^-)^{(r_{m-1})}$$

with $\sum_i r_i \leq m$.

In particular, $dim_{\mathbf{K}}W_q(\boldsymbol{\pi}) < \infty$. $\square$

The modules $W_q(m)$ and $W_q(\boldsymbol{\pi})$, together with their classical analogues, have the following universal properties.

Proposition 3.1. *Let* $\lambda \in P_e^+$.

   (i) *Let* V_q *be any integrable* $\mathbf{U}_q^e$-*module generated by an element* v *of* $(V_q)_m$ *satisfying* $\mathbf{U}_q(>).v = 0$. *Then,* V_q *is a quotient of* $W_q(m)$.

(ii) *Let V_q be a finite-dimensional quotient $\mathbf{U}_q$-module of $W_q(m)$ and let v be the image of w_m in V_q. Assume that $\ker(ev_{\boldsymbol{\pi}}).v = 0$ for some $\boldsymbol{\pi} = (\pi_1, \cdots, \pi_m)$, where the $\pi_i \in \overline{\mathbf{K}}$. Then, V_q is a quotient of $W_q(\boldsymbol{\pi})$.*

(iii) *Let V_q be finite-dimensional $\mathbf{U}_q$-module generated by an element $v \in (V_q)_m$ and such that $\mathbf{U}_q(>).v = 0$ and $\ker(ev_{\boldsymbol{\pi}}).v = 0$ for some $\boldsymbol{\pi}$. Then, V_q is a quotient of $W_q(\boldsymbol{\pi})$.*

Analogous statements hold in the classical case.

Proof. This proposition was proved in [CP4] in the case when $\boldsymbol{\pi} \in \mathbf{K}^m$. The proof in this case is identical, and follows immediately from the defining relations of $W_q(m)$ and $W_q(\boldsymbol{\pi})$. $\qquad\qquad\qquad\qquad\qquad\qquad$ □

One can now deduce the following theorem, which classifies the irreducible finite-dimensional representations of $\mathbf{U}_q$ over $\mathbf{K}$.

Theorem 3. Let $\boldsymbol{\pi} \in \overline{\mathbf{K}}^m$ be as above. Then, $W_q(\boldsymbol{\pi})$ has a unique irreducible quotient $\mathbf{U}_q$-module $V_q(\boldsymbol{\pi})$. Conversely, any irreducible finite-dimensional $\mathbf{U}_q$-module is isomorphic to $V_q(\boldsymbol{\pi})$ for a suitable choice of $\boldsymbol{\pi}$.

Proof. To prove that $W_q(\boldsymbol{\pi})$ has a unique irreducible quotient, it suffices to prove that it has a unique maximal $\mathbf{U}_q$-submodule. For this, it suffices to prove that, if N is any submodule, then

$$N \cap W_q(\boldsymbol{\pi})_m = \{0\}.$$

Since $W_q(\boldsymbol{\pi})_m = \mathbf{U}_q(0).w_{\boldsymbol{\pi}}$ is an irreducible $\mathbf{U}_q(0)$-module, it follows that

$$N \cap W_q(\boldsymbol{\pi})_m \neq \{0\} \implies w_{\boldsymbol{\pi}} \in N,$$

and hence that $N = W_q(\boldsymbol{\pi})$. Conversely, if V is any finite-dimensional irreducible module, one can show as in [CP1], [CP4] that there exists $0 \neq v \in V_m$ such that $\mathbf{U}_q(>).v = 0$ and that $\Lambda_r.v = 0$ if $|r| > m$. This shows that V_m must be an irreducible module for $\mathbf{K}[\Lambda_1, \cdots, \Lambda_m, \Lambda_m^{-1}]$, and the result follows. $\quad$ □

It follows from the preceding discussion that, to prove Lemma 2.1, we must show that, if $\mathbf{F}_{\boldsymbol{\pi}}$ is an extension of $\mathbf{K}$ of degree d, then

(3.1) $$\dim_{\mathbf{K}} W_q(\boldsymbol{\pi}) = 2^m d.$$

Assume from now on that we have a fixed finite extension $\mathbf{F}$ of $\mathbf{K}$ of degree d and an element $\boldsymbol{\pi} \in \mathbf{F}^m$ as above. Given $0 \neq a \in \mathbf{K}$, and $\boldsymbol{\pi} \in \mathbf{F}^m$ where $\mathbf{K} \subset \mathbf{F}$, define

$$\boldsymbol{\pi}_a = (a\pi_1, a^2\pi_2, \cdots, a^m\pi_m).$$

Given any $\mathbf{U}_q$-module M, and $0 \neq a \in \mathbf{K}$, let $\tau_a^* M$ be the $\mathbf{U}_q$-module obtained by pulling back M through the automorphism τ_a defined in Lemma 1.2. The next lemma is immediate from Proposition 3.1.

Lemma 3.2. *We have*

$$\tau_a^* W_q(m) \cong W_q(m), \quad \tau_a^* W_q(\boldsymbol{\pi}) \cong W_q(\boldsymbol{\pi}_a),$$

where the first isomorphism is one of $\mathbf{U}_q^e$-modules and the second is an isomorphism of $\mathbf{U}_q$-modules. $\qquad\qquad\qquad\qquad\qquad\qquad\qquad\qquad$ □

Let $\overline{\mathbf{A}}$ be the integral closure of $\mathbf{A}$ in $\mathbf{F}$. Fix $a \in \mathbf{A}$ such that $\boldsymbol{\pi}_a \in \overline{\mathbf{A}}^m$. By Lemma 3.2, to prove (3.1) it suffices to prove that

$$\dim_{\mathbf{K}} W_q(\boldsymbol{\pi}_a) = 2^m d.$$

Let $\mathbf{L} \supset \mathbf{K}$ be the smallest subfield of $\mathbf{F}$ such that $\boldsymbol{\pi}_a \in \mathbf{L}^m$ and let $\tilde{\mathbf{A}}$ be the integral closure of $\mathbf{A}$ in $\mathbf{L}$. Then, $\tilde{\mathbf{A}}$ is free of rank d as an $\mathbf{A}$-module and

$$\mathbf{L} \cong \tilde{\mathbf{A}} \otimes_{\mathbf{A}} \mathbf{K}.$$

In what follows we write $\boldsymbol{\pi}$ for $\boldsymbol{\pi}_a$. Set

$$W_{\mathbf{A}}(\boldsymbol{\pi}) = \mathbf{U}_{\mathbf{A}} \otimes_{\mathbf{U}_{\mathbf{A}}(0)} \tilde{\mathbf{A}} w_{\boldsymbol{\pi}}.$$

By Lemma 3.1, $W_{\mathbf{A}}(\boldsymbol{\pi})$ is finitely-generated as an $\tilde{\mathbf{A}}$-module, and hence as an $\mathbf{A}$-module. Further,

$$W_q(\boldsymbol{\pi}) \cong W_{\mathbf{A}}(\boldsymbol{\pi}) \otimes_{\mathbf{A}} \mathbf{K}$$

as vector spaces over $\mathbf{K}$. Note, however, that $W_{\mathbf{A}}(\boldsymbol{\pi})$ is not an $\mathbf{U}_{\mathbf{A}}$-module in general, since π_m^{-1} need not be in $\tilde{\mathbf{A}}$. However, $W_{\mathbf{A}}(\boldsymbol{\pi})$ is a $\mathbf{U}_{\mathbf{A}}(+)$-module and

$$W_q(\boldsymbol{\pi}) \cong W_{\mathbf{A}}(\boldsymbol{\pi}) \otimes_{\mathbf{A}} \mathbf{K},$$

as $\mathbf{U}_q(+)$-modules.

Set

$$\mathbf{U}_1(+) = \mathbf{U}_{\mathbf{A}}(+) \otimes_{\mathbf{A}} \mathbf{C}_1.$$

This is essentially the universal enveloping algebra $\mathbf{U}(+)$ of $sl_2 \otimes \mathbf{C}[t]$, and hence

$$\overline{W_q(\boldsymbol{\pi})} = W_{\mathbf{A}}(\boldsymbol{\pi}) \otimes_{\mathbf{A}} \mathbf{C}_1$$

is a module for $\mathbf{U}(+)$.

Since

$$\dim_{\mathbf{K}} W_q(\boldsymbol{\pi}) = \operatorname{rank}_{\mathbf{A}} W_{\mathbf{A}}(\boldsymbol{\pi}) = \dim_{\mathbf{C}} \overline{W_q(\boldsymbol{\pi})},$$

it suffices to prove that

$$\dim_{\mathbf{C}} \overline{W_q(\boldsymbol{\pi})} = 2^m d.$$

Define elements $\Lambda_r \in \mathbf{U}(+)$ in the same way as the elements Λ_r are defined, replacing q by 1.

Lemma 3.3. *With the above notation, there exists a filtration*

$$\overline{W_q(\boldsymbol{\pi})} = W_1 \supset W_2 \supset \cdots \supset W_d \supset W_{d+1} = 0$$

such that, for each $i = 1, \ldots, d$, W_i/W_{i+1} is generated by a non-zero vector v_i such that

$$(3.2) \qquad x_r^+ . v_i = 0, \quad (x_r^-)^{m+1} . v_i = 0 \quad (r \geq 0),$$

$$(3.3) \qquad h_0 . v_i = m v_i, \quad \Lambda_r . v_i = \lambda_{i,r} v_i \quad (r > 0),$$

where the $\lambda_{i,r} \in \mathbf{C}$ and $\lambda_{i,r} = 0$ for $r > m$.

Proof. Let $\overline{W_q(\pi)}_n$ be the eigenspace of h_0 acting on $\overline{W_q(\pi)}$ with eigenvalue $n \in \mathbf{Z}$. Of course,

$$\overline{W_q(\pi)} = \bigoplus_{n=-m}^{m} \overline{W_q(\pi)}_n.$$

We can choose a basis $w_1, w_2, \ldots, w_l$, say, of $\overline{W_q(\pi)}_m$ such that the action of Λ_i, for $i = 1, \ldots, m$, is in upper triangular form. Let W_i be the $\mathbf{U}(+)$-submodule of $\overline{W_q(m)}$ generated by $\{w_i, w_{i+1}, \ldots, w_l\}$. This gives a filtration with the stated properties. To see that $l = d$, note that $W_{\mathbf{A}}(\pi)_m = \tilde{\mathbf{A}}w_m$ is a free $\mathbf{A}$-module of rank d, hence

$$\overline{W_q(\pi)}_m = W_{\mathbf{A}}(\pi)_m \otimes_{\mathbf{A}} \mathbf{C}_1$$

is a vector space of dimension d. $\square$

Lemma 3.4. *Let* $\pi = 1 + \sum_{r=1}^{n} \lambda_r u^r \in \mathbf{C}[u]$ *be a polynomial of degree n, and let $m \geq n$. Let $W_+(\pi, m)$ be the quotient of $\mathbf{U}(+)$ by the left ideal generated by the elements*

$$h - m, \quad \Lambda_r - \lambda_r, \quad x_r^+, \quad (x_r^-)^{m+1},$$

for all $r \geq 0$. Then,

$$dim_{\mathbf{C}} W_+(\pi, m) \leq 2^m.$$

Proof. This is exactly the same as the proof given in [CP5, Sections 3 and 6] that $dim_{\mathbf{C}} W(\pi) \leq 2^{\deg(\pi)}$. We note that the arguments used there only make use of elements of the subalgebra $\mathbf{U}(+)$ of $\mathbf{U}$. $\square$

It follows immediately from this lemma that

$$dim_{\mathbf{C}} \overline{W_q(\pi)} \leq 2^m d.$$

Indeed, each W_i/W_{i+1} in Lemma 3.3 is clearly a quotient of some $W_+(\pi, m)$ satisfying the conditions of Lemma 3.4, and so has dimension $\leq 2^m$.

We have now proved that

$$dim_{\mathbf{K}} W_q(\pi) \leq 2^m d.$$

To prove the reverse inequality, let $\tilde{\mathbf{F}}$ be the splitting field of the polynomial $1 + \sum_{i=1}^{m} \pi_i u^i$ over $\mathbf{F}$, say

$$1 + \sum_{i=1}^{m} \pi_i u^i = \prod_{i=1}^{m} (1 - a_i u),$$

with $a_1, \ldots, a_m \in \tilde{\mathbf{F}}$. Let $V_{\mathbf{F}}(a_i)$ be a two-dimensional vector space over $\tilde{\mathbf{F}}$ with basis $\{v_+, v_-\}$, define an action of $\mathbf{U}_q$ on it by setting $t = a_i$ in the formulas in (2.1), (2.2) and (2.3), and set

$$\tilde{W} = \bigotimes_{i=1}^{m} V_{\tilde{\mathbf{F}}}(a_i).$$

Clearly,

$$dim_{\mathbf{K}} \tilde{W} = 2^m d\tilde{d},$$

where $\tilde{d}$ is the degree of $\tilde{\mathbf{F}}$ over $\mathbf{F}$. If $\{f_1, \ldots, f_{\tilde{d}}\}$ is a basis of $\tilde{\mathbf{F}}$ over $\mathbf{F}$, and if $\tilde{w} = v_+^{\otimes m}$, then

$$\tilde{W} = \bigoplus_{j=1}^{\tilde{d}} \tilde{W}_j,$$

where $\tilde{W}_j$ is the $\mathbf{U}_q$-submodule of $\tilde{W}$ generated by $f_j\tilde{w}$ (see [CP3, Proof of 2.5]). Moreover, the vectors $f_j\tilde{w}$ satisfy the defining relations of $W_q(\boldsymbol{\pi})$, and so are quotients of $W_q(\boldsymbol{\pi})$. It follows that

$$\dim_{\mathbf{K}} W_q(\boldsymbol{\pi}) \geq 2^m d.$$

The proof of Lemma 2.1 is now complete.   □

Turning to Lemma 2.2, set

$$L_{\mathbf{A}}(V) = V \otimes \mathbf{A}[t, t^{-1}].$$

Clearly, $L_{\mathbf{A}}(V)$ is a $\mathbf{U}_{\mathbf{A}}$-module. The map $\mathcal{S}^{(m)}$ takes $L_{\mathbf{A}}(V)^{\otimes m}$ into itself; set

$$\operatorname{im}(\mathcal{S}^{(m)}) = S_q(m), \quad S_{\mathbf{A}}(m) = S_q(m) \cap L_{\mathbf{A}}(V)^{\otimes m}.$$

We have

(3.4) $$S_{\mathbf{A}}^{(m)} \otimes_{\mathbf{A}} \mathbf{K} \cong S_q^{(m)}, \quad S_{\mathbf{A}}^{(m)} \otimes_{\mathbf{A}} \mathbf{C}_1 \cong S^m(L(V)).$$

The first isomorphism above is clear; the second requires the basis constructed in Proposition 2.4. The proof of Proposition 2.3 shows that

$$L_{\mathbf{A}}(V)^{\otimes m} = S_{\mathbf{A}}(m) \oplus (\ker(\mathcal{S}_q^{(m)}) \cap L_{\mathbf{A}}(V)^{\otimes m}).$$

Given $\boldsymbol{\pi} \in \mathbf{F}^m$ such that $\pi_i \in \overline{\mathbf{A}}$, set

$$S_q(\boldsymbol{\pi}) = S_q(m) \otimes_{\mathbf{U}_q(0)} \mathbf{F}, \quad S_{\mathbf{A}}(\boldsymbol{\pi}) = S_{\mathbf{A}} \otimes_{\mathbf{U}_{\mathbf{A}}(0)} \tilde{\mathbf{A}}.$$

Then, $S_q(\boldsymbol{\pi})$ (resp. $S_{\mathbf{A}}(\boldsymbol{\pi})$) is a $\mathbf{U}_q$-module (resp. $\mathbf{U}_{\mathbf{A}}(+)$-module) and

(3.5) $$S_q(\boldsymbol{\pi}) \cong S_{\mathbf{A}}(\boldsymbol{\pi}) \otimes_{\mathbf{A}} \mathbf{K}$$

as $\mathbf{U}_q(+)$-modules. Further, the map $\eta_m : W_q(m) \to S_q(m)$ induces a map $\eta_{\boldsymbol{\pi}} : W_q(\boldsymbol{\pi}) \to S_q(\boldsymbol{\pi})$ that takes $W_{\mathbf{A}}(\boldsymbol{\pi})$ into $S_{\mathbf{A}}(\boldsymbol{\pi})$.

Set $\overline{\mathbf{F}} = \mathbf{F} \otimes_{\mathbf{A}} \mathbf{C}_1$. Let $\overline{\boldsymbol{\pi}} : \mathbf{C}[\Lambda_1, \cdots, \Lambda_m] \to \overline{\mathbf{F}}$ be the homomorphism obtained by sending Λ_i to $\pi_i \otimes 1$ and set

$$S(\overline{\boldsymbol{\pi}}) = S^m(L(V)) \otimes_{\mathbf{U}(0)} \overline{\mathbf{F}}.$$

Now, in [CP4] we proved that $S^m(L(V))$ is a free $\mathbf{C}[\Lambda_1, \cdots, \Lambda_m]$-module of rank 2^m, hence $S(\overline{\boldsymbol{\pi}})$ has dimension $2^m d$. Further, [CP4],

$$W(\overline{\boldsymbol{\pi}}) \cong S(\overline{\boldsymbol{\pi}}) = \mathbf{U}(+).v_+^{\otimes m}.$$

This shows that the induced map $\overline{\eta_{\boldsymbol{\pi}}} : \overline{W_q(\boldsymbol{\pi})} \to \overline{S_q(\boldsymbol{\pi})}$ is surjective and hence, using Lemma 2.1, that it is an isomorphism.

Let $K_q(\boldsymbol{\pi})$ be the kernel of $\eta_{\boldsymbol{\pi}}$ and let $K_{\mathbf{A}}(\boldsymbol{\pi}) = K_q(\boldsymbol{\pi}) \cap W_q(\boldsymbol{\pi})$. Then, $K_{\mathbf{A}}(\boldsymbol{\pi})$ is free $\mathbf{A}$-module and

$$\dim_{\mathbf{K}} K_q(\boldsymbol{\pi}) = \operatorname{rank}_{\mathbf{A}} K_{\mathbf{A}}(\boldsymbol{\pi}).$$

The previous argument shows that

$$\overline{K_q(\pi)} = K_\mathbf{A}(\pi) \otimes_\mathbf{A} \mathbf{C}_1$$

is zero. Hence, $K_q(\pi) = 0$ and the map η_π is an isomorphism for all $\pi \in \overline{A}^m$. But now, by twisting with an automorphism τ_a for $0 \neq a \in \mathbf{K}$, we have a commutative diagram

$$
\begin{array}{ccc}
W_q(\pi_a) & \longrightarrow & S_q(\pi_a) \\
\downarrow & & \downarrow \\
W_q(\pi) & \longrightarrow & S_q(\pi)
\end{array}
$$

for *any* $\pi \in \mathbf{F}^m$, in which the vertical maps are isomorphisms of $\mathbf{U}_q(+)$-modules. If a is such that $\pi_a \in \overline{\mathbf{A}}^m$, the top horizontal map is also an isomorphism, hence so is the bottom horizontal map. Thus, $W_q(\pi) \to S_q(\pi)$ is an isomorphism for *all* $\pi \in \mathbf{F}^m$. It follows from Nakayama's lemma that $\eta_m : W_q(m) \to S_q^{(m)}$ is surjective and the proof of Lemma 2.2 is complete. $\square$

4. THE GENERAL CASE: A CONJECTURE

In this section, we indicate to what extent the results of this paper can be generalized to the higher rank cases, and then state a conjecture in the general case.

Thus, let $\}$ be a finite-dimensional simple Lie algebra of rank n of type A, D or E and let $\hat{\}}$ be the corresponding untwisted affine Lie algebra. Given any dominant integral weight λ for $\mathfrak{g}$, one can define an integrable $\mathbf{U}_q(\hat{\}})$-module $W_q(\lambda)$ on which the centre acts trivially, [CP4]. These modules have a family of finite-dimensional quotients $W_q(\pi)$, where $\pi = (\pi^1, \cdots, \pi^n)$ and the $\pi^i \in \overline{\mathbf{K}}^{\lambda(i)}$. The module $W_q(\pi)$ has a unique irreducible quotient $V_q(\pi)$ and one can prove the analogue of Theorem 3. (The proofs of these statements are the same as in the sl_2 case.)

We make the following

Conjecture. For any π as above,

$$\dim_\mathbf{K} W_q(\pi) = m_\lambda,$$

where $m_\lambda \in \mathbf{N}$ is given by

$$m_\lambda = \prod_{i=1}^n (m_i)^{\lambda_i}, \quad m_i = \dim_\mathbf{K} W_q(i),$$

and $W_q(i)$ is the finite-dimensional module associated to the n-tuple $(\pi^1, \cdots, \pi^n)$ with $\pi^j = \{0\}$ if $j \neq i$ and $\pi^i = \{1\}$. $\square$

In the case of sl_2, the conjecture is established in this paper. It follows from the results in [C] that $W_q(i)$ is in fact an irreducible $\mathbf{U}_q(\hat{\}})$-module and hence [CP2] the values of the m_i are actually known. The results of [C] also establish the conjecture for all π associated to the fundamental weight λ_i of $\}$, for all $i = 1, \cdots, n$.

Using the results in [VV], one can show that

$$\dim_{\mathbf{K}} W_q(\boldsymbol{\pi}) \geq m_\lambda.$$

It suffices to prove the reverse inequality in the case when the $\pi^i \in \overline{\mathbf{A}}^{\lambda(i)}$ for all i. One can prove exactly as in this paper that the $\mathbf{U}_q(+)$-modules $W_q(\boldsymbol{\pi})$ admit an $\mathbf{U}_{\mathbf{A}}(+)$-lattice $W_{\mathbf{A}}(\boldsymbol{\pi})$, so that

$$\dim_{\mathbf{K}} W_q(\boldsymbol{\pi}) = \operatorname{rank}_{\mathbf{A}} W_{\mathbf{A}}(\boldsymbol{\pi}) = \dim_{\mathbf{C}} \overline{W_q(\boldsymbol{\pi})}.$$

Thus, it suffices to prove the conjecture in the classical case, i.e.,

$$\dim_{\mathbf{C}} W(\boldsymbol{\pi}) = m_\lambda,$$

where m_λ is defined above.

REFERENCES

[B]   J. Beck, Braid group action and quantum affine algebras, *Commun. Math. Phys.* **165** (1994), 555-568.

[BCP]   J. Beck, V. Chari and A. Pressley, An algebraic characterization of the affine canonical basis, *Duke Math. J.* **99** (1999), 455-487.

[C]   V. Chari, On the Fermionic formula and a conjecture of Kirillov and Reshetikhin, preprint, qa/0006090.

[CP1]   V. Chari, and A. Pressley, Quantum affine algebras, *Commun. Math. Phys.* **142** (1991), 261-283.

[CP2]   V. Chari, and A. Pressley, Quantum affine algebras and their representations, *Canadian Math. Soc. Conf. Proc.* **16** (1995), 59-78.

[CP3]   V. Chari, and A. Pressley, Quantum affine algebras and integrable quantum systems, *NATO Advanced Science Institute on Quantum Fields and Quantum Space-Time, Series B, Physics, Vol. 364*, ed. G. t'Hooft et al., 1997, Plenum Press, pp. 245-264.

[CP4]   V. Chari and A. Pressley, Weyl modules for classical and quantum affine algebras, preprint, qa/0004174.

[Dr]   V.G. Drinfeld, A new realization of Yangians and quantum affine algebras, *Soviet Math. Dokl.* **36** (1988), 212-216.

[G]   H. Garland, The arithmetic theory of loop algebras, *J. Algebra* **53** (1978), 480-551.

[J]   N. Jing, On Drinfeld realization of quantum affine algebras. The Monster and Lie algebras (Columbus, OH, 1996), pp. 195-206, *Ohio State Univ. Math. Res. Inst. Publ.* **7**, de Gruyter, Berlin, 1998.

[FM]   E. Frenkel and E. Mukhin, Combinatorics of q-characters of finite-dimensional representations of quantum affine algebras, preprint, math.qa/9911112.

[K]   M. Kashiwara, Crystal bases of the modified quantized enveloping algebra, *Duke Math. J.* **73** (1994), 383-413.

[KMS]   M. Kashiwara, T. Miwa and E. Stern, Decomposition of q-deformed Fock spaces, *Selecta Mathematica* **1** (1995), 787-805.

[L1]   G. Lusztig, Quantum deformations of certain simple modules over enveloping algebras, *Adv. Math.* **70** (1988), 237-249.

[L2]   G. Lusztig, *Introduction to quantum groups*, Progress in Mathematics **110**, Birkhäuser, Boston, 1993.

[VV]   M. Varagnolo and E. Vasserot, Standard modules for quantum affine algebras, preprint qa/0006084.

VYJAYANTHI CHARI, UNIVERSITY OF CALIFORNIA, RIVERSIDE

ANDREW PRESSLEY, KINGS COLLEGE, LONDON

Notes on Balanced Categories and Hopf Algebras

Bernhard Drabant

Introduction

In the article we investigate balanced categories and balanced Hopf algebras. The close relation of balanced categories, balanced Hopf algebras and ribbon braids allows the use of diagrammatic morphisms in algebraic calculations for balanced Hopf algebras and categories and to discuss algebraic applications in knot theory.

In the first part we consider balanced categories and balanced Hopf algebras as well as ribbon and sovereign categories and Hopf algebras. Sovereign categories have been introduced in [9], sovereign Hopf algebras have been studied in [2]. From the reconstruction theoretical point of view they are the natural objects in relation with sovereign categories [2].

Strong sovereignity will be introduced and it will be shown that a Hopf algebra is strong sovereign if and only if it is a ribbon Hopf algebra. This result immediately implies the redundancy of the relations $S(\theta) = \theta$ and $\theta^2 = u \cdot S(u)$ for the twist element of a ribbon Hopf algebra (H, R, θ).

For every quasitriangular bialgebra a corresponding balanced bialgebra will be constructed by which we easily find an example of a balanced category related to a category of modules. Another example of a balanced category is the balanced construction out of a given monoidal category. Braided balanced categories with duality and braided sovereign categories are equivalent notations [6, 33, 25]. We provide an elementary proof of this fact using results on balanced categories with duality. Moreover we will show that braided sovereign categories are exactly the braided categories whose classes of left and right duality functors are identical. Using traces on balanced categories with duality we will define balanced Markov-Yang-Baxter operators which generalize the notion of enhanced Yang-Baxter operators [29][1].

In the second part of the article we study the ribbon braid group and ribbon braids. Both are examples of balanced categories which are equivalent

---

[1]Compare the definition of balanced and tortile Yang-Baxter operators in [13].

to the balanced category generated by one object [14, 28]. Similarly as in the classical case of ordinary braids one derives a generalized Markov Theorem for ribbon braids. The Markov Theorem shows that balanced categories are closely related to directed ribbon links. In particular the invariants of directed ribbon links are in one-to-one correspondence with maps on the Markov equivalence classes of ribbon braids. As an example we consider invariants of ribbon links arising from balanced Markov-Yang-Baxter operators.

Preliminaries

Bi- and Hopf Algebras. Recall [7] that a quasitriangular bialgebra (H, R) over a commutative ring (or field) $\Bbbk$ is a bialgebra with a universal R-matrix $R \in H \otimes H$ obeying the identities

$$\Delta^{\mathrm{op}}(x) = R \cdot \Delta(x) \cdot R^{-1} \quad \forall x \in H \,,$$
$$(\Delta \otimes \mathrm{id}_H)(R) = R_{13} \cdot R_{23} \,, \tag{0.1}$$
$$(\mathrm{id}_H \otimes \Delta)(R) = R_{13} \cdot R_{12} \,.$$

where $R_{ij} := \sum \mathbb{1} \otimes \cdots \otimes \underset{i.}{R_1} \otimes \cdots \otimes \underset{j.}{R_2} \otimes \cdots \otimes \mathbb{1}$. It is well known (see for instance [15]) that R fulfills the equations $(\varepsilon \otimes \mathrm{id})(R) = 1 = (\mathrm{id} \otimes \varepsilon)(R)$ and $R_{12} R_{13} R_{23} = R_{23} R_{13} R_{12}$ or $\widehat{R}_{12} \widehat{R}_{23} \widehat{R}_{12} = \widehat{R}_{23} \widehat{R}_{12} \widehat{R}_{23}$ for $\widehat{R} = \tau(R)$. The antipode of a quasitriangular Hopf algebra (H, R) is bijective. Furthermore $(S \otimes \mathrm{id})(R) = R^{-1} = (\mathrm{id} \otimes S^{-1})(R)$. The inverse of the element $u := \mathrm{m} \circ (S \otimes \mathrm{id}) \circ \tau(R)$ is given by $u^{-1} = \mathrm{m} \circ (S^{-1} \otimes \mathrm{id}) \circ \tau(R^{-1}) = \mathrm{m} \circ (\mathrm{id} \otimes S) \circ \tau(R^{-1})$ and the following relations hold.

$$S^2(x) = u \cdot x \cdot u^{-1} \ \forall \ x \in H \,,$$
$$S(u) \cdot u \ \text{ is central in } H \tag{0.2}$$
$$\varepsilon(u) = 1 \,, \quad \Delta(u) = (R_{21} \cdot R)^{-1} \cdot (u \otimes u) \,.$$

The axioms of a coquasitriangular bi- or Hopf algebra can be easily obtained from the definition of quasitriangularity and formally applying categorical duality[2].

Monoidal Categories. We assume that the (braided) monoidal categories are strict, that is the tensor product is associative and the tensor product of an object V with the unit object $\mathbb{1}$ is V. Each monoidal category is equivalent to a strict one by Mac Lane's Coherence Theorem [21, 22]. We use graphical calculus in (strict) monoidal categories [24, 15, 31]. Our conventions are shown in Figure 1.

We will frequently make use of categorical dualization. Suppose a certain categorical statement can be formulated in a category $\mathcal{C}$, in terms $S(\mathcal{C})$, as well as in the dual category $\mathcal{C}^{\mathrm{op}}$, in terms $S(\mathcal{C}^{\mathrm{op}})$. Then we say that the

---

[2]See the next paragraph for details on categorical duality

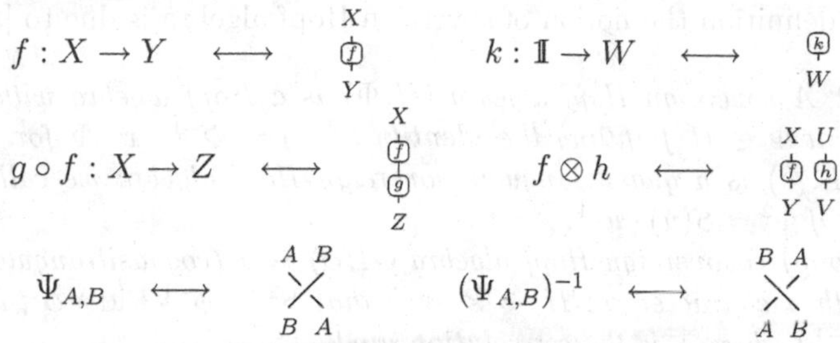

$$f : X \to Y \quad \longleftrightarrow \quad \overset{X}{\underset{Y}{\boxed{f}}} \qquad\qquad k : \mathbb{I} \to W \quad \longleftrightarrow \quad \overset{\textcircled{k}}{\underset{W}{}}$$

$$g \circ f : X \to Z \quad \longleftrightarrow \quad \overset{X}{\underset{Z}{\boxed{f}\boxed{g}}} \qquad\qquad f \otimes h \quad \longleftrightarrow \quad \overset{X\;\;U}{\underset{Y\;\;V}{\boxed{f}\;\boxed{h}}}$$

$$\Psi_{A,B} \quad \longleftrightarrow \quad \overset{A\;B}{\underset{B\;A}{\times}} \qquad (\Psi_{A,B})^{-1} \quad \longleftrightarrow \quad \overset{B\;A}{\underset{A\;B}{\times}}$$

Figure 1: Graphical presentation of a morphism f, a morphism k with unit object domaine, composition $g \circ f$, tensor product $f \otimes g$, braiding Ψ, and inverse braiding Ψ^{-1}.

statement S can be dualized in the categorical sense and call $(S(\mathcal{C}^{\mathrm{op}})^{\mathrm{op}})$ the dual statement of $S(\mathcal{C})$. Observe that there are statements in certain monoidal categories which can not be dualized.

Braids and Links. We discuss topological structures of (ribbon) braids and links in the setting of piecewise linear topology [27]. In particular we freely use notions like continuous map, regular neighborhood, orientation, isotopy, etc. within this piecewise linear framework. We suppose familiarity with elementary knot theory, and we refer henceforth to all the excellent articles on the subject from which [1, 3, 9, 13, 15, 19, 20, 26, 28, 31, 33] is only a small excerpt. In particular the categories of graphs and tangles were invented by Yetter [32], Freyd and Yetter [8, 9], and Turaev [30]. The idea to describe the categories of graphs and tangles via generators and relations is mainly due to Yetter [32] and Turaev [30].

Acknowledgements: I thank Yuri Bespalov, Shahn Majid, Susan Montgomery and Jozef Przytycki for valuable discussions.

1   Balanced Hopf Algebras

Let us briefly recall the definition of balanced and ribbon Hopf algebras (see for instance [26, 15]).

DEFINITION 1.1 *A balanced bialgebra* (H, R, θ) *is a quasitriangular bialgebra with an invertible, central element* $\theta \in H$ *such that* $\Delta(\theta) = (R_{21} \cdot R)^{-1} \cdot (\theta \otimes \theta)$ *and* $\varepsilon(\theta) = 1$. *If* H *is a balanced Hopf algebra and the identity* $S(\theta) = \theta$ *holds then* (H, R, θ) *is called a ribbon Hopf algebra.*

Cobalanced bialgebras and coribbon Hopf algebras are the dual analogues of balanced bialgebras and ribbon Hopf algebras respectively.

In the following definition the notion of sovereign Hopf algebra is due to [2].

DEFINITION 1.2 *A sovereign Hopf algebra* (H, Φ) *is a Hopf algebra with a group like element* $\Phi \in H$ *fulfilling the identity* $S^2(x) = \Phi^{-1} \cdot x \cdot \Phi$ *for all* $x \in H$. *If* (H, R, Φ) *is a quasitriangular sovereign Hopf algebra we call it strong sovereign if* $\Phi^2 = S(u) \cdot u^{-1}$.

Dually, a (strong) cosovereign Hopf algebra (H, ϕ) *is a (coquasitrangular) Hopf algebra with a character* $\phi : H \to \mathbb{k}$ *such that* $S^2 = \phi^{-1} * \mathrm{id} * \phi$ *(and* $\phi^2 = (\mu \circ S) * \mu^{-1}$*) where* $*$ *is the convolution product.*

One easily verifies that the antipode S of a (co-)sovereign Hopf algebra has an inverse $S^{-1} = \Phi \cdot S(\textbf{.}) \cdot \Phi^{-1}$ and $S^{-1} = \phi * S * \phi^{-1}$ respectively. Actually, this is equivalent to the definition of (co-)sovereign Hopf algebras in Definition 1.2.

THEOREM 1.3   1. (H, R, θ) *is a balanced Hopf algebra if and only if* (H, R, Φ) *is a quasitriangular sovereign Hopf algebra. The elements* θ *and* Φ *are related by* $\theta \cdot \Phi = S(u)$.

   2. (H, R, θ) *is a ribbon Hopf algebra if and only if* (H, R, Φ) *is a strong sovereign Hopf algebra.*

PROOF. The first part of Theorem 1.3 has been proven in [2] for the corresponding dual statement of cobalanced and coquasitriangular Hopf algebras.

We only have to verify the second part of the theorem. Observe that for any sovereign Hopf algebra

$$\Phi^{-1} \cdot S(u) = S^3(u) \cdot \Phi^{-1} = S(u) \cdot \Phi^{-1}. \tag{1.1}$$

If in addition H is strong sovereign then

$$S(\theta) = S(S(u) \cdot \Phi^{-1}) = S(\Phi^{-1} \cdot S(u))$$
$$= S(\Phi \cdot u) = S(u) \cdot S(\Phi) = S(u) \cdot \Phi^{-1} = \theta$$

where the definition of θ has been used in the first equation, relation (1.1) in the second identity, the assumption $\Phi^2 = S(u) \cdot u^{-1}$ in the third and $S(\Phi) = \Phi^{-1}$ in the fifth equation. Conversely, if (H, R, θ) is a ribbon Hopf algebra we have to show that $\Phi := \theta^{-1} \cdot S(u)$ obeys the identity $\Phi^2 = S(u) \cdot u^{-1}$. By assumption $\theta = S(\theta)$. Therefore $S(\Phi \cdot S(u^{-1})) = \phi \cdot S(u^{-1})$. From the first statement of the theorem we know that Φ is a sovereign group like element. Then we conclude $S(\Phi \cdot S(u^{-1})) = \Phi \cdot S(u^{-1}) \Leftrightarrow S(\Phi) \cdot S^2(u^{-1}) = \Phi \cdot S(u^{-1}) \Leftrightarrow \Phi^{-1} \cdot u^{-1} = \Phi \cdot S(u^{-1}) \Leftrightarrow \Phi^2 = u^{-1} \cdot S(u) = S(u) \cdot u^{-1}$. Hence, (H, R, Φ) is strong sovereign Hopf algebra.   ■

A corresponding dual result of Theorem 1.3 holds for coquasitriangular Hopf algebras and cobalanced and cosovereign structures. From Theorem 1.3 we derive the following corollary which says that the conditions $S(\theta) = \theta$ and $\theta^2 = u \cdot S(u)$ of the twist element θ of a ribbon Hopf algebra are equivalent.

COROLLARY 1.4 *Suppose that (H, R, θ) is a balanced Hopf algebra. Then (H, R, θ) is ribbon if and only if $\theta^2 = u \cdot S(u)$.*

PROOF. From Theorem 1.3 we know that (H, R, θ) is ribbon if and only if $(H, R, \theta^{-1} \cdot S(u))$ is strong sovereign if and only if $(\theta^{-1} \cdot S(u))^2 = S(u) \cdot u^{-1}$ iff $\theta^2 = S(u) \cdot u$. ∎

The next proposition describes the construction of a balanced bialgebra out of a given quasitriangular bialgebra. This process is a straightforward generalization of the construction in [26] which associates to any quasitriangular Hopf algebra a ribbon Hopf algebra.

PROPOSITION 1.5 *Suppose that H is a quasitriangular bialgebra. Then the Laurent polynomial algebra $H[\theta, \theta^{-1}]$ is a balanced bialgebra with $\Delta_{|H} := \Delta_H$, $\Delta(\theta^{\pm 1}) := (R_{21} \cdot R)^{\mp 1} \cdot (\theta^{\pm 1} \otimes \theta^{\pm 1})$, $\varepsilon_{|H} := \varepsilon_H$, and $\varepsilon(\theta^{\pm 1}) = 1$.*

PROOF. We follow the lines of [26]. Therefore we only sketch the most important arguments. $H[\theta, \theta^{-1}]$ is canonically an algebra. The bialgebra structure on $H[\theta, \theta^{-1}] = \bigoplus_{r \in \mathbb{Z}} H_r$ is given by $\varepsilon(\theta^{\pm 1}) = 1$ and $\Delta(\theta^{\pm 1}) = (R_{21} \cdot R)^{\mp 1} \cdot (\theta^{\pm 1} \otimes \theta^{\pm 1})$. To prove coassociativity the identity

$$(\mathbb{1} \otimes R_{21} \cdot R) \cdot (\mathrm{id} \otimes \Delta)(R_{21} \cdot R) = (R_{21} \cdot R \otimes \mathbb{1}) \cdot (\Delta \otimes \mathrm{id})(R_{21} \cdot R) \quad (1.2)$$

needs to be verified in particular. If H were a quasitriangular Hopf algebra (1.2) could be conveniently derived with the help of (0.2). To prove (1.2) for a quasitriangular bialgebra H we proceed as follows. Using (0.1) we obtain

$$(R_{21} \cdot R) \cdot \Delta(a) = \Delta(a) \cdot (R_{21} \cdot R) \quad \forall \, a \in H. \quad (1.3)$$

We transform the left and right hand side of (1.2) with the help of (0.1).

$$
\begin{aligned}
(\mathbb{1} \otimes R_{21} \cdot R) \cdot (\mathrm{id} \otimes \Delta)(R_{21} \cdot R) &= R_{32} \cdot R_{23} \cdot R_{21} \cdot R_{31} \cdot R_{13} \cdot R_{12}, \\
(R_{21} \cdot R \otimes \mathbb{1}) \cdot (\Delta \otimes \mathrm{id})(R_{21} \cdot R) &= R_{32} \cdot R_{31} \cdot R_{13} \cdot R_{23} \cdot R_{21} \cdot R_{12}.
\end{aligned} \quad (1.4)
$$

Furthermore (0.1) and (1.3) yield .

$$
\begin{aligned}
R_{23} \cdot R_{21} \cdot R_{31} \cdot R_{13} &= (\tau \otimes \mathrm{id})(R_{13} \cdot R_{12}) \cdot (\tau \otimes \mathrm{id})(\mathbb{1} \otimes R_{21} \cdot R) \\
&= (\tau \otimes \mathrm{id})((\mathbb{1} \otimes R_{21} \cdot R) \cdot (R_{13} \cdot R_{12})) = R_{31} \cdot R_{13} \cdot R_{23} \cdot R_{21}.
\end{aligned} \quad (1.5)
$$

Inserting (1.5) into (1.4) yields (1.2).

It is easy to show that R is an R-matrix for $H[\theta, \theta^{-1}]$. Then by definition of $\Delta(\theta)$ it follows that $H[\theta, \theta^{-1}]$ is a balanced bialgebra. ∎

REMARK 1 If H is a quasitriangular Hopf algebra then $H(\theta) := H[\theta]/(\theta^2 - u \cdot S(u))$ is a ribbon Hopf algebra with $S_{|H} := S_H$ and ribbon twist θ [26]. H is a quasitriangular sub-bialgebra of $H[\theta, \theta^{-1}]$ and $H(\theta)$ respectively.

2 Balanced and Sovereign Categories

Braided monoidal categories were introduced in [10, 12]. The definition of a twist in a braided category was given in [28]. Balanced categories have been studied in [14, 28, 9] and sovereign categories in [6, 33, 25].

We recall the definition of (adjoint) duality in a monoidal category [10, 12, 14, 16][3]. Suppose that $(\mathcal{C}, \otimes)$ is a monoidal category. An object X in $\mathcal{C}$ has a left dual $^*X$ if there exist morphisms $\mathrm{ev}_X : {^*X} \otimes X \to \mathbb{1}$ and $\mathrm{coev}_X : \mathbb{1} \to X \otimes {^*X}$ such that

$$
\begin{aligned}
(\mathrm{id}_X \otimes \mathrm{ev}_X) \circ (\mathrm{coev}_X \otimes \mathrm{id}_X) &= \mathrm{id}_X, \\
(\mathrm{ev}_X \otimes \mathrm{id}_{*X}) \circ (\mathrm{id}_{*X} \otimes \mathrm{coev}_X) &= \mathrm{id}_{*X}.
\end{aligned}
\tag{2.1}
$$

The pair $(\mathrm{ev}, \mathrm{coev})$ is an adjunction between the objects $^*X$ and X [11]. If every object in $\mathcal{C}$ has a left dual then $\mathcal{C}$ is a category with left duality. Similar definitions yield the notion of right duality. The left (or right) dual is unique up to canonical isomorphism. The contravariant duality functor $^*(-) : \mathcal{C} \longrightarrow \mathcal{C}$ is given by $^*(X) = {^*X}$ and $^*(f) : {^*Y} \to {^*X}$, where $f : X \to Y$ and $^*(f) := (\mathrm{ev}_Y \otimes \mathrm{id}_{*X}) \circ (\mathrm{id}_{*Y} \otimes f \otimes \mathrm{id}_{*X}) \circ (\mathrm{id}_{*Y} \otimes \mathrm{coev}_X)$. Observe that $^*(-) : (\mathcal{C}, \otimes) \longrightarrow (\mathcal{C}^{\mathrm{op}}, \otimes^{\mathrm{op}})$ is a monoidal functor which is unique up to monoidal isomorphism. Similarly the right duality functor $(-)^* : \mathcal{C} \longrightarrow \mathcal{C}$ will be defined. Left and right duality functors are not isomorphic in general. Henceforth we use the graphical presentation $\mathrm{ev}_X := \smile : {^*X} \otimes X \to \mathbb{1}$ and $\mathrm{coev}_X := \frown : \mathbb{1} \to X \otimes {^*X}$, and similarly for the right duals.

In a braided monoidal category with left duality $(\mathrm{ev}, \mathrm{coev})$ right duality can be defined by $\mathrm{ev}'_X = \mathrm{ev}_X \circ \Psi_{X, *X}$ and $\mathrm{coev}'_X = \Psi^{-1}_{X, *X} \circ \mathrm{coev}_X$ for $X \in \mathrm{Ob}(\mathcal{C})$ [4].

DEFINITION 2.1 *The category* $(\mathcal{C}, \otimes, \Psi, \Theta)$ *is called balanced if* $(\mathcal{C}, \otimes, \Psi)$ *is braided and* $\Theta : \mathrm{id}_{\mathcal{C}} \overset{\bullet}{\cong} \mathrm{id}_{\mathcal{C}}$ *is a natural isomorphism which obeys the identities*

$$
\Theta_{X \otimes Y} = \Psi_{Y, X} \circ \Psi_{X, Y} \circ (\Theta_X \otimes \Theta_Y)
\tag{2.2}
$$

for all $X, Y \in \mathrm{Ob}(\mathcal{C})$. *A balanced category* $\mathcal{C}$ *with left duality* $(\mathrm{ev}, \mathrm{coev})$ *(and hence right duality) is called ribbon category if in addition* $\Theta_{*X} = {^*(\Theta_X)}$ *for all* $X \in \mathrm{Ob}(\mathcal{C})$.

Naturality of Θ and (2.2) imply that $\Theta_{\mathbb{1}} = \mathrm{id}_{\mathbb{1}}$ in any balanced category $(\mathcal{C}, \Theta)$.

REMARK 2 A balanced (ribbon) functor $F : \mathcal{C} \to \mathcal{D}$ is a braided monoidal functor which preserves the twist, $F(\Theta^{\mathcal{C}}_X) = \Theta^{\mathcal{D}}_{F(X)}$ (and is compatible with duality).

---

[3]Adjoint duality may not be confused with the categorical duality discussed in the Introduction.

[4]A category with right and left duality is called autonomous.

Next we are going to discuss the construction of a balanced category out of any monoidal category. This construction interpolates the more general center construction [23, 13] which assigns to every monoidal category a braided category $\mathcal{Z}(\mathcal{C})$, and the more special double construction [15] which assigns to every category with left duality a ribbon category $\mathcal{D}(\mathcal{C})$.

DEFINITION 2.2 *Let $\mathcal{C}$ be a monoidal category. Then the category $\mathcal{B}(\mathcal{C})$ is defined as follows.*

1. *Objects are tuples $(V, \psi_{V,-}, \theta_V)$ such that*

   (a) *V is an object in $\mathcal{C}$,*

   (b) *$\psi_{V,-} : V \otimes (-) \xrightarrow{\cdot} (-) \otimes V$ is a natural isomorphism in $\mathcal{C}$.*

   (c) *$\theta_V : V \to V$ is an automorphism in $\mathcal{C}$*

   and the identities $\psi_{V,X \otimes Y} = (\mathrm{id}_X \otimes \psi_{V,Y}) \circ (\psi_{V,X} \otimes \mathrm{id}_Y)$ and $(\mathrm{id}_X \otimes \theta_V) \circ \psi_{V,X} = \psi_{V,X} \circ (\theta_V \otimes \mathrm{id}_X)$ hold for all objects X, Y in $\mathcal{C}$.

2. *Morphisms $f : (V, \psi_{V,-}, \theta_V) \to (W, \psi_{W,-}, \theta_W)$ are morphisms $f : V \to W$ in $\mathcal{C}$ such that $(\mathrm{id}_X \otimes f) \circ \psi_{V,X} = \psi_{W,X} \circ (f \otimes \mathrm{id}_X)$ and $f \circ \theta_V = \theta_W \circ f$ for all objects X in $\mathcal{C}$.*

Note that $\mathcal{B}(\mathcal{C})$ is bigger than $\mathcal{D}(\mathcal{C})$. For instance, with $(V, \psi_{V,-}, \theta_V)$ also $(V, \psi_{V,-}, \theta_V^n)$ and $(V, \psi_{V,-}, \lambda \cdot \theta_V)$ are objects of $\mathcal{B}(\mathcal{C})$ for any $n \in \mathbb{N}$ and $\lambda \in \mathrm{Aut}(\mathbb{1}_{\mathcal{C}})$.

The next proposition states that the category $\mathcal{B}(\mathcal{C})$ is balanced. Therefore this construction provides a balanced category $\mathcal{B}(\mathcal{C})$ for any monoidal category $\mathcal{C}$.

PROPOSITION 2.3 *If $(\mathcal{C}, \otimes, \mathbb{1})$ is a monoidal category then $\mathcal{B}(\mathcal{C})$ is a balanced category with*

1. *unit object $\mathbb{1}_{\mathcal{B}(\mathcal{C})} := (\mathbb{1}, \psi_{\mathbb{1},-} = \mathrm{id}_{(-)}, \theta_{\mathbb{1}} = \mathrm{id}_{\mathbb{1}})$,*

2. *tensor product $(V, \psi_{V,-}, \theta_V) \otimes (W, \psi_{W,-}, \theta_W) := (V \otimes W, \psi_{V \otimes W,-}, \theta_{V \otimes W})$ where $\theta_{V \otimes W} := (\theta_V \otimes \theta_W) \circ \psi_{W,V} \circ \psi_{V,W}$, and $\psi_{V \otimes W,X} := (\psi_{V,X} \otimes \mathrm{id}_W) \circ (\mathrm{id}_V \otimes \psi_{W,X})$ for any object X in $\mathcal{C}$,*

3. *braiding $\Psi_{(V, \psi_{V,-}, \theta_V), (W, \psi_{W,-}, \theta_W)} := \psi_{V,W}$,*

4. *twist $\Theta_{(V, \psi_{V,-}, \theta_V)} := \theta_V$.*

PROOF. The proof is a straightforward specialization/generalization of the corresponding proofs of the center and double constructions [23, 13, 17] respectively. ∎

For every monoidal category $\mathcal{C}$ there exists a monoidal functor $\Pi_{\mathcal{C}} : \mathcal{B}(\mathcal{C}) \longrightarrow \mathcal{C}$ defined by $(V, \psi_{V,-}, \theta_V) \to V$ and $f \to f$ [17]. The functor $\Pi_{\mathcal{C}}$ is not balanced in general. If $F : \mathcal{C}_1 \longrightarrow \mathcal{C}_2$ is a balanced functor of balanced categories then $\mathcal{B}(F) : \mathcal{C}_1 \longrightarrow \mathcal{B}(\mathcal{C}_2)$ with $\mathcal{B}(F)(V) := (V, \Psi^{\mathcal{B}_2}_{F(V),-}, \Theta^{\mathcal{B}_2}_{F(V)})$, $\mathcal{B}(F)(f) := F(f)$ is a balanced functor, and it holds $\Pi_{\mathcal{B}_2} \circ \mathcal{B}(F) = F$. If $\mathcal{C}_1 \xrightarrow{F} \mathcal{C}_2 \xrightarrow{G} \mathcal{C}_3$ is a sequence of balanced functors then $\mathcal{B}(G) \circ F = \mathcal{B}(G \circ F)$.

Similarly as in [17] one proves the next proposition.

PROPOSITION 2.4 *If $\mathcal{C}$ is a monoidal category, $\mathcal{B}$ is a balanced category and $F : \mathcal{B} \longrightarrow \mathcal{C}$ is a (strict) monoidal functor which is bijective on the objects and surjective on the morphisms, then there exists a unique balanced functor $\mathcal{B}(F) : \mathcal{B} \longrightarrow \mathcal{B}(\mathcal{C})$ such that $F = \Pi_{\mathcal{C}} \circ \mathcal{B}(F)$. In particular the balanced functor $\mathcal{B}(\mathrm{id}_{\mathcal{B}}) : \mathcal{B} \to \mathcal{B}(\mathcal{B})$ is unique such that $\Pi_{\mathcal{B}} \circ \mathcal{B}(\mathrm{id}_{\mathcal{B}}) = \mathrm{id}_{\mathcal{B}}$.* ∎

In the following we will consider balanced categories with duality. In particular we will discuss their relation to so-called sovereign categories.

PROPOSITION 2.5 *Let $(\mathcal{C}, \Theta)$ be a balanced category with left duality $(\mathrm{ev}, \mathrm{coev})$. Then for every object X in $\mathcal{C}$ the inverse $(\Theta_X)^{-1}$ is given by*

$$(\Theta_X)^{-1} = \raisebox{-1em}{\includegraphics[height=3em]{fig1}} = \raisebox{-1em}{\includegraphics[height=3em]{fig2}} \qquad (2.3)$$

The category $\mathcal{C}$ is ribbon if and only if

$$(\Theta_X)^{-2} = \raisebox{-1em}{\includegraphics[height=3em]{fig3}} \qquad (2.4)$$

for every object X in $\mathcal{C}$.

PROOF. We use similar techniques as for instance in [15]. Since Θ is functorial and fulfills (2.2) and $\Theta_{\mathbb{I}} = \mathrm{id}_{\mathbb{I}}$ it holds

$$\mathrm{coev}_X = \mathrm{coev}_X \circ \Theta_{\mathbb{I}} = \Theta_{X \otimes {}^*X} \circ \mathrm{coev}_X = (\Theta_X \otimes \Theta_{*X}) \circ \Psi \circ \Psi \circ \mathrm{coev}_X . \tag{2.5}$$

Applying (2.1) one obtains $\mathrm{id}_X = (\mathrm{id}_X \otimes \mathrm{ev}_X) \circ ((\Theta_X \otimes \Theta_{*X}) \circ \Psi \circ \Psi \circ \mathrm{coev}_X \otimes \mathrm{id}_X)$ from which (2.3) can be derived using functoriality of the braiding Ψ.

Now suppose that $\mathcal{C}$ is ribbon, then ${}^*\Theta_X = \Theta_{*X}$. Therefore $(\mathrm{id} \otimes \Theta_{*X}) \circ \mathrm{coev}_X = (\Theta_X \otimes \mathrm{id}) \circ \mathrm{coev}_X)$, and we can apply $(\Theta_X)^{-1} \circ |$ on both sides of (2.3) to get (2.4). Conversely, if (2.4) holds we apply $| \circ \Theta_X$ on both sides of

(2.4). Taking into account that (2.3) holds in particular for ribbon categories we obtain

$$
\text{[diagram]} = \text{[diagram]} \tag{2.6}
$$

Using (2.1) and functoriality of the braiding one immediately derives from (2.6) the identity $^*\Theta_X = \Theta_{*X}$. Therefore $(\mathcal{C}, \Theta)$ is a ribbon category. ∎

PROPOSITION 2.6 *If $\mathcal{C}$ is a balanced category with left duality* (ev, coev) *then*

$$
\mathrm{ev}^r_X = \mathrm{ev}_X \circ \Psi_{X,*X} \circ (\Theta_X \otimes \mathrm{id}_{*X}), \quad \mathrm{coev}^r_X = (\Theta_{*X} \otimes \mathrm{id}_X) \circ \Psi_{X,*X} \circ \mathrm{coev}_X . \tag{2.7}
$$

defines a right duality on $\mathcal{C}$. Conversely, if (ev, coev) *is a right duality on $\mathcal{C}$ then*

$$
\mathrm{ev}^l_X = \mathrm{ev}_X \circ \Psi_{X^*,X} \circ (\Theta_{X^*} \otimes \mathrm{id}_X), \quad \mathrm{coev}^l_X = (\Theta_X \otimes \mathrm{id}_{X^*}) \circ \Psi_{X^*,X} \circ \mathrm{coev}_X . \tag{2.8}
$$

is a left duality. These structures are inverse to each other in the following sense. Suppose that (ev, coev) *is a left duality of the balanced category $\mathcal{C}$. Then* $((\mathrm{ev}^r)^l, (\mathrm{coev}^r)^l) = (\mathrm{ev}, \mathrm{coev})$.

PROOF. With the help of (2.3) one easily verifies that $(\mathrm{ev}^r, \mathrm{coev}^r)$ according to (2.7) defines a right duality on $\mathcal{C}$.

By definition of $(\mathrm{ev}^r, \mathrm{coev}^r)$ we have $X^* = {}^*X$ where $^*X$ is the left dual w. r. t. (ev, coev). Then we use (2.8), (2.7) and (2.3) to derive

$$
(\mathrm{ev}^r)^l_X = \mathrm{ev}^r_X \circ \Psi_{X^*,X} \circ (\Theta_{X^*} \otimes \mathrm{id}_X)
$$
$$
= \mathrm{ev}_X \circ \Psi \circ (\Theta_X \otimes \Theta_{*X}) \circ \Psi
$$
$$
= \text{[diagram]} = \text{[diagram]} = \mathrm{ev}_X \tag{2.9}
$$

Similarly we derive $(\mathrm{coev}^r)^l_X = \mathrm{coev}_X$. ∎

Next we recall the definition of sovereign categories (see for instance [33, 25]. These are categories whose left and right duality are isomorphic to each other.

DEFINITION 2.7 *Let $\mathcal{C}$ be a monoidal category with left and right duality. If the left and right duality functors of $\mathcal{C}$ are monoidally isomorphic through $\varphi : (.)^* \overset{\bullet}{\cong} {}^*(.)$, then $(\mathcal{C}, \varphi)$ is called sovereign category.*

It will be shown in Theorem 2.10 (see also [6, 33, 25]) that balanced categories with duality are equivalent to braided sovereign categories. Propositions 2.8 and 2.9 are preliminary steps.

PROPOSITION 2.8 *If $(\mathcal{C}, \Theta)$ is a balanced category with duality and if the right dual structure is given by (2.7) then the left duality functor on $\mathcal{C}$ coincides monoidally with the right duality functor. This implies that* id $: (-)^* \xrightarrow{\bullet} {}^*(-)$ *is a natural monoidal isomorphism, and therefore $(\mathcal{C}, \mathrm{id}_{*(-)})$ is a braided sovereign category.*

PROOF. To verify the proposition we have to prove that $\big((-)^*, \rho_{\mathbb{1}} = \mathrm{id}_{\mathbb{1}}, \rho_{X,Y}\big)$ and $\big({}^*(-), \lambda_{\mathbb{1}} = \mathrm{id}_{\mathbb{1}}, \lambda_{X,Y}\big)$ coincide. The natural morphisms $\lambda_{X,Y}$ and $\rho_{X,Y}$ are given by

$$\lambda_{X,Y} = (\mathrm{ev}_{X\otimes Y} \otimes \mathrm{id}_{*Y} \otimes \mathrm{id}_{*X}) \circ (\mathrm{id}_{*(X\otimes Y)} \otimes (\mathrm{id}_X \otimes \mathrm{coev}_Y \otimes \mathrm{id}_{*X}) \circ \mathrm{coev}_X)$$
$$\rho_{X,Y} = (\mathrm{id}_{Y^*} \otimes \mathrm{id}_{X^*} \otimes \mathrm{ev}^r_{X\otimes Y}) \circ ((\mathrm{id}_{Y^*} \otimes \mathrm{coev}^r_X \otimes \mathrm{id}_Y) \circ \mathrm{coev}^r_Y \otimes \mathrm{id}_{(X\otimes Y)^*})$$
$$\tag{2.10}$$

By construction $X^* = {}^*X$ for all objects X in $\mathcal{C}$. For any morphism $f : X \to Y$ we obtain

$$f^* = (\mathrm{id} \otimes \mathrm{ev}^r) \circ (\mathrm{id} \otimes f \otimes \mathrm{id}) \circ (\mathrm{coev}^r \otimes \mathrm{id}) = \quad\cdots\quad = \quad\cdots\quad = {}^*f$$
$$\tag{2.11}$$

where we used definition (2.7), naturality of Θ and the definition of ${}^*f$ in the second equation of (2.11). In the third identity we use functoriality of the braiding Ψ and (2.1). Eventually, (2.3) yields the final identity. Therefore the functors $(-)^*$ and ${}^*(-)$ coincide.

It remains to show that $\rho_{X,Y} = \lambda_{X,Y}$ for all objects X, Y in $\mathcal{C}$. We will use the notation

$$\mathrm{ev}_{X\otimes Y} = \quad {}^*(X\otimes Y) \;\; X \;\; Y \qquad\qquad\tag{2.12}$$

Then we use (2.7) to rewrite $\rho_{X,Y}$, and with the help of (2.2) and (2.5) we obtain

$$\rho_{X,Y} = \qquad = \qquad = $$

$$= \qquad = \lambda_{X,Y}$$

In particular this implies that $(\mathcal{C}, \mathrm{id}_{(-)^*})$ is a braided sovereign category. ∎

The converse of Proposition 2.8 holds as well.

PROPOSITION 2.9 *Let $(\mathcal{C}, \varphi)$ be a braided sovereign category and define*

$$\Theta_X := (\mathrm{id}_X \otimes \mathrm{ev}'_X) \circ (\Psi_{X,X} \otimes \varphi_X^{-1}) \circ (\mathrm{id}_X \otimes \mathrm{coev}_X) \quad \forall\, X \in \mathrm{Ob}(\mathcal{C}) \quad (2.13)$$

where $(\mathrm{ev}, \mathrm{coev})$ *and* $(\mathrm{ev}', \mathrm{coev}')$ *are the left and right duality respectively. Then $(\mathcal{C}, \Theta)$ is a balanced category with duality.*

PROOF. We only have to prove that the above defined Θ is a twist of a balanced category. It is rather simple to show that Θ_X is an isomorphism. For any morphism $f : X \to Y$ we have $\mathrm{ev}'_Y \circ (f \otimes \mathrm{id}) = \mathrm{ev}'_X \circ (\mathrm{id} \otimes f^*)$, $f^* \circ \varphi_Y^{-1} = \varphi_X^{-1} \circ {}^*f$ and $(\mathrm{id} \otimes {}^*f) \circ \mathrm{coev}_Y = (f \otimes \mathrm{id}) \circ \mathrm{coev}_X$. Hence $\Theta_Y \circ f = f \circ \Theta_X$. If $\mathrm{\ominus}_i, i \in \{1, 2\}$, are suitable morphisms then the identity

$$= \qquad\qquad (2.14)$$

holds. Since φ is monoidal, i.e. $\varphi_{X \otimes Y} \cong \varphi_Y \otimes \varphi_X$, the identity (2.2) easily follows from (2.13) and (2.14) ∎

The first part of the following theorem has been proven in [6, 33] and reformulated in [25]. Using the previous results we will provide an elementary proof of that.

THEOREM 2.10 *Suppose that $\mathcal{C}$ is a braided category with duality. Then $\mathcal{C}$ is balanced if and only if $\mathcal{C}$ is sovereign. In this case there exists for any left duality a right duality such that the sovereign structure coincides with the identical natural transformation.*

PROOF. If $\mathcal{C}$ is balanced with duality then by Proposition 2.8 $\mathcal{C}$ is braided sovereign. Conversely, Proposition 2.9 states that there exists a twist on every braided sovereign category. Then we use again Proposition 2.8 to verify that every sovereign structure φ on $\mathcal{C}$ has the form $\varphi = \mathrm{id} \circ \delta = \delta$ where $\delta : (-)^* \overset{\bullet}{\to} (-)^{*'}$ is the canonical monoidal isomorphism of the functors $(-)^*$ and $(-)^{*'}$. ∎

Therefore the twist of a balanced category with duality corresponds in a very natural way to sovereign structures which are identities of a given duality functor.

COROLLARY 2.11 *Let $\mathcal{C}$ be a braided category with duality. Then $\mathcal{C}$ is sovereign if and only if its class of left duality functors coincides monoidally with its class of right duality functors. The twist on $\mathcal{C}$ is given by $\Theta_V :=$ $(\mathrm{id}_V \otimes \mathrm{ev}'_V) \circ (\Psi_{V,V} \otimes \mathrm{id}_{*V}) \circ (\mathrm{id}_V \otimes \mathrm{coev}_V)$ where $(\mathrm{ev}, \mathrm{coev})$ and $(\mathrm{ev}', \mathrm{coev}')$ are the respective left and right dual structure of the duality functor.* ∎

Every sovereign category admits a trace [25]. In particular every balanced category $\mathcal{C}$ with (left) duality has a trace map $\mathrm{tr}_{A,B;C} : \mathrm{Hom}_{\mathcal{C}}(A \otimes C, B \otimes C) \to \mathrm{Hom}_{\mathcal{C}}(A, B)$ on its Hom-sets, defined by

$$\mathrm{tr}_{A,B;C}(u) := \qquad\qquad\qquad (2.15)$$

Standard methods [20, 25] will be used to prove the following identities.

A. $\mathrm{tr}_{A',B;C}(u \circ (f \otimes \mathrm{id}_C)) = \mathrm{tr}_{A,B;C}(u) \circ f$,

B. $\mathrm{tr}_{A,B';C}((g \otimes \mathrm{id}_C) \circ u) = g \circ \mathrm{tr}_{A,B;C}(u)$,

C. $\mathrm{tr}_{A,B;C'}(w \circ (\mathrm{id}_A \otimes h)) = \mathrm{tr}_{A,B;C}((\mathrm{id}_B \otimes h) \circ w)$,

D. $\mathrm{tr}_{A \otimes A',B \otimes B';C}(k \otimes l) = k \otimes \mathrm{tr}_{A',B';C}(l)$,

E. $\mathrm{tr}_{A,B;C \otimes D}(v) = \mathrm{tr}_{A,B;C}(\mathrm{tr}_{A \otimes C,B \otimes C;D}(v))$

where $u : A \otimes C \to B \otimes C$, $v : A \otimes C \otimes D \to B \otimes C \otimes D$, $w : A \otimes C \to B \otimes C'$, $f : A' \to A$, $g : B \to B'$, $h : C' \to C$, $k : A \to B$, $l : A' \otimes C \to B' \otimes C$. If $q : C \to C$ we denote the trace $\mathrm{tr}_{\mathbb{1},\mathbb{1};C}(q)$ either by $\mathrm{tr}_C(q)$ or $\mathrm{tr}(q)$ if it is clear from the context.

In [13, 14] balanced Yang-Baxter operators have been defined. A balanced Yang-Baxter operator $(\widehat{R}, \varphi)$ on a functor $F : \mathcal{V} \longrightarrow \mathcal{C}$ from a category $\mathcal{V}$ to a (strict) monoidal category $\mathcal{C}$ is a pair of natural isomorphisms $\widehat{R} : \otimes \circ (F, F) \overset{\bullet}{\to} \otimes^{\mathrm{op}} \circ (F, F)$ and $\varphi : F \overset{\bullet}{\to} F$ obeying the identities $(\widehat{R} \otimes \mathrm{id}) \circ (\mathrm{id} \otimes \widehat{R}) \circ (\widehat{R} \otimes \mathrm{id}) = (\mathrm{id} \otimes \widehat{R}) \circ (\widehat{R} \otimes \mathrm{id}) \circ (\mathrm{id} \otimes \widehat{R})$ and $\widehat{R}^{\pm 1} \circ (\varphi \otimes \mathrm{id}) = (\mathrm{id} \otimes \varphi) \circ \widehat{R}^{\pm 1}$.

We will define balanced Markov-Yang-Baxter operators or MYB operators. They are generalizations of the enhanced Yang-Baxter operators defined in [29].

DEFINITION 2.12 *Let $\mathcal{C}$ be a balanced category with duality and the trace on $\mathcal{C}$ be defined as in (2.15). Then the tuple $R = (\widehat{R}, \varphi, \boldsymbol{\mu})$ is called a balanced Markov-Yang-Baxter operator if $(\widehat{R}, \varphi)$ is a balanced Yang-Baxter operator on a functor $F : \mathcal{V} \longrightarrow \mathcal{C}$, and*

1. *$\boldsymbol{\mu} : F \overset{\bullet}{\to} F$ is a natural endomorphism,*

2. *$\boldsymbol{\mu} \circ \varphi = \varphi \circ \boldsymbol{\mu}$ and $\widehat{R} \circ (\boldsymbol{\mu} \otimes \boldsymbol{\mu}) = (\boldsymbol{\mu} \otimes \boldsymbol{\mu}) \circ \widehat{R}$,*

3. *$\mathrm{tr}_{F(V), F(V); F(V)} \left(\widehat{R}_{V, V}^{\pm 1} \circ (\boldsymbol{\mu}_V \otimes \boldsymbol{\mu}_V) \right) = \varphi_V^{\pm 1} \circ \boldsymbol{\mu}_V$ for all objects V in $\mathcal{V}$.*

REMARK 3 1. If $\mathcal{C}$ is a ribbon category and $F : \mathcal{V} \longrightarrow \mathcal{C}$ a functor then one uses (2.4) to show that $(\Psi^{\mathcal{C}} \circ (F, F), \Theta^{\mathcal{C}} \circ F, \mathrm{id}_F)$ is a balanced Markov-Yang-Baxter operator on F.

2. A Markov-Yang-Baxter operator or enhanced Yang-Baxter operator $R = (\widetilde{R}, \boldsymbol{\mu})$ on a functor F [29] is a balanced Markov-Yang-Baxter operator on F for which $\varphi = \mathrm{id}_F$. Then we can define balanced Markov-Yang-Baxter operators in terms of Markov-Yang-Baxter operators. $(\widehat{R}, \varphi, \boldsymbol{\mu})$ is a balanced MYB operator if and only if $(\widetilde{R}, \boldsymbol{\mu})$ is an MYB operator and there exists an automorphism φ of F obeying the identities $(\varphi \otimes \mathrm{id}) \circ \widetilde{R}^{\pm 1} = \widetilde{R}^{\pm 1} \circ (\mathrm{id}_V \otimes \varphi)$ and $\varphi \circ \boldsymbol{\mu} = \boldsymbol{\mu} \circ \varphi$. In this case $\widetilde{R} = (\varphi^{-1} \otimes \mathrm{id}) \circ \widehat{R}$.

Given a balanced Yang-Baxter operator $R = (\widehat{R}, \varphi)$ on an object V in $\mathcal{C}$. The group $G_n^R \subset \mathrm{Aut}(V^{\otimes n})$ is defined by the generators $\psi_i := \underset{1.}{\mathrm{id}_V} \otimes \cdots \otimes \underset{i-1.}{\mathrm{id}_V} \otimes \widehat{R} \otimes \underset{i+2.}{\mathrm{id}_V} \otimes \cdots \otimes \underset{n.}{\mathrm{id}_V}$ and $\phi_j := \underset{1.}{\mathrm{id}_V} \otimes \cdots \otimes \underset{i-1.}{\mathrm{id}_V} \otimes \varphi \otimes \underset{i+1.}{\mathrm{id}_V} \otimes \cdots \otimes \underset{n.}{\mathrm{id}_V}$ for $i \in \{1, \ldots, n-1\}$ and $j \in \{1, \ldots, n\}$.

PROPOSITION 2.13 *If $R = (\widehat{R}, \varphi, \boldsymbol{\mu})$ is a balanced Markov-Yang-Baxter operator on an object V in $\mathcal{C}$ then*

$$\mathrm{tr}(c \circ b \circ \boldsymbol{\mu}^{\otimes n}) = \mathrm{tr}(b \circ c \circ \boldsymbol{\mu}^{\otimes n}),$$
$$\mathrm{tr}(\phi_n^{\pm 1} \circ \psi_n^{\mp 1} \circ (b \otimes \mathrm{id}_V) \circ \boldsymbol{\mu}^{\otimes n+1}) = \mathrm{tr}(b \circ \boldsymbol{\mu}^{\otimes n}) \tag{2.16}$$

for any $b, c \in G_n^R$.

PROOF. The first identity of (2.16) follows from property (C) of the trace tr and Definition 2.12.2 since $b, c \in G_n^R$ by assumption. To verify the second identity of (2.16) we use throughout the commutativity of μ according to Definition 2.12.2. Then we apply successively the Properties (C), (E), (D), (D) and Definition 2.12.3 to obtain the following series of equations.

$$\mathrm{tr}\left(\phi_n^{\pm 1} \circ \psi_n^{\mp 1} \circ (b \otimes \mathrm{id}_V) \circ \mu^{\otimes n+1}\right)$$
$$= \mathrm{tr}\left((b \circ \phi_n^{\pm 1} \circ \mu^{\otimes n} \otimes \mathrm{id}_V) \circ (\mathrm{id}_{V^{\otimes n}} \otimes \mu) \circ \psi_n^{\mp 1}\right)$$
$$= \mathrm{tr}\left(\mathrm{tr}_{V^{\otimes n}, V^{\otimes n}; V}\left((b \circ \phi_n^{\pm 1} \circ (\mu^{\otimes n-1} \otimes \mathrm{id}_V) \otimes \mathrm{id}_V) \circ (\mathrm{id}_{V^{\otimes n-1}} \otimes \mu \otimes \mu) \circ \psi_n^{\mp 1}\right)\right)$$
$$= \mathrm{tr}\left(b \circ \phi_n^{\pm 1} \circ (\mu^{\otimes n-1} \otimes \mathrm{id}_V) \circ \mathrm{tr}_{V^{\otimes n}, V^{\otimes n}; V}\left((\mathrm{id}_{V^{\otimes n-1}} \otimes \mu \otimes \mu) \circ \psi_n^{\mp 1}\right)\right)$$
$$= \mathrm{tr}\left(b \circ \phi_n^{\pm 1} \circ (\mu^{\otimes n-1} \otimes \mathrm{id}_V) \circ (\mathrm{id}_{V^{\otimes n-1}} \otimes \mathrm{tr}_{V, V; V}\left(\mu \otimes \mu\right) \circ \psi^{\mp 1}\right)\right)$$
$$= \mathrm{tr}\left(b \circ \phi_n^{\pm 1} \circ (\mu^{\otimes n-1} \otimes \mathrm{id}_V) \circ (\mathrm{id}_{V^{\otimes n-1}} \otimes \phi^{\mp 1} \circ \mu)\right)$$
$$= \mathrm{tr}(b \circ \mu^{\otimes n})$$

■

Examples

The categories of modules and comodules over certain types of bi- and Hopf algebras over a field $\Bbbk$ are a source of the different kinds of categories discussed above. In the simplest case it is known that the category of modules (or comodules) over a bialgebra is monoidal.

Let H-mod be the category of modules over a bi- or Hopf algebra H and by H-comod its comodule category. The subscript $(.)_f$ denotes the corresponding finite-dimensional sub-categories. The following results are either well known or may be derived straightforwardly.

Suppose that H is a bialgebra. Then

1. (H, R) is a quasitriangular bialgebra if and only if H-mod is braided.

2. H is a balanced bialgebra if and only if H-mod is balanced.

3. H is a Hopf algebra with bijective antipode if and only if H-comod$_f$ has left and right duality. If $S^2 = \mathrm{id}_H$ then the induced duality functors are involutive, i.e. $(V^*)^* = V$ and $(f^*)^* = f$. If H is finite-dimensional then involutivity of the duality functor $(-)^*$ implies $S^2 = \mathrm{id}_H$.

4. If H is a ribbon Hopf algebra then H-mod$_f$ is a ribbon category. If H is a finite-dimensional quasitriangular Hopf algebra and H-mod$_f$ is a ribbon category then H is a ribbon Hopf algebra.

5. If H is a Hopf algebra with bijective antipode, then H is a cosovereign Hopf algebra if and only if H-comod$_f$ is a sovereign category.

Next we prove the equivalence of the balanced categories $D(H)[\theta^{\pm 1}]$-mod and $\mathcal{B}(H$-mod), where H is a finite-dimensional Hopf algebra with bijective antipode and $D(H)$ is its quantum double [7]. This result extends the corresponding outcomes in [23] and [17] on the center construction and the double construction respectively.

PROPOSITION 2.14 *Let H be a finite-dimensional Hopf algebra with bijective antipode, and denote by $D(H)$ the quantum double of H. Then the categories $D(H)[\theta, \theta^{-1}]$-mod and $\mathcal{B}(H$-mod) are equivalent balanced categories.*

PROOF. The quantum double $D(H)$ is a quasitriangular Hopf algebra, and therefore the categories $D(H)[\theta, \theta^{-1}]$-mod and $\mathcal{B}(H$-mod) are balanced.

Given a quasitriangular bialgebra B we define the category $\mathcal{D}(B)$ with objects (V, θ_V) where V is a left B-module and $\theta_V : V \to V$ is a B-module isomorphism. The morphisms $f : (V, \theta_V) \to (W, \theta_W)$ are B-module morphisms such that $f \circ \theta_V = \theta_W \circ f$. Then every object (V, θ_V) becomes a $B[\theta, \theta^{-1}]$-module through $\theta \triangleright v := \theta_V(v)$ for $v \in V$. Conversely for every $B[\theta, \theta^{-1}]$-module $(V, \triangleright)$ the morphism $\theta_V := \theta \triangleright (-)$ renders (V, θ_V) an object in $\mathcal{D}(B)$. Then any morphism in one of the categories is morphism in the other category, and one easily verifies that the categories $B[\theta, \theta^{-1}]$-mod and $\mathcal{D}(B)$ are isomorphic.

With the help of this fact one can prove the proposition using exactly the same techniques as in the proof of Theorem 5.4.1.(iii) in [17]. We will not go into details. ∎

3   Ribbon Braids and Ribbon Links

In this section we consider ribbon braids and ribbon links. In the first subsection we discuss the algebraic aspects of ribbon braid groups. In the second subsection we discuss ribbon braids and ribbon links and a ribbon version of the Markov Theorem which implies a one-to-one correspondence of ribbon links and certain equivalence classes of ribbon braids.

Ribbon Braid Group

Ribbon braid groups are defined as follows (compare [14]).

DEFINITION 3.1 *Let $n \in \mathbb{N}$. Then the group RB_n is the group generated by* *the elements $\{\phi_i\}_{i=1}^n$ and $\{\varsigma_j\}_{j=1}^{n-1}$ and the relations*

$$\varsigma_j \cdot \varsigma_k = \varsigma_k \cdot \varsigma_j \quad \forall\, j,k \in \{1, \ldots, n-1\},\ |j-k| > 1,$$
$$\varsigma_i \cdot \varsigma_{i+1} \cdot \varsigma_i = \varsigma_{i+1} \cdot \varsigma_i \cdot \varsigma_{i+1} \quad \forall\, i \in \{1, \ldots, n-2\},$$
$$\phi_i \cdot \phi_j = \phi_j \cdot \phi_i,$$
$$\phi_i \cdot \varsigma_i = \varsigma_i \cdot \phi_{i+1}, \tag{3.1}$$
$$\phi_{i+1} \cdot \varsigma_i = \varsigma_i \cdot \phi_i,$$
$$\phi_i \cdot \varsigma_j = \varsigma_j \cdot \phi_i \quad \text{for } |j-i| > 1 \text{ or } j = i+1.$$

The group RB_n will be called the n.th ribbon braid group. For $n = 0$ we define $RB_0 := \{1\}$.

Because RB_n is canonically embedded in RB_m for $m \geq n$ the direct limit of all RB_n exists and is called the ribbon braid group RB_∞.

LEMMA 3.2 *Every element $b \in RB_n$ can be written in the form $b = \phi_1^{m_1} \cdot \ldots \cdot \phi_n^{m_n} \cdot \omega(\varsigma_j^{\pm 1})$ where $m_i \in \mathbb{Z}$ for all $i \in \{1, \ldots, n\}$ and ω is a word in $(\varsigma_j^{\pm 1})_{j=1}^{n-1}$.* ∎

The ribbon braid group RB_n is a semidirect product (see also [14]).

PROPOSITION 3.3 *The n.th ribbon braid group is isomorphic to the semidirect product $\mathbb{Z}^n {}_\alpha\!\rtimes B_n$ where the action $\alpha : B_n \to \mathrm{Aut}(\mathbb{Z}^n)$ is the canonical action of the underlying symmetric group on the n-tuples $(z_1, \ldots, z_n) \in \mathbb{Z}^n$. Explicitely, the group isomorphism $\chi : RB_n \to \mathbb{Z}^n {}_\alpha\!\rtimes B_n$ is given by $\chi(\phi_i) = (e_i, \mathbb{1})$, $\chi(\varsigma_j) = (0, \varsigma_j)$, where $e_i = (0, \ldots, 1, \ldots, 0)$ is the canonical i.th basis element in $\mathbb{Z}^n$.*

PROOF. Using the defining relations (3.1) of RB_n one easily verifies that χ is a group morphism. The inverse of χ is given by $\chi^{-1} : \mathbb{Z}^n {}_\alpha\!\rtimes B_n \to RB_n$, $\chi^{-1}(\mathbf{z}, \varsigma) := g_1(\mathbf{z}) \cdot g_2(\varsigma)$ where $g_1 : \mathbb{Z}^n \to RB_n$, $g_1(e_i) = \phi_i$ and $g_2 : B_n \to RB_n$, $g_2(\varsigma_j) = \varsigma_j$. ∎

The usual braid group B_∞ (or on B_n) has a linear ordering which is compatible with right multiplication. It is defined as follows [4, 5].

An element $b \in B_n$ is said to be positive/neutral/negative if there exists some $k \in \{1, \ldots, n-1\}$ and a word ω in the generators $\{\varsigma_i\}_{i=1}^{n-1}$ such that respectively

$$b = \begin{cases} \omega(\varsigma_k, \varsigma_{k+1}^{\pm 1}, \ldots, \varsigma_{n-1}^{\pm 1}) & \text{(positive)} \\ \omega(\varsigma_{k+1}^{\pm 1}, \ldots, \varsigma_{n-1}^{\pm 1}) & \text{(neutral)} \\ \omega(\varsigma_k^{-1}, \varsigma_{k+1}^{\pm 1}, \ldots, \varsigma_{n-1}^{\pm 1}) & \text{(negative)} \end{cases} \tag{3.2}$$

Then the linear ordering "$\prec$" is defined as $b \prec b'$ if $b' \cdot b^{-1}$ is positive. It is compatible with right multiplication.

COROLLARY 3.4 *The ribbon braid group RB_n has a linear ordering "$\prec$" which is compatible with right multiplication. It is defined by $(\mathbf{z}, b) \prec (\mathbf{z}', b')$ if and only if $\sum_{i=1}^{n} z_i \leq \sum_{i=1}^{n} z_i'$ and $b \prec b'$.*

PROOF. 1.) Suppose $(\mathbf{z}_1, b_1) \prec (\mathbf{z}_2, b_2) \prec (\mathbf{z}_3, b_3)$. Then $\sum_i z_{1,i} \leq \sum_i z_{2,i} \leq \sum_i z_{3,i}$ and $b_1 \prec b_2 \prec b_3$. Therefore $(\mathbf{z}_1, b_1) \prec (\mathbf{z}_3, b_3)$. 2.) If $(\mathbf{z}_1, b_1) \prec (\mathbf{z}_2, b_2)$ then obviously $(\mathbf{z}_1, b_1) \neq (\mathbf{z}_2, b_2)$ and $(\mathbf{z}_2, b_2) \not\prec (\mathbf{z}_1, b_1)$. 3.) Let $(\mathbf{z}_1, b_1) \prec (\mathbf{z}_2, b_2)$, and $(\mathbf{z}_3, b_3) \in RB_n$ be arbitrary. Then $(\mathbf{z}_j, b_j) \cdot (\mathbf{z}_3, b_3) = (\mathbf{z}_j \cdot (b_j \rhd \mathbf{z}_3), b_j \cdot b_3)$, and $\sum_i (\mathbf{z}_1 \cdot (b_1 \rhd \mathbf{z}_3))_i = \sum_i z_{1,i} + \sum_i (b_1 \rhd \mathbf{z}_3)_i = \sum_i z_{1,i} + \sum_i z_{3,i} \leq \sum_i z_{2,i} + \sum_i z_{3,i} = \sum_i (\mathbf{z}_2 \cdot (b_2 \rhd \mathbf{z}_3))_i$. Hence $(\mathbf{z}_1, b_1) \cdot (\mathbf{z}_3, b_3) \prec (\mathbf{z}_2, b_2) \cdot (\mathbf{z}_3, b_3)$. ∎

LEMMA 3.5 *RB_∞ is torsionfree, i.e. the order of every element $g \neq \mathbb{1}$ is infinite.* ∎

REMARK 4 The ribbon braid group RB_∞ is a strict balanced category in a canonical way. The objects are the non-negative integers $n \in \mathbb{N}_0$, and 0 is the unit object. The morphisms $b : n \to n$ are the elements $b \in RB_n$. If $m \neq n$ then $\hom(n, m) := \emptyset$. Composition in the category is the group multiplication. The tensor product is given by $b \otimes b' = b \cdot \mathrm{sh}^{|b|}(b')$, where $\mathrm{sh} : RB_\infty \to RB_\infty$ is the canonical shift homomorphism assigning $\varsigma_j \mapsto \varsigma_{j+1}$, $\phi_i \mapsto \phi_{i+1}$, and $|b|$ is the minimal integer such that $b \in RB_{|b|}$. The braiding $\Psi_{n,m} : n \otimes m \to m \otimes n$ and the twist $\Theta_n : n \to n$ will be defined using $\Psi_{1,1} := \sigma_1$, $\Theta_1 := \phi_1$ and the defining relations of a balanced category. Thus the category RB_∞ is equivalent to the free balanced category generated by one object [14].

We recall the definition of the braid groups $B_{1,n}$ of type B. It turns out that the ribbon braid group RB_n is a certain quotient of $B_{1,n}$.

DEFINITION 3.6 *Denote by $B_{1,n}$ the n.th braid group of type B. This is the group generated by $\{\varsigma_i\}_{i=1}^{n-1}$ and ϕ_1 obeying the relations[5]*

$$\varsigma_j \cdot \varsigma_k = \varsigma_k \cdot \varsigma_j \ \forall j, k \in \{1, \ldots, n-1\}, \ |j - k| > 1,$$
$$\varsigma_i \cdot \varsigma_{i+1} \cdot \varsigma_i = \varsigma_{i+1} \cdot \varsigma_i \cdot \varsigma_{i+1} \ \forall i \in \{1, \ldots, n-2\},$$
$$\varsigma_1 \cdot \phi_1 \cdot \varsigma_1 \cdot \phi_1 = \phi_1 \cdot \varsigma_1 \cdot \phi_1 \cdot \varsigma_1,$$
$$\phi_1 \cdot \varsigma_i = \varsigma_i \cdot \phi_1 \ \forall i > 1.$$

$$(3.3)$$

Define $\phi_j := \varsigma_{j-1} \cdots \varsigma_1 \cdot \phi_1 \cdot \varsigma_1^{-1} \cdots \varsigma_{j-1}^{-1}$ for all $j \in \{2, \ldots, n\}$. The direct limit of the groups $(B_{1,n})_{n \in \mathbb{N}_0}$ will be denoted by $B_{1,\infty}$.

---

[5]It can be shown that $B_{1,n}$ is isomorphic to the semidirect product $\langle x_1, \ldots, x_n \rangle \rtimes B_n$ where $\langle x_1, \ldots, x_n \rangle$ is the free group generated by n generators (see e.g. [18]).

PROPOSITION 3.7 *There exists a unique group epimorphism* $P_{1,n}^B : B_{1,n} \to RB_n$ *such that* $P_{1,n}^B(\varsigma_i) = \varsigma_i$ *and* $P_{1,n}^B(\phi_j) = \phi_j$ *for all* $i \in \{1, \ldots, n-1\}$ *and* $j \in \{1, \ldots, n\}$. *Explicitly*

$$RB_n = B_{1,n} / \langle \varsigma_i \cdot \phi_i \cdot \varsigma_i^{-1} = \varsigma_i^{-1} \cdot \phi_i \cdot \varsigma_i, \ i \in \{1, \ldots, n-1\} \rangle . \qquad (3.4)$$

PROOF. Obviously the mapping

$$\mathrm{p} : \overline{B_{1,n}} := B_{1,n} / \langle \varsigma_i \cdot \phi_i \cdot \varsigma_i^{-1} = \varsigma_i^{-1} \cdot \phi_i \cdot \varsigma_i, \ i \in \{1, \ldots, n-1\} \rangle \to RB_n$$

which is given by $\mathrm{p}(\varsigma_i) = \varsigma_i$ and $\mathrm{p}(\phi_j) = \phi_j$ is a group morphism. Conversely, the generators $\{\varsigma_i\}_i$ in $\overline{B_{1,n}}$ fulfill the braid identities corresponding to the first and second equation of (3.1). By definition of $B_{1,n}$ it holds $\phi_1 \cdot \varsigma_j = \varsigma_j \cdot \phi_1$ $\forall\, j > 1$. Using the identities (in $\overline{B_{1,n}}$) $\phi_1 \cdot \varsigma_1 = \varsigma_1 \cdot \phi_2$, $\phi_1 \cdot \varsigma_1 \cdot \phi_1 \cdot \varsigma_1^{-1} = \varsigma_1 \cdot \phi_1 \cdot \varsigma_1^{-1} \cdot \phi_1$ and $\phi_{i+1} = \varsigma_i^{-1} \cdot \phi_i \cdot \varsigma_i$ the remaining identities in (3.1) can be proven for the generators $\{\varsigma_i\}_i$ and $\{\phi_j\}_j$ in $\overline{B_{1,n}}$. Since RB_n is universal as semidirect product there is a group morphism $\mathrm{q} : RB_n \to \overline{B_{1,n}}$ with $\mathrm{q}(\varsigma_i) = \varsigma_i$, $\mathrm{q}(\phi_j) = \phi_j$, implying that $\mathrm{q} = \mathrm{p}^{-1}$ and therefore $RB_n \cong \overline{B_{1,n}}$. ∎

Ribbon Braids and Ribbon Links

We will use the notations and results of [26, 31, 14, 33] to define ribbon braids and related topological constructions. We will not go into details and assume that the reader is adequately familiar with this matter. We denote the category of ribbon tangles by R-Tang. In contrast to [26, 31] we use the "up-to-down" composition for tangles in R-Tang.

DEFINITION 3.8 *An oriented directed ribbon n-braid is an oriented downward directed* (n, n)-*ribbon tangle* Ω *in* R-Tang *whose cores form an ordinary n-braid. Two ribbon braids are equivalent if there is an isotopy of ribbon tangles which is an isotopy of braids of the corresponding cores of the ribbon braids.*

REMARK 5 Ribbon links and ribbon tangles can be described equivalently by so-called double links and double tangles and their corresponding double link diagrams and double tangle diagrams [14, 28]. A double link (or double tangle) $\mathcal{L}$ will be denoted by its two components, $\mathcal{L} = (L, L')$. We will henceforth use either the description in terms of ribbon links/tangles or double links/tangles depending on which is the most suitableour purposes.

The above definition and the results of [26, 31, 14] imply that the category R-Braid is a balanced subcategory of R-Tang generated by one object. The special ribbon braids

$$\sigma^- = \qquad\qquad \sigma = \qquad\qquad \varphi^- = \qquad\qquad \varphi = \qquad\qquad \mathrm{id} =$$

generate the morphisms of R-Braid monoidally. Moreover, for given $n \in \mathbb{N}$ the ribbon n-braids R-Braid$_n$ are a group generated by $(\sigma_j)_{j=1}^{n-1}$ and $(\varphi_i)_{i=1}^{n}$ where (in R-Tang)

$$\sigma_j := \mathrm{id}_1. \otimes \ldots \otimes \mathrm{id}_{(j-1)}. \otimes \sigma \otimes \mathrm{id}_{(j+2)}. \otimes \ldots \otimes \mathrm{id}_n.$$

$$\varphi_i := \mathrm{id}_1. \otimes \ldots \otimes \mathrm{id}_{(i-1)}. \otimes \varphi \otimes \mathrm{id}_{(i+1)}. \otimes \ldots \otimes \mathrm{id}_n.$$

These generators obey identities (3.1) if ς_i will be replaced by σ_i and ϕ_j by φ_j. Hence, by Lemma 3.2 and Proposition 3.3 every morphism $\beta \in$ R-Braid$_n$ can be written in the form $\beta = \varphi_1^{m_1} \cdot \ldots \cdot \varphi_n^{m_n} \cdot \omega(\sigma_j^{\pm 1})$ where $m_i \in \mathbb{Z}$ for all $i \in \{1, \ldots, n\}$ and ω is a word in $(\sigma_j^{\pm 1})_{j=1}^{n-1}$. Using standard arguments (see [1, 3, 14, 28, 15]) one concludes that if two ribbon braids $\beta_1 = \varphi_1^{m_1} \cdot \ldots \cdot \varphi_n^{m_n} \cdot \omega_1(\sigma_j^{\pm 1})$ and $\beta_2 = \varphi_1^{k_1} \cdot \ldots \cdot \varphi_n^{k_n} \cdot \omega_2(\sigma_j^{\pm 1})$ are equivalent then $(m_1, \ldots, \mathrm{m}_n) = (k_1, \ldots, k_n)$ and $\omega_1(\sigma_j^{\pm 1}) \cong \omega_2(\sigma_j^{\pm 1})$. Hence, the n.th ribbon braid group RB_n and the group of ribbon n-braids R-Braid$_n$ are canonically isomorphic. Therefore R-Braid is equivalent to the free balanced category generated by one object.

REMARK 6 If a ribbon braid diagram may be represented in standard position [31] we will occasionally denote it by the underlying diagram of the core braid. For example $\sigma^- = \times$.

The closure of a ribbon braid or ribbon tangle had been defined in [32].

DEFINITION 3.9 *Let β be a ribbon n-braid. Then the closure $\widehat{\beta}$ is obtained from β by attaching n oriented bands $s_i = I \times I$ which are directed along the core $I \times \{1/2\}$ such that*

1. *The subset $\{0\} \times I$ of the boundary of the band s_i will be attached to the i.th segment at the bottom of the ribbon braid and $\{1\} \times I$ will be glued to the i.th segment at the top of the ribbon braid in such a way that the orientations and directions are compatible.*

2. *The linking number of the canonically directed boundary strands $I \times \{0\}$ and $I \times \{1\}$ of the band s_i is 0 while the segments $\{0\} \times I$ and $\{1\} \times I$ are kept fixed.*

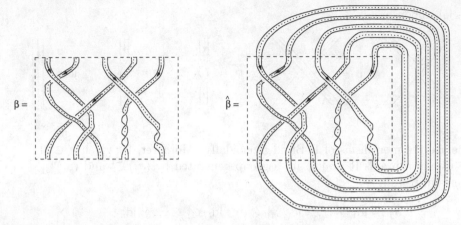

Figure 2: A directed ribbon braid β and its directed closure $\widehat{\beta}$.

3. The bands $(s_i)_{i=1}^{n}$ are pairwise disjoint and intersect with the ribbon braid β only in its segments in $\mathbb{R} \times \{0\}$ and $\mathbb{R} \times \{1\}$ according to (1).

4. Along their directed cores the bands s_i proceed from the bottom to the top of the ribbon braid in a right handed way with respect to the vectors $(1, 0, 0)$ and $(0, 0, 1)$ [6].

An example of a ribbon braid and its closure is shown in Figure 2. Obviously the closure is uniquely defined up to equivalence of ribbon links.

PROPOSITION 3.10 *The closure of a ribbon braid is an oriented directed ribbon link. Conversely every oriented directed ribbon link is the closure of a ribbon braid.*

PROOF. Obviously the closure of a ribbon braid is an oriented directed ribbon link. On the other hand the oriented directed ribbon link $\mathcal{L}$ can be described as a double link (L, L') according to Remark 5. Using standard arguments and results (see [3, 14, 28, 31, 15]) we proceed as follows. We perform a (combinatorial) isotopy I on the directed link L which leads to a closure $\widehat{b}$ (of a downward directed braid b) which is combinatorially equivalent to L [3]. The regular neighbourhood of any component k of L and the accompanying knot k' of k will be mapped homeomorphically into a regular neighbourhood of the transformed knot in $\widehat{b}$ and an accompanying knot in this transformed regular neighbourhood. Hence, $\mathcal{L}$ is ambient isotopic to a ribbon link $\widetilde{\mathcal{L}} = (I(b) = \widehat{b}, I(L'))$ which has a double link diagram where the projection of the knot l' of any component (l, l') of $\widetilde{\mathcal{L}}$ intersects with the projection of any knot of the double link $\widetilde{\mathcal{L}}$ other than l only in the rectangle neighbourhood

---

[6]The vectors $(1, 0, 0)$ and $(0, 0, 1)$ are defined in $\mathbb{R}^2 \times I$ containing the ribbon braid b.

containing the braid b. Application of another appropriate isotopy transforms all selfcrossings of any component (l, l') of $\tilde{\mathcal{L}}$ into the interior of the rectangle neighbourhood containing b and yields a closure of a double braid β which is isotopic to $\mathcal{L}$. ∎

The Markov Theorem for ordinary braids [3] can be adapted to ribbon braids as follows.

THEOREM 3.11 *Let $\mathcal{L}$ and $\tilde{\mathcal{L}}$ be two directed ribbon links represented as closure of double braids, $\mathcal{L} \cong (\widehat{(b, b')}, n)$ and $\tilde{\mathcal{L}} \cong (\widehat{(\tilde{b}, \tilde{b}')}, \tilde{n})$ where $|b| = n$ and $|\tilde{b}| = \tilde{n}$. Then $\mathcal{L}$ is equivalent to $\tilde{\mathcal{L}}$ as directed ribbon link if and only if the ribbon braids (b, b') and $(\tilde{b}, \tilde{b}')$ are connected by a finite sequence of directed ribbon braids $((b, b'), n) \cong ((b_1, b'_1), n_1) \to \dots \to ((b_m, b'_m), n_m) \cong ((\tilde{b}, \tilde{b}'), n')$ such that each ribbon braid $((b_i, b'_i), n_i)$ can be obtained from $((b_{i-1}, b'_{i-1}), n_{i-1})$ by one of the two following moves or their inverses in R-Braid.*

1. $(\beta, r) \to (\gamma \cdot \beta \cdot (\gamma)^{-1}, r)$.

2. $(\beta, r) \to (\varphi_r^{\mp 1} \cdot \sigma_r^{\pm 1} \cdot \beta, r + 1)$.

PROOF. Of course, closures of directed ribbon braids which differ by a finite sequence of the above mentioned moves are isotopic.

Conversely, suppose that $\mathcal{L} \cong \tilde{\mathcal{L}}$ are isotopic ribbon links, and let $\mathcal{L} \cong \widehat{(b, b')}$ and $\tilde{\mathcal{L}} \cong \widehat{(\tilde{b}, \tilde{b}')}$. Then in particular $b \cong b'$ and according to the ordinary Markov Theorem [3] there exists a sequence of ordinary Markov moves $(b, n) \cong (b_1, n_1) \to \dots \to (b_m, n_m) \cong (\tilde{b}, \tilde{n})$ on ordinary braids such that each braid (b_i, n_i) can be obtained from (b_{i-1}, n_{i-1}) by one of the ordinary Markov moves or their inverses. These ordinary moves will be used to design ribbon moves as follows. An ordinary conjugation gives rise to a ribbon braid conjugation where the ordinary generators Ψ_j of the braid group will be replaced by the generators σ_j of the ribbon braid group R-Braid. An ordinary Markov move $(c, r) \to (\Psi_r \cdot c, r + 1)$ will be replaced by a ribbon move $(\beta, r) \to (\varphi_r^{\mp 1} \cdot \sigma_r^{\pm 1} \cdot \beta, r+1)$ of the corresponding ribbon braid β. Then by construction, this sequence of ribbon Markov moves $((b, b'), n) \cong ((b_1, b'_1), n_1) \to \dots \to ((b_m, b'_m), n_m)$ yields a double braid $((b_m, b'_m), n_m)$ whose closure is isotopic to the closure of $((b, b'), n)$ and to $\mathcal{L}$ and $\tilde{\mathcal{L}}$. Since the underlying braid b_m of the double braid (b_m, b'_m) and the braid $\tilde{b}$ of the double braid $(\tilde{b}, \tilde{b}')$ are isotopic by construction, one concludes with the help of Lemma 3.2 that the ribbon braids (b_m, b'_m) and $(\tilde{b}, \tilde{b}')$ differ at most by $(\tilde{b}, \tilde{b}') = \varphi_1^{m_1} \cdot \dots \cdot \varphi_{\tilde{n}}^{m_{\tilde{n}}} \cdot (b_m, b'_m)$. Furthermore the internal linking numbers of each double knot component of the closure of $(\tilde{b}, \tilde{b}')$ and its corresponding counterpart in the closure of (b_m, b'_m) coincide. Hence, the cycle decomposition sums of the exponents $(m_i)_{i=1}^{\tilde{n}}$ vanish. That is, if $\pi = (k_1^1, \dots, k_{s_1}^1) \cdot \dots \cdot (k_1^r, \dots, k_{s_r}^r)$ is the maximal-length cycle decomposition of the underlying permutation π of the ribbon braids (b_m, b'_m) and $(\tilde{b}, \tilde{b}')$ then $\sum_{i=1}^{s_j} m_{k_i^j} = 0$ for all $j \in \{1, \dots, r\}$. Since

R-Braid$_n \cong \mathbb{Z}^n_\alpha \rtimes B_n$ one derives $(b_m, b'_m) = \chi_r^{-1} \cdot \ldots \cdot \chi_1^{-1} \cdot (\tilde{b}, \tilde{b}') \cdot \chi_1 \cdot \ldots \chi_r$ where $\chi_j = \varphi_{k_1^j}^{m_{k_1^j}} \cdot \ldots \cdot \varphi_{k_{s_j-1}^j}^{(\sum_{i=1}^{s_j-1} m_{k_i^j})}$ for $j \in \{1, \ldots, r\}$. This means that (b_m, b'_m) and $(\tilde{b}, \tilde{b}')$ differ by conjugation in R-Braid$_{\tilde{n}}$.   ∎

Like in [3] one argues that a map $f : RB_\infty \cong$ R-Braid$_\infty \to S$ induces an isotopy invariant $\hat{f}$ in the set S if f is invariant under the moves (1) and (2) of Theorem 3.11. And conversely, a ribbon link invariant $\hat{f}$ in S defines a map $f : RB_\infty \cong$ R-Braid$_\infty \to S$, $f(\beta) := \hat{f}(\hat{\beta})$ which is invariant under the moves (1) and (2) of Theorem 3.11.

DEFINITION 3.12 *A ribbon Markov trace* Tr *is a map* Tr : R-Braid$_\infty \to S$ *which obeys the relations* $\mathrm{Tr}(\beta_1 \cdot \beta_2) = \mathrm{Tr}(\beta_2 \cdot \beta_1)$ *and* $\mathrm{Tr}(\varphi_n^{\mp 1} \cdot \sigma_n^{\pm 1} \cdot \beta_1) = \mathrm{Tr}(\beta_1)$ *for any* $\beta_1, \beta_2 \in$ R-Braid$_n$ *and* $n \in \mathbb{N}_0$. *Hence,* Tr *induces a ribbon link invariant* Tr.

In the remainder of this section we lean on Turaev's work [29]. We will define a special class of ribbon Markov traces arising from balanced Markov-Yang-Baxter operators on a balanced category with duality. Suppose that $R = (\hat{R}, \varphi)$ is a balanced Yang-Baxter operator on an object V. Then the previously defined group G_n^R is generated by $(\psi_i)_{i=1}^{n-1}$ and $(\phi_j)_{j=1}^n$ which obey the identities (3.1). Therefore a unique group homomorphism $v_n^R : RB_n \cong$ R-Braid$_n \to G_n^R$ exists such that $v_n^R(\sigma_i) = \psi_i$ and $v_n^R(\varphi_j) = \phi_j$.

PROPOSITION 3.13 *Let* $\mathcal{C}$ *be a balanced category with duality and* $R = (\hat{R}, \varphi, \mu)$ *a balanced Markov-Yang-Baxter operator on an object* V *in* $\mathcal{C}$. *Then* Tr^R : R-Braid$_\infty \to \mathrm{End}_\mathcal{C}(\mathbb{1})$, *defined by* $\mathrm{Tr}^R(\beta) := \mathrm{tr}(v_{|\beta|}^R(\beta) \circ \mu^{\otimes |\beta|})$ *for* $\beta \in$ R-Braid$_{|\beta|}$, *is a ribbon Markov trace, and therefore* $\widehat{\mathrm{Tr}^R}$ *is a ribbon link invariant in* $\mathrm{End}_\mathcal{C}(\mathbb{1})$.

PROOF. Since the $(v_n^R)_{n \in \mathbb{N}_0}$ are group morphisms and $v_{n+1}^R(\beta) = v_n^R(\beta) \otimes \mathrm{id}_V$ for $\beta \in$ R-Braid$_n$ we can use Proposition 2.13 to complete the proof.   ∎

Let henceforth $\mathcal{C}$ be the category of finite-dimensional vector spaces over a field $\Bbbk$ (or more general the finite rank modules over a commutative ring $\Bbbk$). Then any (balanced Markov-)Yang-Baxter operator can be rescaled, $R = (\hat{R}, \varphi, \mu) \to R' = (\lambda \cdot \hat{R}, \lambda \cdot \varphi, \mu)$, $\lambda \in \Bbbk^*$.

Suppose that $R = (\hat{R}, \varphi, \mu)$ is a balanced Markov-Yang-Baxter operator on a vector space V and $\sum_i \alpha_i \cdot \hat{R}^i = 0$. It follows from the properties of R that $(\sum_{i:\mathrm{odd}} \alpha_i \cdot \hat{R}^i) \circ (\mathrm{id}_V \otimes \varphi - \varphi \otimes \mathrm{id}_V) = 0$ and $(\sum_{i:\mathrm{even}} \alpha_i \cdot \hat{R}^i) \circ (\mathrm{id}_V \otimes \varphi - \varphi \otimes \mathrm{id}_V) = 0$. If one of the partial sums $\sum_{i \text{ odd}} \alpha_i \cdot \hat{R}^i$ or $\sum_{i \text{ even}} \alpha_i \cdot \hat{R}^i$ is an isomorphism the linear morphism φ is a multiple of the identity, $\lambda \cdot \mathrm{id}_V$. Then, rescaling with this paramter λ yields an enhanced Yang-Baxter operator [29]. The so obtained ribbon link invariant of a ribbon link $\mathcal{L} = (L, L')$ is the link

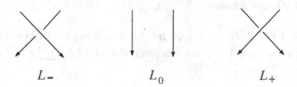

Figure 3: Conway triple (L_-, L_0, L_+)

invariant of L multiplied by λ^m where $m = \sum_{(k,k') \in \mathcal{L}} \mathrm{lk}(k, k')$ is the sum of the internal linking numbers (or generalized twists) of the particular components (k, k') of the ribbon link $\mathcal{L}$. Especially the polynomial invariants related to certain algebras allowing algorithmic computation of the invariants in terms of skein relations on Conway triples (see Figure 3) only admit such kind of MYB invariants of ribbon links. Examples are the invariants coming from all sorts of Hecke algebras, like the Jones or the HOMFLYPT polynomials. More interesting with this respect are therefore balanced MYB operators with higher order minimal polynomials yielding more subtle annihilation relations – of course, such MYBs need to be constructed explicitly. Similarly as in [29] we define two sorts of annihilation of ribbon link invariants. Suppose that T is an invariant of (directed) ribbon links in a $\Bbbk$-module S. Let $f \in \Bbbk[t, t^{-1}]$ be a Laurent polynomial, $f = \sum_{i=p}^{q} k_i \cdot t^i$. Then we say that f annihilates T with respect to the braiding, $f *_\sigma T = 0$, if for any tuple of oriented ribbon links $(L_p, L_{p+1}, \dots, L_q)$ whose diagrams coincide outside some small disk D and differ by

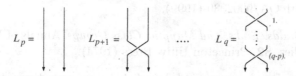

$$(3.5)$$

within the disk D, the equation $\sum_{i=p}^{q} k_i \cdot T(L_i) = 0$ holds.

A Laurent polynomial $g = \sum_{i=r}^{s} l_i \cdot t^i$ annihilates the ribbon link invariant T with respect to the twisting, $g *_\varphi T = 0$, if for any tuple of oriented ribbon links $(L_r, L_{r+1}, \dots, L_s)$ whose diagrams coincide outside some small disk D and differ by

$$L_r = \quad L_{r+1} = \quad \dots \quad L_s = \qquad (3.6)$$

within the disk D, the equation $\sum_{i=r}^{s} l_i \cdot T(L_i) = 0$ holds.

PROPOSITION 3.14 *Let* $R = (\widehat{R}, \varphi, \mu)$ *be a balanced Markov-Yang-Baxter operator on a* $\Bbbk$*-module* V *either obeying the identity* $\sum_{i=p}^{q} k_i \cdot \widehat{R}^i = 0$ *or* $\sum_{j=r}^{s} l_j \cdot \phi^j = 0$. *Then*

$$\left(\sum_{i=p}^{q} k_i \cdot t^i \right) *_{\sigma} \widehat{\mathrm{Tr}^R} = 0 \quad and \quad \left(\sum_{j=r}^{s} l_j \cdot t^j \right) *_{\varphi} \widehat{\mathrm{Tr}^R} = 0 \qquad (3.7)$$

hold respectively.

PROOF. Let $(L_k)_k$ be ribbon links whose diagrams differ in a sufficiently small disk D either by (3.5) or (3.6). Then the identities (3.7) can be derived in a similar way as the corresponding results in [29].   ∎

REMARK 7 As in [29] one argues that for any ribbon EYB operator $R :=$ $(\widehat{R}, \Phi, \mu)$ over a finite-dimensional vector space $\Bbbk$ there exist nontrivial polynomials f and g of degree $\leq \dim(V)^2$ and $\leq \dim(V)$ respectively such that $f *_{\sigma} T^R = 0$ and $g *_{\varphi} T^R = 0$.

References

[1] E. Artin, *Theory of Braids*, Ann. of Math., **48**, No. 2, 101–126 (1947).

[2] J. Bichon, *Cosovereign Hopf Algebras*, Dept. Math. Univ. Montpelier, Preprint, math.QA/9902030 (1999).

[3] J. Birman, *Braids, Links, and Mapping Class Groups*, Annals of Mathematics Studies, **82**, Princeton Univ. Press (1974).

[4] P. Dehornoy, *Braid groups and left distributive operations*, Trans. Amer. Math. Soc., **345**, No. 1, 115–151 (1994).

[5] P. Dehornoy, *Strange questions about braids*, J. Knot Theory and its Ramifications, **8**, No. 5, 589–620 (1999).

[6] P. Deligne, *Catégories tannakiennes*, The Grothendieck Festschrift, 111–195, **2**, Birkhäuser (1990).

[7] V. G. Drinfel'd, *Quantum groups*, Proceedings of the International Congress of Mathematicians, A. Gleason ed., 798–820 (1987).

[8] P. Freyd and D. Yetter, *Braided compact closed categories with applications to low dimensional topology*, Adv. Math., **77**, 156–182, (1989).

[9] P. Freyd and D. Yetter, *Coherence Theorems via Knot Theory*, J. Pure Appl. Algebra, **78**, 49–76 (1992).

[10] A. Joyal and R. Street, *Braided tensor categories*, Macquarie Math. Reports, **860081** (1986).

[11] A. Joyal and R. Street, *An introduction to Tannaka duality and quantum groups*, Lecture Notes in Mathematics, **1488**, 412–492, Category Theory Proceedings, A. Carboni et al eds. (1990).

[12] A. Joyal and R. Street, *The geometry of tensor calculus*, Adv. Math., **88**, 55–112, (1991).

[13] A. Joyal and R. Street, *Tortile Yang-Baxter operators in tensor categories*, J. Pure Appl. Algebra, **71**, 43–51 (1991).

[14] A. Joyal and R. Street, *Braided tensor categories*, Adv. Math., **102**, 20–78, (1993).

[15] C. Kassel, *Quantum Groups*, GTM **155**, Springer (1995).

[16] G. M. Kelly and M. L. Laplaza, *Coherence for compact closed categories*, J. Pure Appl. Algebra, **19**, 193–213, (1980).

[17] C. Kassel and V. G. Turaev, *Double construction for monoidal categories*, Acta Math., **175**, 1–48, (1995).

[18] S. Lambropoulou, *Knot theory related to generalized and cyclotomic Hecke algebras of type B*, J. Knot Theory and its Ramifications, **8**, No. 5, 621–658, (1999).

[19] W. B. R. Lickorish, *An Introduction to Knot Theory*, GTM **175**, Springer (1997).

[20] V. V. Lyubashenko, *Tangles and Hopf algebras in braided categories*, J. Pure Appl. Algebra, **98**, 245 (1995).

[21] S. Mac Lane, *Natural associativity and commutativity*, Rice University Studies, **49**, 28 (1963).

[22] S. Mac Lane, *Categories. For the Working Mathematician*, GTM **5**, Springer (1972).

[23] S. Majid, *Representations, Duals and Quantum Doubles of Monoidal Categories*, Rend. Circ. Math. Palermo, **26**, No. 2, 197–206, (1991).

[24] S. Majid, *Algebras and Hopf algebras in braided categories*, Advances in Hopf Algebras, Lecture Notes in Pure and Appl. Math., Dekker, **158**, 55 (1994).

[25] G. Maltsiniotis, *Traces dans les catégories monoïdales, dualité et catégories monoïdales fibrées*, Cahiers Topologie Géom. Différentielle Catég., **36**, No. 3, 195–288 (1995).

[26] N. Yu. Reshetikhin and V. G. Turaev, *Ribbon Graphs and Their Invariants Derived from Quantum Groups*, Commun. Math. Phys., **127**, 1 (1990).

[27] C. P. Rourke and B. J. Sanderson, *Introduction to Piecewise-Linear Topology*, Ergebnisse der Mathematik und ihrer Grenzgebiete **69**, Springer (1972).

[28] M. C. Shum, *Tortile tensor categories*, J. Pure Appl. Algebra **93**, 57–110, (1994).

[29] V. G. Turaev, *The Yang-Baxter equation and invariants of links*, Inventiones Math. **92**, 527–553, (1988).

[30] V. G. Turaev, *Operator invariants of tangles and R-matrices*, Math. USSR Izvestia **35**, 411–444, (1990).

[31] V. G. Turaev, *Quantum Invariants of Knots and 3-Manifolds*, Studies in Mathematics **18**, Walter de Gruyter (1994).

[32] D. Yetter, *Markov Algebras*, Contemporary Mathematics **78**, 705–730 (1988).

[33] D. Yetter, *Framed Tangles and a Theorem of Deligne on Braided Deformations of Tannakian Categories*, Contemporary Mathematics **134**, 325–350 (1992).

Mail to: DAMTP, UNIVERSITY OF CAMBRIDGE, SILVER STREET, CAMBRIDGE CB3 9EW, UK. **E-mail:** b.drabant@damtp.cam.ac.uk

Lectures on the dynamical Yang-Baxter equations

Pavel Etingof and Olivier Schiffmann

1 Introduction

This paper arose from a minicourse given by the first author at MIT in the Spring of 1999, when the second author extended and improved his lecture notes of this minicourse. It contains a systematic and elementary introduction to a new area of the theory of quantum groups – the theory of the classical and quantum dynamical Yang-Baxter equations.

The quantum dynamical Yang-Baxter equation is a generalization of the ordinary quantum Yang-Baxter equation. It first appeared in physical literature in the work of Gervais and Neveu [GN], and was first considered from a mathematical viewpoint by Felder [F], who attached to every solution of this equation a quantum group, and an interesting system of difference equations, - the quantum Knizhnik-Zamolodchikov-Bernard (qKZB) equation. Felder also considered the classical analogue of the quantum dynamical Yang-Baxter equation – the classical dynamical Yang-Baxter equation. Since then, this theory was systematically developed in many papers, some of which are listed below. By now, the theory of the classical and quantum dynamical Yang-Baxter equations and their solutions has many applications, in particular to integrable systems and representation theory. To discuss this theory and some of its applications is the goal of this paper.

The structure of the paper is as follows.

In Section 2 we consider the exchange construction, which is a natural construction in classical representation theory that leads one to discover the quantum dynamical Yang-Baxter equation and interesting solutions of this equation (dynamical R-matrices). In this section we define the main objects of the paper – the fusion and exchange matrices for Lie algebras and quantum groups, and compute them for the Lie algebra sl_2 and quantum group $U_q(sl_2)$.

In Section 3 we define the quantum dynamical Yang-Baxter equation, and see that the exchange matrices are solutions of this equation. We also study the quasiclassical limit of the quantum dynamical Yang-Baxter equation – the classical dynamical Yang-Baxter equation. We conjecture that any solution of this equation can be quantized. We compute classical limits of exchange matrices, which provides interesting examples of solutions of the classical dynamical Yang-Baxter equation, which we call basic solutions.

In Section 4 we give a classification of solutions of the classical dynamical Yang-Baxter equation for simple Lie algebras defined on a Cartan subalgebra, satisfying the unitarity condition. The result is, roughly, that all such solutions can be obtained from the basic solutions.

In Section 5 we discuss the geometric interpretation of solutions of the classical dynamical Yang-Baxter equation, which generalizes Drinfeld's geometric interpretation of solutions of the classical Yang-Baxter equation via Poisson-Lie groups. This interpretation is in terms of Poisson-Lie groupoids introduced by Weinstein.

In Section 6 we give a classification of solutions of the quantum dynamical Yang-Baxter equation for the vector representation of gl_N, satisfying the Hecke condition. As in the classical case, the result states that all such solutions can be obtained from the basic solutions which arise from the exchange construction.

In Section 7 we discuss the "noncommutative geometric" interpretation of solutions of the quantum dynamical Yang-Baxter equation, which generalizes the interpretation of solutions of the quantum Yang-Baxter equation via quantum groups. This interpretation is in terms of quantum groupoids (or, more precisely, H-Hopf algebroids).

In Section 8 we give a defining equation satisfied by the universal fusion matrix – the Arnaudon-Buffenoir-Ragoucy-Roche (ABRR) equation, and prove it in the Lie algebra case. We give applications of this equation to computing the quasiclassical limit of the fusion matrix, and to computation of the fusion matrix itself for sl_2.

In Section 9 we discuss the connection of solutions of the quantum dynamical Yang-Baxter equation to integrable systems and special functions, in particular to Macdonald's theory. Namely, we consider weighted traces of intertwining operators between representations of quantum groups, and give difference equations for them which in a special case reduce to Macdonald-Ruijsenaars difference equations.

Appendix A contains the classification of solutions of the classical dynamical Yang-Baxter equation for simple Lie algebras defined on subspaces of the Cartan subalgebra.

Appendix B contains a proof of the ABRR equation in the quantum case.

At the end we review some of the existing literature that is relevant to the theory of the dynamical Yang-Baxter equations.

To keep these lectures within bounds, we do not discuss dynamical Yang-Baxter equations with spectral parameter. These equations are related to affine Lie algebras and quantum affine algebras just like the equations without spectral parameter are related to finite dimensional Lie algebras and quantum groups. Most of the definitions and results of these lectures can be carried over to this case, which gives rise to a more interesting but also more complicated theory than the theory described here. A serious discussion of this theory would require a separate course of lectures.

Acknowledgements. We thank the participants of the minicourse at MIT and of the "Quantum groups" conference in Durham (July 1999) for interesting remarks and discussions. We are grateful to IHES and Harvard University for hospitality. The work of P.E. was partially supported by the NSF grant DMS-9700477, and was partly done while he was employed by the Clay Mathematics Institute as a CMI Prize Fellow.

2 Intertwining operators, fusion and exchange matrices.

2.1. The exchange construction. We start by giving a simple and natural construction in classical representation theory which leads to discovery of the quantum dynamical Yang-Baxter equation.

Let $\mathfrak{g}$ be a simple complex Lie algebra, $\mathfrak{h} \subset \mathfrak{g}$ a Cartan subalgebra and $\Delta \subset \mathfrak{h}^*$ the associated root system. Let Π be a set of simple roots, $\Delta^+ \subset \Delta$ the associated system of positive roots. Let $\mathfrak{g} = \mathfrak{n}_- \oplus \mathfrak{h} \oplus \mathfrak{n}_+$ be the corresponding polarization of $\mathfrak{g}$ and let $\mathfrak{g}_\alpha$ be the root subspaces of $\mathfrak{g}$. Let $\langle \, , \, \rangle$ be the nondegenerate invariant symmetric form on $\mathfrak{g}$ normalized by the condition $\langle \alpha, \alpha \rangle = 2$ for long roots. Finally, for each $\alpha \in \Delta$, choose some $e_\alpha \in \mathfrak{g}_\alpha$ in such a way that $\langle e_\alpha, e_{-\alpha} \rangle = 1$.

For $\lambda \in \mathfrak{h}^*$, let $\mathbb{C}_\lambda$ be the one-dimensional $(\mathfrak{h} \oplus \mathfrak{n}_+)$-module such that $\mathbb{C}_\lambda = \mathbb{C} x_\lambda$ with $h.x_\lambda = \lambda(h)x_\lambda$ for $h \in \mathfrak{h}$ and $\mathfrak{n}_+.x_\lambda = 0$. The Verma module of highest weight λ is the induced module

$$M_\lambda = \mathrm{Ind}_{\mathfrak{h} \oplus \mathfrak{n}_+}^{\mathfrak{g}} \, \mathbb{C}_\lambda.$$

Notice that M_λ is a free $U(\mathfrak{n}_-)$-module and can be identified with $U(\mathfrak{n}_-)$ as a linear space by the map $U(\mathfrak{n}_-) \xrightarrow{\sim} M_\lambda$, $u \mapsto u.x_\lambda$.

Define a partial order on $\mathfrak{h}^*$ by putting $\mu < \nu$ if there exist $\alpha_1, \ldots \alpha_r \in \Delta^+$, $r > 0$, such that $\nu = \mu + \alpha_1 + \ldots + \alpha_r$. Let $M_\lambda = \bigoplus_{\mu \leq \lambda} M_\lambda[\mu]$ denote the decomposition of M_λ into weight subspaces.

The following proposition is standard.

Proposition 2.1. *The module M_λ is irreducible for generic values of λ.*

Define also the dual Verma module M_λ^* to be the graded dual vector space $\bigoplus_\mu M_\lambda[\mu]^*$ equipped with the following $\mathfrak{g}$-action:

$$(a.u)(v) = -u(a.v) \; \forall a \in \mathfrak{g}, \; u \in M_\lambda^*, \; v \in M_\lambda.$$

Let x_λ^* be the lowest weight vector of M_λ^* satisfying $\langle x_\lambda, x_\lambda^* \rangle = 1$.

Now let V be a finite-dimensional $\mathfrak{g}$-module. Let $V = \bigoplus_{\nu \in \mathfrak{h}^*} V[\nu]$ be its decomposition into weight subspaces. Let $\lambda, \mu \in \mathfrak{h}^*$ and let us consider $\mathfrak{g}$-module intertwining operators

$$\Phi : \ M_\lambda \to M_\mu \otimes V.$$

If Φ is such an intertwining operator, define its "expectation value" by

$$\langle \Phi \rangle = \langle \Phi.x_\lambda, x_\mu^* \rangle \in V[\lambda - \mu].$$

Remark. This definition is similar to the notion of expectation value in quantum field theory.

Proposition 2.2. *Let M_μ be irreducible. Then the map*

$$\mathrm{Hom}_{\mathfrak{g}}(M_\lambda, M_\mu \otimes V) \to V[\lambda - \mu], \ \ \Phi \mapsto \langle \Phi \rangle$$

is an isomorphism.

Proof. By Frobenius reciprocity, we have

$$\mathrm{Hom}_{\mathfrak{g}}(M_\lambda, M_\mu \otimes V) = \mathrm{Hom}_{\mathfrak{h} \oplus \mathfrak{n}_+}(\mathbb{C}_\lambda, M_\mu \otimes V) = \mathrm{Hom}_{\mathfrak{h} \oplus \mathfrak{n}_+}(\mathbb{C}_\lambda \otimes M_\mu^*, V).$$

Moreover, since M_μ is irreducible, we have $M_\mu^* = \mathrm{Ind}_{\mathfrak{h}}^{\mathfrak{h} \oplus \mathfrak{n}_+} \mathbb{C}_{-\mu}$ as an $\mathfrak{h} \oplus \mathfrak{n}_+$-module. In particular,

$$\mathrm{Hom}_{\mathfrak{h} \oplus \mathfrak{n}_+}(\mathbb{C}_\lambda \otimes M_\mu^*, V) = \mathrm{Hom}_{\mathfrak{h}}(\mathbb{C}_\lambda \otimes \mathbb{C}_{-\mu}, V) = V[\lambda - \mu].$$

■

This proposition can be reformulated as follows: for any $v \in V[\lambda - \mu]$ there exists a unique intertwining operator $\Phi_\lambda^v : \ M_\lambda \to M_\mu \otimes V$ such that

$$\Phi_\lambda^v(x_\lambda) \in x_\mu \otimes v + \bigoplus_{\nu < \mu} M_\mu[\nu] \otimes V.$$

Notice that Φ_λ^v (for fixed v) is defined only for generic values of λ. Identifying the Verma modules M_λ and M_μ with $U(\mathfrak{n}_+)$, we can view Φ_λ^v as a linear map $U(\mathfrak{n}_+) \to U(\mathfrak{n}_+) \otimes V$. It is easy to see that the coefficients of this map (in any basis) are rational functions of λ.

We would now like to consider the "algebra" of such intertwining operators. Let us denote by $\mathrm{wt}(u) \in \mathfrak{h}^*$ the weight of any homogeneous vector u in a $\mathfrak{g}$-module. Let V, W be two finite-dimensional $\mathfrak{g}$-modules, and let $v \in V$, $w \in W$ be two homogeneous vectors. Let $\lambda \in \mathfrak{h}^*$ and consider the composition

$$\Phi_\lambda^{w,v} : \ M_\lambda \xrightarrow{\Phi_\lambda^v} M_{\lambda - \mathrm{wt}(v)} \otimes V \xrightarrow{\Phi_{\lambda - \mathrm{wt}(v)}^w} M_{\lambda - \mathrm{wt}(v) - \mathrm{wt}(w)} \otimes W \otimes V.$$

(Here and below we abuse notations and write Φ instead of $\Phi \otimes 1$). Then $\Phi_\lambda^{w,v} \in \mathrm{Hom}_{\mathfrak{g}}(M_\lambda, M_{\lambda - \mathrm{wt}(v) - \mathrm{wt}(w)} \otimes W \otimes V)$. Hence by Proposition 1.2, for generic λ there exists a unique element $u \in V \otimes W[\mathrm{wt}(v) + \mathrm{wt}(w)]$ such that $\Phi_\lambda^u = \Phi_\lambda^{w,v}$. It is clear that the assignment $(v, w) \mapsto u$ is bilinear, and defines an $\mathfrak{h}$-linear map

$$J_{WV}(\lambda) : W \otimes V \to W \otimes V,$$
$$w \otimes v \mapsto \langle \Phi_\lambda^{w,v} \rangle$$

Definition. We call the operator $J_{WV}(\lambda)$ the *fusion matrix* of V and W.

We will now list some fundamental properties of fusion matrices. First let us introduce an important piece of notation to be used throughout this text. If $A_1, \ldots A_r$ are semisimple $\mathfrak{h}$-modules and $F(\lambda) : A_1 \otimes \ldots \otimes A_r \to A_1 \otimes \ldots \otimes A_r$ is a linear operator depending on $\lambda \in \mathfrak{h}^*$ then, for any homogeneous $a_1, \ldots a_r$ we set

$$F(\lambda - h^{(i)})(a_1 \otimes \ldots \otimes a_r) := F(\lambda - \mathrm{wt}(a_i))(a_1 \otimes \ldots \otimes a_r).$$

Proposition 2.3. *Let V, W be finite-dimensional $\mathfrak{g}$-modules. Then*

1. *$J_{WV}(\lambda)$ is a rational function of λ.*

2. *$J_{WV}(\lambda)$ is strictly lower triangular, i.e. $J = 1 + N$ where*

$$N(W[\nu] \otimes V[\mu]) \subset \bigoplus_{\tau < \nu, \mu < \sigma} W[\tau] \otimes V[\sigma].$$

In particular, $J_{WV}(\lambda)$ is invertible.

3. *Let U, V, W be finite-dimensional $\mathfrak{g}$-modules. Then the fusion matrices satisfy the following dynamical 2-cocycle condition:*

$$J_{U \otimes W, V}(\lambda)(J_{UW}(\lambda - h^{(3)}) \otimes 1) = J_{U, W \otimes V}(\lambda)(1 \otimes J_{WV}(\lambda)).$$

on $U \otimes V \otimes W$.

Proof. Statements 1. and 2. follow from the definitions and from the fact that the intertwining operators Φ_λ^v are rational functions of λ. To prove statement 3., let $u \in U$, $v \in V$, $w \in W$ be homogeneous elements and consider the composition

$$M_\lambda \xrightarrow{\Phi_\lambda^v} M_{\lambda - \mathrm{wt}(v)} \otimes V \xrightarrow{\Phi_{\lambda - \mathrm{wt}(v)}^w} M_{\lambda - \mathrm{wt}(v) - \mathrm{wt}(w)} \otimes W \otimes V$$
$$\xrightarrow{\Phi_{\lambda - \mathrm{wt}(v) - \mathrm{wt}(w)}^u} M_{\lambda - \mathrm{wt}(u) - \mathrm{wt}(v) - \mathrm{wt}(w)} \otimes U \otimes W \otimes V.$$

The dynamical 2-cocycle condition follows from the associativity relation

$$\Phi_{\lambda - \mathrm{wt}(v) - \mathrm{wt}(w)}^u \circ (\Phi_{\lambda - \mathrm{wt}(v)}^w \circ \Phi_\lambda^v) = (\Phi_{\lambda - \mathrm{wt}(v) - \mathrm{wt}(w)}^u \circ \Phi_{\lambda - \mathrm{wt}(v)}^w) \circ \Phi_\lambda^v$$

and from the definition of the fusion matrices. ∎

The fusion matrices can be viewed as the structure constants for multiplication in the "algebra" of intertwining operators. We now turn to the structure constants for "commutation relations". Let V, W be two finite-dimensional $\mathfrak{g}$-modules. Let us define

$$R_{VW}(\lambda) = J_{VW}(\lambda)^{-1} J_{WV}^{21}(\lambda) \in \mathrm{Hom}_{\mathfrak{h}}(V \otimes W, V \otimes W),$$

where $J^{21} = PJP$ with $P(x \otimes y) = y \otimes x$. The above definition can be rephrased in terms of intertwining operators as follows: $R_{VW}(\lambda)(v \otimes w) = \sum_i v_i \otimes w_i$ where $\Phi_\lambda^{w,v} = P \sum_i \Phi_\lambda^{v_i, w_i}$.

Definition. The operator $R_{VW}(\lambda)$ is called the *exchange matrix* of V and W.

Proposition 2.4. *Let U, V, W be three finite-dimensional $\mathfrak{g}$-modules. Then the exchange matrices satisfy the following relation*

$$R_{VW}(\lambda - h^{(3)})R_{VU}(\lambda)R_{WU}(\lambda - h^{(1)}) = R_{WU}(\lambda)R_{VU}(\lambda - h^{(2)})R_{VW}(\lambda) \quad (2.1)$$

in the algebra $\mathrm{Hom}_{\mathfrak{h}}(V \otimes W \otimes U, V \otimes W \otimes U)$.

Proof. Let $u \in U$, $v \in V$, $w \in W$ be homogeneous elements and, as in Proposition 2.3, consider the composition $\Phi_\lambda^{u,w,v} = \Phi_{\lambda-\mathrm{wt}(v)-\mathrm{wt}(w)}^{u} \circ \Phi_{\lambda-\mathrm{wt}(v)}^{w} \circ \Phi_\lambda^{v}$. The proof of relation (2.1) is obtained by rewriting $\Phi_\lambda^{u,w,v}$ as $\sum \sigma \Phi_\lambda^{v_i, w_i, u_i}$ where $\sigma : U \otimes W \otimes V \to V \otimes W \otimes U$, $x \otimes y \otimes z \mapsto z \otimes y \otimes x$, using exchange matrices in two different ways according to the following hexagon

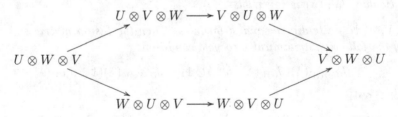

Remark. One can also deduce this proposition from Part 3. of Proposition 2.3. Namely, one can show that if $J(\lambda)$ is any element of the completion of $U(\mathfrak{g}) \otimes U(\mathfrak{g})$ which satisfies the dynamical 2-cocycle condition (where $J_{VW}(\lambda)$ denotes the projection of $J(\lambda)$ to the product $V \otimes W$ of finite dimensional modules V, W) then the element $R(\lambda) = J(\lambda)^{-1} J^{21}(\lambda)$ satisfies the quantum dynamical Yang-Baxter equation.

Example 1. Let us evaluate the fusion and exchange matrices in the simplest example. Namely, take $\mathfrak{g} = \mathfrak{sl}_2 = \mathbb{C}e \oplus \mathbb{C}h \oplus \mathbb{C}f$ and $V = \mathbb{C}^2 = \mathbb{C}v_+ \oplus \mathbb{C}v_-$ with

$$h.v_\pm = \pm v_\pm, \qquad e.v_- = v_+, \qquad e.v_+ = 0, \qquad f.v_- = 0, \qquad f.v_+ = v_-.$$

Let us compute the fusion matrix $J_{VV}(\lambda)$. By the triangularity property of $J_{VV}(\lambda)$, we have

$$J_{VV}(\lambda)(v_\pm \otimes v_\pm) = v_\pm \otimes v_\pm, \qquad J_{VV}(\lambda)(v_- \otimes v_+) = v_- \otimes v_+,$$

so it remains to compute $J_{VV}(\lambda)(v_- \otimes v_+)$. Consider the intertwiner $\Phi_\lambda^{v_-}$: $M_\lambda \to M_{\lambda+1} \otimes V$. By definition, $\Phi_\lambda^{v_-}(x_\lambda) = x_{\lambda+1} \otimes v_- + y(\lambda)fx_{\lambda+1} \otimes v_+$. To determine the function $y(\lambda)$, we use the intertwining property:

$$\begin{aligned}
0 = \Phi_\lambda^{v_-}(ex_\lambda) = (e \otimes 1 + 1 \otimes e)\Phi_\lambda^{v_-}(x_\lambda) &= x_{\lambda+1} \otimes v_+ + y(\lambda)efx_{\lambda+1} \otimes v_+ \\
&= x_{\lambda+1} \otimes v_+ + y(\lambda)(h + fe)x_{\lambda+1} \otimes v_+ \\
&= x_{\lambda+1} \otimes v_+ + (\lambda+1)y(\lambda)x_{\lambda+1} \otimes v_+
\end{aligned}$$

Hence $y(\lambda) = -\frac{1}{\lambda+1}$. A similar computation shows that $\Phi_{\lambda+1}^{v_+}(x_{\lambda+1}) = x_\lambda \otimes v_+$. Thus

$$\Phi_\lambda^{v_+,v_-}(x_\lambda) = \Phi_{\lambda+1}^{v_+}\Phi_\lambda^{v_-}(x_\lambda) = x_\lambda \otimes (v_+ \otimes v_- - \frac{1}{\lambda+1}v_- \otimes v_+) + \text{lower weight terms}.$$

Therefore $J_{VV}(\lambda)(v_+ \otimes v_-) = v_+ \otimes v_- - \frac{1}{\lambda+1}v_- \otimes v_+$, and

$$J_{VV}(\lambda) = \begin{pmatrix} 1 & 0 & 0 & 0 \\ 0 & 1 & 0 & 0 \\ 0 & -\frac{1}{\lambda+1} & 1 & 0 \\ 0 & 0 & 0 & 1 \end{pmatrix}$$

The exchange matrix is now easily computed. In the basis $(v_+ \otimes v_+, v_+ \otimes v_-, v_- \otimes v_+, v_- \otimes v_-)$ it is given by

$$R_{VV}(\lambda) = \begin{pmatrix} 1 & 0 & 0 & 0 \\ 0 & 1 & -\frac{1}{\lambda+1} & 0 \\ 0 & \frac{1}{\lambda+1} & 1 - \frac{1}{(\lambda+1)^2} & 0 \\ 0 & 0 & 0 & 1 \end{pmatrix}.$$

2.2. Generalization to quantum groups.

The construction of intertwining operators, fusion and exchange matrices admit natural quantum analogues. Let $U_q(\mathfrak{g})$ be the quantum universal enveloping algebra associated to $\mathfrak{g}$, as defined in [CP], Chapter 6., and for each $\lambda \in \mathfrak{h}^*$, let M_λ be the Verma module of highest weight λ. Then Proposition 2.2 and the definition of the fusion matrices $J_{W,V}(\lambda)$ are identical to the classical case. In this situation,

Proposition 2.3, parts 2., 3. hold. However, the fusion matrices are no longer rational functions of λ, but rather trigonometric functions (i.e rational functions of $q^{<\lambda,\alpha>}$, $\alpha \in \Delta$).

Let $\mathcal{R} \in U_q(\mathfrak{g}) \hat{\otimes} U_q(\mathfrak{g})$ be the universal R-matrix of $U_q(\mathfrak{g})$. Let V, W be two finite-dimensional $U_q(\mathfrak{g})$-modules. The exchange matrix is defined as

$$R_{VW}(\lambda) = J_{VW}^{-1}(\lambda) \mathcal{R}_{VW}^{21} J_{W,V}^{21}(\lambda)$$

where $\mathcal{R}_{VW}^{21}$ is the evaluation of $\mathcal{R}^{21}$ on $V \otimes W$.

In terms of intertwining operators, the exchange matrix has the following interpretation. Recall that if V and W are any two $U_q(\mathfrak{g})$-modules then $P\mathcal{R}_{VW} : V \otimes W \to W \otimes V$ is a $U_q(\mathfrak{g})$-intertwiner. Then $R_{VW}(\lambda)(v \otimes w) = \sum_i v_i \otimes w_i$ where $P\mathcal{R}_{WV}\Phi_\lambda^{w,v} = \sum_i \Phi_\lambda^{v_i,w_i}$.

With this definition, Proposition 2.4 is satisfied. The quantum analogues of the fusion and exchange matrices in example 1 are

$$J_{VV}(\lambda) = \begin{pmatrix} 1 & 0 & 0 & 0 \\ 0 & 1 & 0 & 0 \\ 0 & \frac{q^{-1}-q}{q^{2(\lambda+1)}-1} & 1 & 0 \\ 0 & 0 & 0 & 1 \end{pmatrix},$$

$$R_{VV}(\lambda) = \begin{pmatrix} q & 0 & 0 & 0 \\ 0 & 1 & \frac{q^{-1}-q}{q^{2(\lambda+1)}-1} & 0 \\ 0 & \frac{q^{-1}-q}{q^{-2(\lambda+1)}-1} & \frac{(q^{2(\lambda+1)}-q^2)(q^{2(\lambda+1)}-q^{-2})}{(q^{2(\lambda+1)}-1)^2} & 0 \\ 0 & 0 & 0 & q \end{pmatrix}.$$

3   The dynamical Yang-Baxter equations.

3.1. Proposition 2.4 motivates the following definition. Let $\mathfrak{h}$ be a finite-dimensional abelian Lie algebra and let V be a semisimple $\mathfrak{h}$-module. Let us denote by M the field of meromorphic functions on $\mathfrak{h}^*$. Let us equip M with the trivial $\mathfrak{h}$-module structure.

Definition. Let $R : V \otimes V \otimes M \to V \otimes V \otimes M$ be an $\mathfrak{h}$-invariant and M-linear map. Then the *quantum dynamical Yang-Baxter equation* (QDYBE) is the following equation with respect to R:

$$R^{12}(\lambda - h^{(3)})R^{13}(\lambda)R^{23}(\lambda - h^{(1)}) = R^{23}(\lambda)R^{13}(\lambda - h^{(2)})R^{12}(\lambda).$$

A *quantum dynamical R-matrix* is an invertible solution of this equation.

It follows from Proposition 2.4 that for any simple complex Lie algebra $\mathfrak{g}$ and for any finite-dimensional $\mathfrak{g}$-module V, the exchange matrix $R_{VV}(\lambda)$ is a quantum dynamical R-matrix. The same is true if we replace the Lie algebra $\mathfrak{g}$ by the quantum group $U_q(\mathfrak{g})$.

Remarks. 1. The usual quantum Yang-Baxter equation is recovered from the quantum dynamical Yang-Baxter equation when $\mathfrak{h} = 0$.

2. A constant solution of the quantum dynamical Yang-Baxter equation is the same thing as a solution of the ordinary quantum Yang-Baxter equation which is $\mathfrak{h}$-invariant.

3. In physical literature, the variable λ is called a dynamical variable. This gave rise to the name "dynamical R-matrix".

Replacing λ by $\frac{\lambda}{\gamma}$ in the QDYBE yields the following equation

$$\tilde{R}^{12}(\lambda - \gamma h^{(3)})\tilde{R}^{13}(\lambda)\tilde{R}^{23}(\lambda - \gamma h^{(1)}) = \tilde{R}^{23}(\lambda)\tilde{R}^{13}(\lambda - \gamma h^{(2)})\tilde{R}^{12}(\lambda), \quad (3.1)$$

which is called the quantum dynamical Yang-Baxter equation with step γ.

Proposition 3.1. *Let $\mathfrak{h}$ be an abelian Lie algebra. Let V be a finite dimensional semisimple $\mathfrak{h}$-module and let $R : \mathfrak{h}^* \to \mathrm{End}_{\mathfrak{h}}(V \otimes V)[[\gamma]]$ be a series of meromorphic functions of the form $R = 1 - \gamma r + O(\gamma^2)$. If R satisfies the quantum dynamical Yang-Baxter equation with step γ then r satisfies the following classical analogue of the quantum dynamical Yang-Baxter equation:*

$$\sum_i \left(x_i^{(1)} \frac{\partial r^{23}(\lambda)}{\partial x^i} - x_i^{(2)} \frac{\partial r^{13}(\lambda)}{\partial x^i} + x_i^{(3)} \frac{\partial r^{12}(\lambda)}{\partial x^i} \right) +$$
$$[r^{12}(\lambda), r^{13}(\lambda)] + [r^{12}(\lambda), r^{23}(\lambda)] + [r^{13}(\lambda), r^{23}(\lambda)] = 0 \quad (3.2)$$

where (x_i) is a basis of $\mathfrak{h}$ and (x^i) is the dual basis of $\mathfrak{h}^*$.

This leads to the following definition:

Definition. Let $\mathfrak{g}$ be a finite-dimensional Lie algebra and let $\mathfrak{h} \subset \mathfrak{g}$ be a Lie subalgebra. The *classical dynamical Yang-Baxter equation* (CDYBE) is equation (3.2) with respect to a holomorphic, $\mathfrak{h}$-invariant function $r : U \to \mathfrak{g} \otimes \mathfrak{g}$, where $U \subset \mathfrak{h}^*$ is an open region. A solution to this equation is called a *classical dynamical r-matrix*.

Remarks. 1. The ordinary classical Yang-Baxter equation is recovered from the classical dynamical Yang-Baxter equation when $\mathfrak{h} = 0$.

2. A constant solution of the classical dynamical Yang-Baxter equation is the same thing as an $\mathfrak{h}$-invariant solution of the ordinary classical Yang-Baxter equation.

We will now consider asymptotic behavior of fusion and exchange matrices, and obtain solutions to the CDYBE. Let $\mathfrak{g}$ be a simple complex Lie algebra. Let V, W be two finite-dimensional $\mathfrak{g}$-modules and let $J_{VW}(\lambda)$ and $R_{VW}(\lambda)$ be the fusion and exchange matrices of V and W.

Proposition 3.2 ([EV3]). *1. The function $J_{VW}(\frac{\lambda}{\gamma})$ is regular at $\gamma = 0$ for generic values of λ.*

2. Set $J_{VW}(\frac{\lambda}{\gamma}) = 1 + \gamma j_{VW}(\lambda) + O(\gamma^2)$. Then $j_{VW}(\lambda)$ is the evaluation on $V \otimes W$ of the element

$$j(\lambda) = -\sum_{\alpha > 0} \frac{e_{-\alpha} \otimes e_{\alpha}}{\langle \alpha, \lambda \rangle} \in \mathfrak{n}_- \otimes \mathfrak{n}_+.$$

Corollary 3.1. *We have $R_{VW}(\frac{\lambda}{\gamma}) = 1 - \gamma r(\lambda)_{|V \otimes W} + O(\gamma^2)$ where*

$$r(\lambda) = j(\lambda) - j^{21}(\lambda) = \sum_{\alpha > 0} \frac{e_{\alpha} \otimes e_{-\alpha} - e_{-\alpha} \otimes e_{\alpha}}{\langle \alpha, \lambda \rangle}. \qquad (3.3)$$

A proof of Proposition 3.2, which is based on computing the asymptotics of intertwining operators at $\lambda \to \infty$, is given in [EV3]. Later we will give another proof of this Proposition.

It follows from Proposition 3.1 that $r(\lambda)$ in (3.3) is a classical dynamical r-matrix. Let us call it the *basic rational dynamical r-matrix*.

Proposition 3.2 and Corollary 3.1 have natural quantum analogues. Let $U_q(\mathfrak{g})$ be the quantum group associated to $\mathfrak{g}$ with quantum parameter $q = e^{-\varepsilon\gamma/2}$ for some fixed $\varepsilon \in \mathbb{C}$ and formal parameter γ. Let V, W be two finite-dimensional $U_q(\mathfrak{g})$-modules and let $R_{VW}(\lambda)$ be the exchange matrix. Set $\tilde{R}_{VW}(\lambda) = R_{VW}(\frac{\lambda}{\gamma})$.

Proposition 3.3 ([EV3]). *We have*

$$\tilde{R}_{VW}(\lambda) = 1 - \gamma r^{\varepsilon}_{VW}(\lambda) + O(\gamma^2)$$

where $r^{\varepsilon}_{VW} : \mathfrak{h}^* \to \mathrm{End}_{\mathfrak{h}}(V \otimes W)$ is the evaluation on $V \otimes W$ of the following universal element:

$$r^{\varepsilon}(\lambda) = \frac{\varepsilon}{2}\Omega + \sum_{\alpha > 0} \frac{\varepsilon}{2}\coth\left(\frac{\varepsilon}{2}\langle \alpha, \lambda \rangle\right)(e_{\alpha} \otimes e_{-\alpha} - e_{-\alpha} \otimes e_{\alpha}) \in \mathfrak{g} \otimes \mathfrak{g} \quad (3.4)$$

where $\Omega \in S^2\mathfrak{g}$ is the inverse element to the form $(\,,\,)$ (the Casimir element).

It follows from Proposition 3.1 that $r^{\varepsilon}(\lambda)$ is a solution of the CDYBE. Let us call it the *basic trigonometric dynamical r-matrix*.

3.2. Quantization and quasiclassical limit. Let $\mathfrak{h}$ be an abelian Lie algebra and let V be a finite-dimensional semisimple $\mathfrak{h}$-module. Let $r : \mathfrak{h}^* \to \mathrm{End}_{\mathfrak{h}}(V \otimes V)$ be a classical dynamical r-matrix. Suppose that $R : \mathfrak{h}^* \to \mathrm{End}_{\mathfrak{h}}(V \otimes V)[[\gamma]]$ is of the form $R = 1 - \gamma r + O(\gamma^2)$ and satisfies the QDYBE.

Definition. R is called a *quantization* of r. Conversely, r is called the *quasiclassical limit* of R.

For instance, the exchange matrix $\tilde{R}_{VV}(\lambda)$ constructed from a Lie algebra $\mathfrak{g}$ is a quantization of the evaluation on $V \otimes V$ of the basic rational dynamical r-matrix. Similarly, exchange matrices constructed from quantum groups provide quantization of the basic trigonometric dynamical r-matrix.

Conjecture: Any classical dynamical r-matrix admits a quantization.

Notice that when $\mathfrak{h} = 0$, this conjecture reduces to the conjecture of Drinfeld [Dr] about quantization of classical (non-dynamical) r-matrices, which was proved in [EK]. In the skew-symmetric case, the conjecture was recently proved in [Xu2], using the theory of Fedosov quantization.

3.3. Unitarity conditions.

Recall the following notions introduced by Drinfeld. A classical r-matrix $r \in \mathfrak{g} \otimes \mathfrak{g}$ is a *quasitriangular* structure on a Lie algebra $\mathfrak{g}$ if $r + r^{21} \in (S^2\mathfrak{g})^{\mathfrak{g}}$. It is a *triangular* structure on $\mathfrak{g}$ if $r + r^{21} = 0$.

This definition is natural in the theory of Lie bialgebras. Namely, a classical r-matrix $r \in \mathfrak{g} \otimes \mathfrak{g}$ defines a Lie bialgebra structure on $\mathfrak{g}$ by $\delta : \mathfrak{g} \to \Lambda^2\mathfrak{g}$, $x \mapsto [1 \otimes x + x \otimes 1, r]$ if and only if $r + r^{21} \in (S^2\mathfrak{g})^{\mathfrak{g}}$. In the case of a simple Lie algebra $\mathfrak{g}$ we have $(S^2\mathfrak{g})^{\mathfrak{g}} = \mathbb{C}\Omega$, so that a classical r-matrix r is quasitriangular if $r + r^{21} = \varepsilon\Omega$ for some $\varepsilon \in \mathbb{C}$, and it is triangular if moreover $\varepsilon = 0$.

This leads one to make the following definition:

Definition. A classical dynamical r-matrix $r : \mathfrak{h}^* \to (\mathfrak{g} \otimes \mathfrak{g})^{\mathfrak{h}}$ has *coupling constant* ε if

$$r + r^{21} = \varepsilon\Omega. \tag{3.5}$$

Equation (3.5) is called the *unitarity condition*. Notice that the basic rational dynamical r-matrix $r(\lambda)$ and the basic trigonometric dynamical r-matrix $r^\varepsilon(\lambda)$ have coupling constants 0 and ε respectively.

4 Classification of classical dynamical r-matrices.

In this section, we give the classification of all dynamical r-matrices $r : \mathfrak{h}^* \to \mathfrak{g} \otimes \mathfrak{g}$ which have coupling constant $\varepsilon \in \mathbb{C}$.

4.1. Gauge transformations.

Consider the following operations on mero-morphic maps $r : \mathfrak{h}^* \to (\mathfrak{g} \otimes \mathfrak{g})^{\mathfrak{h}}$.

1. $r(\lambda) \mapsto r(\lambda) + \sum_{i<j} C_{ij}(\lambda) x_i \wedge x_j$,, where $\sum_{i,j} C_{ij}(lambda) d\lambda_i \wedge d\lambda_j$ is a closed meromorphic 2-form.

2. $r(\lambda) \mapsto r(\lambda - \nu)$, where $\nu \in \mathfrak{h}^*$.

3. $r(\lambda) \mapsto (A \otimes A) r(A^* \lambda)$, where $A \in W$, the Weyl group of $\mathfrak{g}$.

Lemma 4.1. *Transformations 1-3 preserve the set of classical dynamical r-matrices.*

The proof is straightforward.

Two classical dynamical r-matrices which can be obtained one from the other by a sequence of such transformations will be called *gauge-equivalent*.

4.2. Classification of dynamical r-matrices with zero coupling constant.

Let $\mathfrak{l} \supset \mathfrak{h}$ be a reductive Lie subalgebra of $\mathfrak{g}$. Define

$$r^{\mathfrak{l}}(\lambda) = \sum_{\substack{\alpha > 0 \\ e_\alpha \in \mathfrak{l}}} \frac{e_\alpha \otimes e_{-\alpha} - e_{-\alpha} \otimes e_\alpha}{(\lambda, \alpha)}. \tag{4.1}$$

It is clear that this is the image of the basic rational dynamical r-matrix of $\mathfrak{l}$ under the embedding $\mathfrak{l} \subset \mathfrak{g}$.

Theorem 4.1 ([EV1]). *Any classical dynamical r-matrix $r : \mathfrak{h}^* \to (\mathfrak{g} \otimes \mathfrak{g})^{\mathfrak{h}}$ with zero coupling constant is gauge-equivalent to $r^{\mathfrak{l}}(\lambda)$ for some $\mathfrak{l}$.*

4.3. Classification of dynamical r-matrices with coupling constant

$\varepsilon \in \mathbb{C}^*$. Let $X \subset \Pi$, and denote by $\langle X \rangle \subset \Delta$ the set of all roots which are linear combinations of elements in $X \cup -X$. For any $\alpha \in \Delta$ introduce a meromorphic function $\varphi_\alpha : \mathfrak{h}^* \to \mathbb{C}$ by the following rule. Set $\varphi_\alpha(\lambda) = \frac{\varepsilon}{2}$ if $\alpha \in \Delta^+ \backslash \langle X \rangle$, $\varphi_\alpha(\lambda) = -\frac{\varepsilon}{2}$ if $\alpha \in \Delta^- \backslash \langle X \rangle$ and

$$\varphi_\alpha(\lambda) = \frac{\varepsilon}{2} \mathrm{cotanh} \left(\frac{\varepsilon}{2} (\lambda, \alpha) \right)$$

if $\alpha \in \langle X \rangle$.

Theorem 4.2 ([EV1]). *Let $X \subset \Pi$. Set*

$$r_X(\lambda) = \frac{\varepsilon}{2} \Omega + \sum_{\alpha \in \Delta} \varphi_\alpha(\lambda) e_\alpha \otimes e_{-\alpha}.$$

Then $r_X^\varepsilon(\lambda)$ is a classical dynamical r-matrix with coupling constant ε. Moreover, any classical dynamical r-matrix with coupling constant ε is gauge-equivalent to $r_X^\varepsilon(\lambda)$ for a suitable $X \subset \Pi$.

Remarks. 1. The basic trigonometric dynamical r-matrix $r^\varepsilon(\lambda)$ is obtained when we take $X = \Pi$. Moreover, the r-matrix $r_X^\varepsilon(\lambda)$ is equal to a limit of $r^\varepsilon(\lambda - \nu)$ when ν tends to infinity in $\mathfrak{h}^*$ in an appropriate direction. In other words, every classical dynamical r-matrix with nonzero coupling constant ε is a limiting case of the basic trigonometric r-matrix.

2. Let W be the Weyl group of $\mathfrak{g}$, and let $w \in W$. Let $\lambda \in \mathfrak{h}^*$ tend to infinity in a generic way in the Weyl chamber associated to w. Then

$$\lim r^1(\lambda) = \frac{1}{2}\sum_i x_i \otimes x_i + \sum_{\alpha \in w(\Delta^+)} e_\alpha \otimes e_{-\alpha},$$

which is the standard classical r-matrix corresponding to the polarization of $\mathfrak{g}$ associated to w. Hence the basic trigonometric dynamical r-matrix $r^1(\lambda)$ interpolates all $\mathfrak{h}$-invariant classical (non-dynamical) r-matrices r satisfying $r + r^{21} = \Omega$, (up to the addition of a skew 2-form in $\Lambda^2\mathfrak{h}$).

A classification of all classical dynamical r-matrices $r : \mathfrak{l}^* \to (\mathfrak{g} \otimes \mathfrak{g})^{\mathfrak{l}}$ where $\mathfrak{g}$ is a simple Lie algebra and $\mathfrak{l} \subset \mathfrak{h}$ is given in [S]. This classification generalizes both the above classification (when $\mathfrak{l} = \mathfrak{h}$) and the Belavin-Drinfeld classification of classical r-matrices (when $\mathfrak{l} = 0$) (see Appendix A).

5 Classical dynamical r-matrices and Poisson-Lie groupoids

In this section we give a geometric interpretation of the CDYBE. Let us first briefly recall the relationship between the classical Yang-Baxter equation and the theory of Poisson-Lie groups, developed by Drinfeld.

5.1. Poisson-Lie groups. Let G be a (complex or real) Lie group, let $\mathfrak{g}$ be its Lie algebra and let $\mathcal{O}(G)$ be the algebra of regular functions on G. Let $\{\,,\,\} : \mathcal{O}(G) \times \mathcal{O}(G) \to \mathcal{O}(G)$ be a Poisson structure on G. Let Π be the Poisson bivector field, defined by the relation $\{f, g\} = df \otimes dg(\Pi)$. Recall that $(G, \{\,,\,\})$ is called a Poisson-Lie group if the multiplication map $m : G \times G \to G$ is a Poisson map.

Let $\rho \in \Lambda^2\mathfrak{g}$ and consider the following bivector field:

$$\Pi_\rho = R_\rho - L_\rho,$$

where R_ρ (resp. L_ρ) is the left-invariant (resp. right-invariant) bivector field satisfying $(R_\rho)_{|e} = \rho$ (resp. $(L_\rho)_{|e} = \rho$); in other words, R_ρ, L_ρ stand for the translates of ρ by right and left shifts respectively.

Proposition 5.1 (Drinfeld). *The bivector Π_ρ defines a Poisson-Lie group structure on G if and only if*

$$[\rho_{12}, \rho_{13}] + [\rho_{13}, \rho_{23}] + [\rho_{12}, \rho_{23}] \in (\Lambda^3\mathfrak{g})^\mathfrak{g}.$$

When this is the case, G is called a *coboundary* Poisson-Lie group. Two cases are of special interest:

1. The exists $T \in (S^2 \mathfrak{g})^{\mathfrak{g}}$ such that $[\rho_{12}, \rho_{13}] + [\rho_{13}, \rho_{23}] + [\rho_{12}, \rho_{23}] = \frac{1}{4}[T_{12}, T_{23}]$. This implies that $r = \rho + \frac{1}{2}T$ satisfies the classical Yang-Baxter equation, and $\Pi_\rho = R_r - L_r$. In this case G is called a *quasitriangular* Poisson-Lie group.

2. We have $[\rho_{12}, \rho_{13}] + [\rho_{13}, \rho_{23}] + [\rho_{12}, \rho_{23}] = 0$. In this case, G is called a *triangular* Poisson-Lie group.

5.2. Poisson-Lie groupoids.

It turns out that, in order to generalize this correspondence to the dynamical case, groups must be replaced by groupoids. Recall that a groupoid is a (small) category where all morphisms are isomorphisms. It is equivalent to the following data: two sets X and P (the set of morphisms, or the groupoid itself, and the set of objects, or the base, respectively), two surjective maps $s, t : X \rightarrow P$ (the source and target maps), an injective map $E : P \rightarrow X$ (the identity morphisms), a map $m : \{(a, b) \in X \times X \mid t(a) = s(b)\} \rightarrow X$ ($m(a, b) = b \circ a$, the composition of morphisms), and an involution $i : X \rightarrow X$ such that $s(i(x)) = t(x)$, $t(i(x)) = s(x)$, $m(i(x), x) = \mathrm{Id}_{s(x)}$ and $m(x, i(x)) = \mathrm{Id}_{t(x)}$ for all $x \in X$, satisfying some obvious axioms. One can visualize elements of X as arrows $s(a) \xrightarrow{a} t(a)$.

Note that when $|P| = 1$, the notion of a groupoid coincides with the notion of a group.

A Lie groupoid is a groupoid with a smooth structure (in particular, the sets of objects and the sets of morphisms are smooth manifolds and the structure maps are smooth, see [M]).

Now we would like to generalize the notion of a Poisson-Lie group to groupoids. The usual definition does not generalize directly since if X is a Lie groupoid and a Poisson manifold then the set of points $(a, b) \in X^2$ for which the multiplication is defined is not necessarily a Poisson submanifold, so we cannot require that the multiplication map be Poisson. But this difficulty can be bypassed using the following observation:

Proposition 5.2. *Let X, Y be two Poisson manifolds and let $f : X \rightarrow Y$ be a smooth map. Consider the graph $\Gamma_f = \{(x, f(x))\} \subset X \times \overline{Y}$, where $\overline{Y}$ is the manifold Y, with the opposite Poisson structure $\{,\}_{\overline{Y}} = -\{,\}_Y$. Then f is a Poisson map if and only if Γ_f is a coisotropic submanifold of $X \times \overline{Y}$, i.e if and only if for any $z \in \Gamma_f$, $(T_z \Gamma_f)^\perp \subset T_z^*(X \ltimes \overline{Y})$ is an isotropic subspace with respect to the Poisson form Π on $T_z(X \times \overline{Y})^*$.*

This gives rise to the following notion of a Poisson-Lie groupoid, first introduced by Weinstein [W].

Definition. A Lie groupoid X with a Poisson structure is called a *Poisson-Lie groupoid* if $\Gamma_m \subset X \times X \times \overline{X}$ is a coisotropic submanifold.

We now restrict ourselves to a particular class of Lie groupoids. Let G be a Lie group, let $\mathfrak{g}$ be its Lie algebra, $\mathfrak{h} \subset \mathfrak{g}$ a subalgebra and H a Lie subgroup of G with Lie algebra $\mathfrak{h}$. Let $U \subset \mathfrak{h}^*$ be an open set. Consider the following groupoid: $X = U \times G \times U$, $P = U$ with $s(u_1, g, u_2) = u_1$, $t(u_1, g, u_2) = u_2$. The composition $m((u_1, g, u_2), (u_3, g', u_4))$ is defined only when $u_2 = u_3$ and $m((u_1, g, v), (v, g', u_4)) = (u_1, gg', u_4)$. If a is a function on U we set $a_1 = s^*(a) \in \mathcal{O}(X)$ and $a_2 = t^*(a) \in \mathcal{O}(X)$. Let $\rho : U \to \Lambda^2 \mathfrak{g}$ be a regular function.

The group H^2 acts on X by

$$(h_1, h_2)(u_1, g, u_2) = (\mathrm{Ad}^*(h_1)u_1, h_1 g h_2^{-1}, \mathrm{Ad}^*(h_2)u_2).$$

We want to define a Poisson structure on X for which $(-s, t)$ is a moment map for this action. This forces the following relations

$$\{a_1, b_1\} = -[a, b]_1, \qquad \{a_2, b_2\} = [a, b]_2, \qquad \{a_1, b_2\} = 0, \tag{5.1}$$
$$\{a_1, f\} = R_a f, \qquad \{a_2, f\} = L_a f.$$

We try to complete the definition of the Poisson structure on X by adding the relation

$$\{f, g\} = (df \otimes dg)(R_{\rho(u_1)} - L_{\rho(u_2)}), \tag{5.2}$$

where f, g are any functions on X pulled back from G and a, b are linear functions on U.

Proposition 5.3 ([EV1]). *Formulae (5.1) and (5.2) define a Poisson-Lie groupoid structure on X if and only if*

1.

$$\sum_i \left(x_i^{(1)} \frac{\partial \rho^{23}}{\partial x^i} - x_i^{(2)} \frac{\partial \rho^{13}}{\partial x^i} + x_i^{(3)} \frac{\partial \rho^{12}}{\partial x^i} \right) + [\rho^{12}, \rho^{13}] + [\rho^{12}, \rho^{23}] + [\rho^{13}, \rho^{23}]$$

is a constant $\mathfrak{g}$-invariant element of $\Lambda^3 \mathfrak{g}$, and

2. *ρ is $\mathfrak{h}$-invariant.*

When this is the case, X is called a *coboundary* dynamical Poisson-Lie groupoid, which will be denoted by X_r. Two cases are of special interest:

1. The exists $T \in (S^2 \mathfrak{g})^\mathfrak{g}$ such that

$$\frac{1}{4}[T_{12}, T_{23}] = \sum_i \left(x_i^{(1)} \frac{\partial r^{23}(\lambda)}{\partial x^i} - x_i^{(2)} \frac{\partial r^{13}(\lambda)}{\partial x^i} + x_i^{(3)} \frac{\partial r^{12}(\lambda)}{\partial x^i} \right)$$
$$+ [\rho_{12}, \rho_{13}] + [\rho_{13}, \rho_{23}] + [\rho_{12}, \rho_{23}].$$

This implies that $r = \rho + \frac{1}{2}T$ satisfies the classical dynamical Yang-Baxter equation. In this case X_r is called a *quasitriangular* dynamical Poisson-Lie groupoid.

2. We have

$$\sum_i \left(x_i^{(1)} \frac{\partial r^{23}(\lambda)}{\partial x^i} - x_i^{(2)} \frac{\partial r^{13}(\lambda)}{\partial x^i} + x_i^{(3)} \frac{\partial r^{12}(\lambda)}{\partial x^i} \right)$$

$$+ [\rho_{12}, \rho_{13}] + [\rho_{13}, \rho_{23}] + [\rho_{12}, \rho_{23}] = 0.$$

In this case, X_r is called a *triangular* dynamical Poisson-Lie groupoid.

Thus, the basic rational solution defined above gives rise to a triangular dynamical Poisson-Lie groupoid, and the basic trigonometric solution gives rise to a quasitriangular one.

6 Classification of quantum dynamical R-matrices

In this section we give the classification of all quantum dynamical R-matrices $R : \mathfrak{h}^* \to \mathrm{End}_{\mathfrak{h}}(V \otimes V)$, where $\mathfrak{h}$ is the Cartan subalgebra of $\mathfrak{gl}(n, \mathbb{C})$ consisting of diagonal matrices, and $V = \mathbb{C}^n$ is the vector representation, which satisfy an additional *Hecke condition*, a quantum analogue of the unitarity condition.

6.1. Hecke condition. Let $\mathfrak{h}$ be the abelian Lie algebra of diagonal N by N matrices, and let V be the standard N-dimensional $\mathfrak{h}$-module. Let $h_1, \ldots h_n$ be the standard basis of $\mathfrak{h}$, $\lambda_1, \ldots \lambda_n$ be the corresponding coordinate functions on $\mathfrak{h}^*$, and V_i, $i = 1, \ldots n$ be the (one-dimensional) weight subspaces of V of weight ω_i where $\langle \omega_i, h_j \rangle = \delta_{ij}$.

Consider the $\mathfrak{h}$-module $V \otimes V$. Its weight subspaces are $V_a \otimes V_b \oplus V_b \otimes V_a$ and $V_a \otimes V_a$.

Definition. An operator $R : \mathfrak{h}^* \to \mathrm{End}_{\mathfrak{h}}(V \otimes V)$ satisfies the *Hecke condition* with parameter $q \in \mathbb{C}^*$ if the eigenvalues of PR (where P is the permutation matrix) are 1 on $V_a \otimes V_a$ and $1, -q$ on $V_a \otimes V_b \oplus V_b \otimes V_a$.

This condition can be thought of as a quantum analogue of the unitarity condition for classical r-matrices, since it is easy to show that the quasiclassical limit of an operator satisfying the Hecke condition satisfies the unitarity condition. In particular if R satisfies the Hecke condition with $q = 1$ then $RR^{21} = 1$, which can be thought of as a quantization of the relation $r + r^{21} = 0$.

The terminology comes from the following remark: if R is a λ-independent solution of the quantum dynamical Yang-Baxter equation satisfying the Hecke condition with parameter q then $\check{R}$ defines a representation of the Hecke algebra H_p of type A_{p-1} on the space $V^{\otimes p}$ for any $p > 1$. Similar representations can be defined for dynamical R-matrices (see [EV2] and Section 7).

6.2. Gauge transformations.

Let $R(\lambda)$ be a quantum dynamical R-matrix satisfying Hecke condition with parameter q. The weight-zero and Hecke conditions imply that

$$R(\lambda) = \sum_a E_{aa} \otimes E_{aa} + \sum_{a \neq b} \alpha_{ab}(\lambda) E_{aa} \otimes E_{bb} + \sum_{a \neq b} \beta_{ab}(\lambda) E_{ab} \otimes E_{ba} \quad (6.1)$$

where E_{ij} is the elementary matrix, and α_{ab}, β_{ab} are meromorphic functions $\mathfrak{h}^* \to \mathbb{C}$. So it is enough to look for solutions of this form.

As in the classical case, we will give the classification of solutions up to some group of transformations.

Definition. A *multiplicative 2-form* on V is a collection meromorphic functions $\{\varphi_{ab} : \mathfrak{h}^* \to \mathbb{C}\}_{a,b=1}^n$ satisfying $\varphi_{ab}\varphi_{ba} = 1$ for all a, b. A multiplicative 2-form $\{\varphi_{ab}(\lambda)\}$ is *closed* if for all a, b, c,

$$\frac{\varphi_{ab}(\lambda)}{\varphi_{ab}(\lambda - \omega_c)} \frac{\varphi_{bc}(\lambda)}{\varphi_{bc}(\lambda - \omega_a)} \frac{\varphi_{ca}(\lambda)}{\varphi_{ca}(\lambda - \omega_b)} = 1.$$

Consider the following operations on meromorphic weight-zero maps $R : \mathfrak{h}^* \to \mathrm{End}_{\mathfrak{h}}(V \otimes V)$ of the form (6.1):

1.

$$R(\lambda) \mapsto \sum_a E_{aa} \otimes E_{aa} + \sum_{a \neq b} \varphi_{ab}(\lambda)\alpha_{ab}(\lambda) E_{aa} \otimes E_{bb} + \sum_{a \neq b} \beta_{ab}(\lambda) E_{ab} \otimes E_{ba},$$

where $\{\varphi_{ab}(\lambda)\}$ is a closed multiplicative 2-form on V,

2. $R(\lambda) \mapsto R(\lambda - \nu)$ where ν is a pseudoconstant, i.e. a meromorphic function $\mathfrak{h}^* \to \mathfrak{h}^*$ such that $\nu(\lambda + \omega_i) = \nu(\lambda)$ for all i (for example, a constant),

3. $R(\lambda) \mapsto (\sigma \otimes \sigma)R(\sigma^{-1}\lambda)(\sigma^{-1} \otimes \sigma^{-1})$ where $\sigma \in \mathfrak{S}_n$ acts on V and $\mathfrak{h}^*$ by permutation of coordinates.

Remark. Here we allow to perform transformation 2 only if the answer is meromorphic.

Lemma 6.1. *Tranformations $1 - 3$ preserve the set of quantum dynamical R-matrices.*

Two R-matrices which can be obtained one from the other by a sequence of such transformations are said to be *gauge-equivalent*.

6.3. Classification for $q = 1$. Let X be a subset of $\{1, \dots n\}$ and write $X = X_1 \cup \dots \cup X_k$ where $X_i = \{a_i \dots b_i\}$ are disjoint intervals. Set

$$R_X(\lambda) = \sum_{a,b=1}^{n} E_{aa} \otimes E_{bb} + \sum_{l=1}^{k} \sum_{\substack{a,b \in X_l \\ a \neq b}} \frac{1}{\lambda_a - \lambda_b} (E_{aa} \otimes E_{bb} + E_{ba} \otimes E_{ab}).$$

Theorem 6.1 ([EV2]). *Let $X \subset \{1, \dots n\}$. Then $R_X(\lambda)$ is a quantum dynamical R-matrix satisfying the Hecke condition with $q = 1$. Moreover, any dynamical R-matrix $R : \mathfrak{h}^* \to \mathrm{End}_{\mathfrak{h}}(V \otimes V)$ is gauge-equivalent to $R_X(\lambda)$ for a unique subset $X \subset \{1, \dots n\}$.*

Remark. The function $R_X(\lambda/\gamma)$ is, up to a gauge transformation, a quantization in the sense of Section 3 of the rational classical dynamical r-matrix (4.1) corresponding to the reductive subalgebra of $\mathfrak{gl}(n)$ spanned by root subspaces $\mathfrak{g}_\alpha, \mathfrak{g}_{-\alpha}$ for $\alpha \in X$.

The most interesting solution R_X corresponds to the case when $X = \{1, \dots, n\}$. We will call it *the basic rational solution* of the QDYBE.

6.4. Classification for $q \neq 1$. Let $\varepsilon \notin 2i\pi\mathbb{Z}$ and set $q = e^\varepsilon$. Let X be a subset of $\{1, \dots n\}$ and again write $X = X_1 \cup \dots \cup X_k$ where $X_i = \{a_i \dots b_i\}$ are disjoint intervals. Set

$$R_X^\varepsilon(\lambda) = \sum_a E_{aa} \otimes E_{aa} + \sum_{a \neq b} \alpha_{ab}(\lambda) E_{aa} \otimes E_{bb} + \sum_{a \neq b} \beta_{ab}(\lambda) E_{ab} \otimes E_{ba},$$

where $\alpha_{ab}(\lambda) = q + \beta_{ab}(\lambda)$ and where $\beta_{ab}(\lambda)$ is defined as follows: $\beta_{ab} = \frac{q-1}{q^{\lambda_a - \lambda_b} - 1}$ if $a, b \in X_l$ for some $1 \leq l \leq k$, $\beta_{ab}(\lambda) = 1 - q$ otherwise if $a > b$ and $\beta_{ab}(\lambda) = 0$ otherwise if $a < b$.

Theorem 6.2 ([EV2]). *Let $X \subset \{1, \dots n\}$. Then $R_X^\varepsilon(\lambda)$ is a quantum dynamical R-matrix satisfying the Hecke condition with $q = e^\varepsilon$. Moreover, any dynamical R-matrix $R : \mathfrak{h}^* \to \mathrm{End}_{\mathfrak{h}}(V \otimes V)$ is gauge-equivalent to $R_X(\lambda)$ for a unique subset $X \subset \{1, \dots n\}$.*

Remark. It can be checked that the $R_X^\varepsilon(\lambda)$ yield (again up to gauge transformations) quantizations of the trigonometric classical dynamical r-matrices with coupling constant ε appearing in Theorem 4.2.

The most interesting solution R_X^ε corresponds to the case when $X = \{1, \dots, n\}$. We will call it *the basic trigonometric solution* of the QDYBE.

6.5. The fusion and exchange matrices for the vector representation of classical and quantum gl_n. The above classification can be applied to compute the fusion and exchange matrices for the vector representation. Namely, we have:

Theorem 6.3 ([EV3]). *1. Let $\mathfrak{g} = \mathfrak{gl}_n$ and let $V = \mathbb{C}^n$ be the vector representation. Then*

$$J_{VV}(\lambda) = 1 + \sum_{a<b} \frac{1}{\lambda_b - \lambda_a + a - b} E_{ba} \otimes E_{ab}$$

$$R_{VV}(\lambda) = \sum_{a=1}^{n} E_{aa} \otimes E_{aa} + \sum_{a \neq b} \frac{1}{\lambda_a - \lambda_b + b - a} E_{ba} \otimes E_{ab} + \sum_{a<b} E_{aa} \otimes E_{bb}$$

$$- \sum_{a>b} \frac{(\lambda_b - \lambda_a + a - b - 1)(\lambda_b - \lambda_a + a - b + 1)}{(\lambda_b - \lambda_a + a - b)^2} E_{aa} \otimes E_{bb}.$$

2. Let $V = \mathbb{C}^n$ be the representation of $U_q(\mathfrak{gl}_N)$ which is the q-analog of the vector representation. Then

$$J_{VV}(\lambda) = 1 + \sum_{a<b} \frac{q^{-1} - q}{q^{2(\lambda_a - \lambda_b + b - a)} - 1} E_{ba} \otimes E_{ab}$$

$$R_{VV}(\lambda) = q \sum_{a=1}^{n} E_{aa} \otimes E_{aa} + \sum_{a \neq b} \frac{q^{-1} - q}{q^{2(\lambda_a - \lambda_b + b - a)} - 1} E_{ba} \otimes E_{ab} + \sum_{a<b} E_{aa} \otimes E_{bb}$$

$$+ \sum_{a>b} \frac{(q^{2(\lambda_b - \lambda_a + a - b)} - q^{-2})(q^{2(\lambda_b - \lambda_a + a - b)} - q^2)}{(q^{2(\lambda_b - \lambda_a + a - b)} - 1)^2} E_{aa} \otimes E_{bb}.$$

Proof. The proof relies on explicit computations and on the classification of quantum dynamical R-matrices (Theorems 6.1 and 6.2). More precisely, it is possible to compute explicitly the coefficients of J corresponding to simple roots, and all the other coefficients are then uniquely determined by Theorems 6.1 and 6.2. ∎

Remark. The matrix coefficients of $J_{VV}(\lambda)$ for nonsimple roots are not as easily computed directly as those for simple roots. The above approach allows one to avoid this calculation.

7 Quantum dynamical R-matrices and quantum groupoids

In this section we will give a "noncommutative geometric" interpretation of the QDYBE which is analogous to the geometric interpretation of the CDYBE given above. More precisely, solutions of the QDYBE we will associate, following [F],[EV2], a kind of quantum group, more precisely a *Hopf algebroid* (or *quantum groupoid*).

The general notion of a Hopf algebroid was introduced in [Lu]. However, here it will be sufficient to use a less general notion, that of an H-Hopf algebroid, which was introduced in [EV2]. Our exposition will follow [EV2, EV3].

7.1. H-bialgebroids. Let H be a commutative and cocommutative finitely generated Hopf algebra over $\mathbb{C}$, $T = \operatorname{Spec} H$ the corresponding commutative affine algebraic group. Assume that T is connected. Let M_T denote the field of meromorphic functions on T. Let us introduce the following definitions.

Definition. *An H-algebra* is an associative algebra A over $\mathbb{C}$ with 1, endowed with an T-bigrading $A = \oplus_{\alpha,\beta \in T} A_{\alpha\beta}$ (called the weight decomposition), and two algebra embeddings $\mu_l, \mu_r : M_T \to A_{00}$ (the left and the right moment maps), such that for any $a \in A_{\alpha\beta}$ and $f \in M_T$, we have

$$\mu_l(f(\lambda))a = a\mu_l(f(\lambda+\alpha)), \quad \mu_r(f(\lambda))a = a\mu_r(f(\lambda+\beta)). \tag{7.1}$$

A morphism $\varphi : A \to B$ of two H-algebras is an algebra homomorphism, preserving the moment maps.

Example 1. Let D_T be the algebra of difference operators $M_T \to M_T$, i.e. the operators of the form $\sum_{i=1}^n f_i(\lambda)T_{\beta_i}$, where $f_i \in M_T$, and for $\beta \in T$ we denote by T_β the field automorphism of M_T given by $(T_\beta f)(\lambda) = f(\lambda+\beta)$.

The algebra D_T is an example of an H-algebra if we define the weight decomposition by $D_T = \oplus(D_T)_{\alpha\beta}$, where $(D_T)_{\alpha\beta} = 0$ if $\alpha \neq \beta$, and $(D_T)_{\alpha\alpha} = \{f(\lambda)T_\alpha^{-1} : f \in M_T\}$, and the moment maps $\mu_l = \mu_r : M_T \to (D_T)_{00}$ to be the tautological isomorphism.

Example 2. This is a generalization of Example 1. Let W be a diagonalizable H-module, $W = \oplus_{\lambda \in T} W[\lambda]$, $W[\lambda] = \{w \in W \mid aw = \lambda(a)w, \text{for all } a \in H\}$, and let $D_{T,W}^\alpha \subset \operatorname{Hom}_{\mathbb{C}}(W, W \otimes D_T)$ be the space of all difference operators on T with coefficients in $\operatorname{End}_{\mathbb{C}}(W)$, which have weight $\alpha \in T$ with respect to the action of H in W.

Consider the algebra $D_{T,W} = \oplus_\alpha D_{T,W}^\alpha$. This algebra has a weight decomposition $D_{T,W} = \oplus_{\alpha,\beta}(D_{T,W})_{\alpha\beta}$ defined as follows: if $g \in \operatorname{Hom}_{\mathbb{C}}(W, W \otimes M_T)$ is an operator of weight $\beta - \alpha$, then $gT_\beta^{-1} \in (D_{T,W})_{\alpha\beta}$.

Define the moment maps $\mu_l, \mu_r : M_T \to (D_{T,W})_{00}$ by the formulas $\mu_r(f(\lambda)) = f(\lambda)$, $\mu_l(f(\lambda)) = f(\lambda-h)$ where $f(\lambda-h)w = f(\lambda-\mu)w$ if $w \in W[\mu]$, $\mu \in T$. The algebra $D_{T,W}$ equipped with this weight decomposition and these moment maps is an H-algebra.

Now let us define the tensor product of H-algebras. Let A, B be two H-algebras and $\mu_l^A, \mu_r^A, \mu_l^B, \mu_r^B$ their moment maps. Define their *matrix tensor product*, $A \widetilde{\otimes} B$, which is also an H-algebra. Let

$$(A \widetilde{\otimes} B)_{\alpha\delta} := \oplus_\beta A_{\alpha\beta} \otimes_{M_T} B_{\beta\delta}, \tag{7.2}$$

where $\otimes_{M_T}$ means the usual tensor product modulo the relation $\mu_r^A(f)a \otimes b = a \otimes \mu_l^B(f)b$, for any $a \in A, b \in B, f \in M_T$. Introduce a multiplication in $A \widetilde{\otimes} B$

by the rule $(a \otimes b)(a' \otimes b') = aa' \otimes bb'$. It is easy to check that the multiplication is well defined. Define the moment maps for $A \widetilde{\otimes} B$ by $\mu_l^{A \widetilde{\otimes} B}(f) = \mu_l^A(f) \otimes 1$, $\mu_r^{A \widetilde{\otimes} B}(f) = 1 \otimes \mu_r^B(f)$.

For any H-algebra A, the algebras $A \widetilde{\otimes} D_T$ and $D_T \widetilde{\otimes} A$ are canonically isomorphic to A. In particular, D_T is canonically isomorphic to $D_T \widetilde{\otimes} D_T$. Thus the category of H-algebras equipped with the product $\widetilde{\otimes}$ is a monoidal category, where the unit object is D_T.

Now let us define the notions of a coproduct and a counit on an H-algebra.

Definition. A *coproduct* on an H-algebra A is a homomorphism of H-algebras $\Delta : A \to A \widetilde{\otimes} A$.

A *counit* on an H-algebra A is a homomorphism of H-algebras $\epsilon : A \to D_T$.

Finally, we can define the notions of an H-bialgebroid and an H-Hopf algebroid.

Definition. An *H-bialgebroid* is an H-algebra A equipped with a coassociative coproduct Δ (i.e. such that $(\Delta \otimes \mathrm{Id}_A) \circ \Delta = (\mathrm{Id}_A \otimes \Delta) \circ \Delta$, and a counit ϵ such that $(\epsilon \otimes \mathrm{Id}_A) \circ \Delta = (\mathrm{Id}_A \otimes \epsilon) \circ \Delta = \mathrm{Id}_A$.

Let A be an H-algebra. A linear map $S : A \to A$ is called *an antiautomorphism* of H-algebras if it is an antiautomorphism of algebras and $\mu_r \circ S = \mu_l$, $\mu_l \circ S = \mu_r$. From these conditions it follows that $S(A_{\alpha\beta}) = A_{-\beta,-\alpha}$.

Let A be an H-bialgebroid, and let Δ, ϵ be the coproduct and counit of A. For $a \in A$, let

$$\Delta(a) = \sum_i a_i^1 \otimes a_i^2. \tag{7.3}$$

Definition. An *antipode* on the H-bialgebroid A is an antiautomorphism of H-algebras $S : A \to A$ such that for any $a \in A$ and any presentation (7.3) of $\Delta(a)$, one has

$$\sum_i a_i^1 S(a_i^2) = \mu_l(\epsilon(a)1), \quad \sum_i S(a_i^1) a_i^2 = \mu_r(\epsilon(a)1),$$

where $\epsilon(a)1 \in M_T$ is the result of the application of the difference operator $\epsilon(a)$ to the constant function 1.

An H-bialgebroid with an antipode is called *an H-Hopf algebroid*.

Remarks.   1. If $H = \mathbb{C}$ then the notions of H-algebra, H-bialgebroid, H-Hopf algebroid are the familiar notions of an algebra, bialgebra, and Hopf algebra.

2. It is easy to see that D_T is an H-bialgebroid where $\Delta : D_T \to D_T \tilde{\otimes} D_T$ is the canonical isomorphism and $\epsilon = \mathrm{Id}$. Furthermore, it is an H-Hopf algebroid with $S(D) = D^*$, where D^* is the formal adjoint to the difference operator D (i.e. $(f(\lambda)\mathcal{T}_\alpha)^* = \mathcal{T}_\alpha^{-1} f(\lambda)$). This H-Hopf algebroid is an analog of the 1-dimensional Hopf algebra in the category of Hopf algebras.

3. One can define the notions of an H-algebra, H-bialgebroid, H-Hopf algebroid if the group T is not connected (for example, a finite group), in essentially the same way as above. More precisely, since in this case the algebra M_T of meromorphic functions on T is not a field but a direct sum of finitely many copies of a field, one should introduce an additional axiom requiring that $A_{\alpha\beta}$ is a free module over $\mu_l(M_T)$ and $\mu_r(M_T)$. Similarly, one can make all the above definitions in the case when M_T is replaced with another algebra of functions on T (rational functions, regular functions on some open set, etc.)

7.2. Dynamical representations of H-bialgebroids.   One of the reasons H-bialgebroids are good analogs of bialgebras is that their representations, like representations of bialgebras, form a tensor category. However, these representations are not the usual representations but rather new objects which we call dynamical representations, and which we will now define.

Definition.   A *dynamical representation* of an H-algebra A is a diagonalizable H-module W endowed with a homomorphism of H-algebras $\pi_W : A \to D_{T,W}$, where $D_{T,W}$ is defined in Example 2.

Definition.   A *homomorphism* of dynamical representations $\varphi : W_1 \to W_2$ is an element of $\mathrm{Hom}_\mathbb{C}(W_1, W_2 \otimes M_T)$ such that $\varphi \circ \pi_{W_1}(x) = \pi_{W_2}(x) \circ \varphi$ for all $x \in A$.

Example.   If A has a counit, then A has *the trivial representation*: $W = \mathbb{C}$, $\pi = \epsilon$.

For diagonalizable H-modules W, U, let $f \in \mathrm{Hom}(W, W \otimes M_T)$ and $g \in \mathrm{Hom}(U, U \otimes M_T)$. Define $f \tilde{\otimes} g \in \mathrm{Hom}(W \otimes U, W \otimes U \otimes M_T)$ as

$$f \tilde{\otimes} g(\lambda) = f^{(1)}(\lambda - h^{(2)})(1 \otimes g(\lambda)) \qquad (7.5)$$

where $f^{(1)}(\lambda - h^{(2)})(1 \otimes g(\lambda)) \, w \otimes u = f(\lambda - \mu)w \otimes g(\lambda)u$ if $g(\lambda)u \in U[\mu]$.

Lemma 1 ([EV2]).   *There is a natural embedding of H-algebras* $\theta_{WU} : D_{T,W} \tilde{\otimes} D_{T,U} \to D_{T,W \otimes U}$ *(an isomorphism if W, U are finite dimensional), given by the formula* $f\mathcal{T}_\beta \otimes g\mathcal{T}_\delta \to (f \tilde{\otimes} g)\mathcal{T}_\delta$.

Now let us define the tensor product of dynamical representations for H-bialgebroids. If A is an H-bialgebroid, and W and U are two dynamical representations of A, then we endow the H-module $W \otimes U$ with the structure of a dynamical representation via $\pi_{W \otimes U}(x) = \theta_{WU} \circ (\pi_W \otimes \pi_U) \circ \Delta(x)$. If $f : W_1 \to W_2$ and $g : U_1 \to U_2$ are homomorphisms of dynamical representations, then so is $f \bar{\otimes} g : W_1 \otimes U_1 \to W_2 \otimes U_2$. Thus, dynamical representations of A form a monoidal category $\text{Rep}(A)$, whose identity object is the trivial representation.

Remark. If A is an H-Hopf algebroid and V is a dynamical representation, then one can define the left and right dual dynamical representations $^*V$ and V^*. We will not discuss this notion here and refer the reader to [EV2].

7.3. The H-bialgebroid associated to a function $R : T \to \text{End}(V \otimes V)$.

Now, following [EV2], let us define an H-bialgebroid $\bar{A}_R$ associated to a meromorphic, zero weight function $R : T \to \text{End}(V \otimes V)$, where V is a finite dimensional diagonalizable H-module (we assume that $R(\lambda)$ is non-degenerate for generic λ). This is the dynamical analogue of the Faddeev-Reshetikhin-Takhtajan-Sklyanin construction of a bialgebra from an element $R \in \text{End}(V \otimes V)$, where V is a vector space.

By definition, the algebra $\bar{A}_R$ is generated by two copies of M_T (embedded as subalgebras) and matrix elements of the operator $L \in \text{End}(V) \otimes \bar{A}_R$. We denote the elements of the first copy of M_T by $f(\lambda^1)$ and of the second copy by $f(\lambda^2)$, where $f \in M_T$. We denote by $L_{\alpha\beta}$ the weight components of L with respect to the natural T-bigrading on $\text{End}(V)$, so that $L = (L_{\alpha\beta})$, where $L_{\alpha\beta} \in \text{Hom}_{\mathbb{C}}(V[\beta], V[\alpha]) \otimes \bar{A}_R$.

Introduce the moment maps for $\bar{A}_R$ by $\mu_l(f) = f(\lambda^1)$, $\mu_r(f) = f(\lambda^2)$, and define the weight decomposition by

$$f(\lambda^1), f(\lambda^2) \in (\bar{A}_R)_{00}, \qquad L_{\alpha\beta} \in \text{Hom}_{\mathbb{C}}(V[\beta], V[\alpha]) \otimes (\bar{A}_R)_{\alpha\beta}.$$

The defining relations for $\bar{A}_R$ are:

$$f(\lambda^1)L_{\alpha\beta} = L_{\alpha\beta}f(\lambda^1 + \alpha); \ f(\lambda^2)L_{\alpha\beta} = L_{\alpha\beta}f(\lambda^2 + \beta); [f(\lambda^1), g(\lambda^2)] = 0;$$

and the dynamical Yang-Baxter relation

$$R^{12}(\lambda^1)L^{13}L^{23} =: L^{23}L^{13}R^{12}(\lambda^2) : . \tag{7.6}$$

Here the $::$ sign means that the matrix elements of L should be put on the right of the matrix elements of R. Thus, if $\{v_a\}$ is a homogeneous basis of V, and $L = \sum E_{ab} \otimes L_{ab}$, $R(\lambda)(v_a \otimes v_b) = \sum R_{cd}^{ab}(\lambda)v_c \otimes v_d$, then (7.6) has the form

$$\sum R_{ac}^{xy}(\lambda^1)L_{xb}L_{yd} = \sum R_{xy}^{bd}(\lambda^2)L_{cy}L_{ax},$$

where we sum over repeated indices.

Define the coproduct on $\bar{A}_R$, $\Delta : \bar{A}_R \to \bar{A}_R \tilde{\otimes} \bar{A}_R$, and the counit of $\bar{A}_R$ by

$$\Delta(L) = L^{12}L^{13}, \epsilon(L_{\alpha\beta}) = \delta_{\alpha\beta} Id_{V[\alpha]} \otimes T_\alpha^{-1},$$

where $Id_{V[\alpha]} : V[\alpha] \to V[\alpha]$ is the identity operator.

Proposition 7.1 ([EV2]). $(\bar{A}_R, \Delta, \epsilon)$ *is an H-bialgebroid.*

Example. Suppose that R is the basic trigonometric solution of the QDYBE (see Section 6). Then the defining relations for $\bar{A}_R$ look like

$$f(\lambda^1)L_{bc} = L_{bc}f(\lambda^1 + \omega_b),$$
$$f(\lambda^2)L_{bc} = L_{bc}f(\lambda^2 + \omega_c),$$
$$L_{as}L_{at} = \frac{\alpha_{st}(\lambda^2)}{1 - \beta_{ts}(\lambda^2)}L_{at}L_{as}, s \neq t,$$
$$L_{bs}L_{as} = \frac{\alpha_{ab}(\lambda^1)}{1 - \beta_{ab}(\lambda^1)}L_{as}L_{bs}, a \neq b,$$
$$\alpha_{ab}(\lambda_1)L_{as}L_{bt} - \alpha_{st}(\lambda_2)L_{bt}L_{as} = (\beta_{ts}(\lambda_2) - \beta_{ab}(\lambda_1))L_{bs}L_{at}, a \neq b, s \neq t,$$

where $\alpha_{ab}(\lambda) = \frac{q^{\lambda_a - \lambda_b + 1} - 1}{q^{\lambda_a - \lambda_b} - 1}$, $\beta_{ab}(\lambda) = \frac{q - 1}{q^{\lambda_a - \lambda_b} - 1}$.

We note that we don't need any special properties of R (like the dynamical Yang-Baxter equation or Hecke condition) to define the H-bialgebroid $\bar{A}_R$. However, if we take a "randomly chosen" function R, the H-bialgebroid $\bar{A}_R$ will most likely have rather bad properties; i.e. it will be rather small and will not have interesting dynamical representations. The simplest way to ensure the existence of at least one interesting dynamical representation is to require that R satisfies the QDYBE. This is so because of the following proposition.

If (W, π_W) is a dynamical representation of an H-algebra A, we denote $\pi_W^0 : A \to \text{Hom}(W, W \otimes M_T)$ the map defined by $\pi_W^0(x)w = \pi_W(x)w, w \in W$ (the difference operator $\pi_W(x)$ restricted to the constant functions). It is clear that π_W is completely determined by π_W^0.

Proposition 7.2. *If R satisfies the QDYBE then $\bar{A}_R$ has a dynamical representation realized in the space V, with $\pi_V^0(\lambda) = R(\lambda)$.*

This representation is called *the vector representation*.

However, even if R satisfies the QDYBE, the H-bialgebroid $\bar{A}_R$ may not be completely satisfactory. In particular, one may ask the following question: does $\bar{A}_R$ define a "good quantum matrix algebra"? More precisely, does the Hilbert series of $\bar{A}_R$ equal to that of the function ring on the matrix algebra, i.e. $(1-t)^{-dim(V)^2}$? In general, the answer is no, even if the quantum dynamical Yang-Baxter equation is satisfied.

In fact, here is the place where the Hecke condition comes handy. Namely, we have the following proposition, which is a generalization of a well known proposition in the theory of quantum groups (due to Faddeev, Reshetikhin, Takhtajan).

Proposition 7.3 ([EV2]). *Suppose that R satisfies the QDYBE and the Hecke condition with q not equal to a nontrivial root of 1. Then the space $\bar{A}^m_R$ of polynomials of degree m in generators $L_{\alpha\beta}$ in $\bar{A}_R$ is a free $M_T \otimes M_T$-module, and the ranks of these modules are given by*

$$\sum_{m \geq 0} rk(\bar{A}^m_R) t^m = (1 - t)^{-n^2}.$$

Remark. The proof of this proposition, like the proof of its nondynamical analog, is based on the fact that under the assumptions of the Proposition, $\bar{A}^m_R$ is a representation of the Hecke algebra. This justifies the name "Hecke condition".

7.4. The H-Hopf algebroid A_R. Suppose now that R satisfies the QDYBE and the Hecke condition where q is not a nontrivial root of 1. In this case, it turns out that, analogously to the nondynamical case, a suitable localization A_R of $\bar{A}_R$ is actually a Hopf algebroid. Namely, define A_R by adjoining to $\bar{A}_R$ a new element L^{-1}, with the relation $LL^{-1} = L^{-1}L = 1$. It is easy to see that the structure of an H-bialgebroid on $\bar{A}_R$ naturally extends to A_R, and it can be shown that A_R admits a unique antipode S such that $S(L) = L^{-1}$. This antipode equips A_R with a structure of an H-Hopf algebroid. This H-Hopf algebroid is a quantization of the group GL_n in the same sense as the H-bialgebroid $\bar{A}_R$ is a quantization of the matrix algebra Mat_n.

7.5. Quasiclassical limit. In conclusion of the section, we would like to explain why the H-Hopf algebroid A_R considered here (for the basic rational or trigonometric solution R of the QDYBE) should be regarded as a quantization of the Poisson groupoid X_r corresponding to the basic rational, respectively trigonometric, solution r of the CDYBE (for the definition of X_r, see Section 5).

To see this, consider the H-Hopf algebroid A_R with $T = \mathfrak{h}^*$ for some finite dimensional abelian Lie algebra $\mathfrak{h}$, and M_T replaced with the ring of regular functions on some open subset U of $\mathfrak{h}^*$. Introduce a formal parameter γ (like in Section 3), and make a change of variable $\lambda \to \lambda/\gamma$ in the defining relations for A_R. It is easy to see that the resulting algebra A^γ_R over $\mathbb{C}[[\gamma]]$ is a deformation of a commutative algebra. The above result about the Hilbert series implies that this deformation is flat, so the quotient algebra $A^0_R :=$ $A^\gamma_R/(\gamma)$ obtains a Poisson structure. Let X be the spectrum of A^0_R; it is an algebraic Poisson manifold. It is not difficult to show that the the moment maps, coproduct, counit, and antipode of A_R define maps s, t, m, E, i (see Section 5) for X, which equips X with the structure of a Poisson groupoid with base U. Moreover, it is easy to check that the Poisson groupoid X is naturally isomorphic to X_r.

8   The universal fusion matrix and the Arnaudon-Buffenoir-Ragoucy-Roche equation

8.1. The ABRR equation. In [ABRR], Arnaudon, Buffenoir, Ragoucy and Roche give a general method for constructing the universal fusion matrix $J(\lambda)$, which lives in some completion of $U_q(\mathfrak{g})^{\otimes 2}$, i.e the unique element satisfying $J_{VW}(\lambda) = J(\lambda)_{|V \otimes W}$ for all V, W. A similar approach is suggested in [JKOS], based on the method of [Fr]

Let $U'(\mathfrak{b}_\pm)$ be the kernel of the projection $U(\mathfrak{b}_\pm) \to U(\mathfrak{h})$. We use the same notations with the index q for the quantum analogs of these objects. We set $\theta(\lambda) = \lambda + \rho - \frac{1}{2} \sum_i x_i^2 \in U\mathfrak{h}$ where as usual $\rho = \frac{1}{2} \sum_{\alpha \in \Delta^+} h_\alpha$ and (x_i) is an orthonormal basis of $\mathfrak{h}$. Set $\mathcal{R}_0 = \mathcal{R} q^{-\sum x_i \otimes x_i}$. It is known that $\mathcal{R}_0 \in 1 + U'_q(\mathfrak{b}_+) \otimes U'_q(\mathfrak{b}_-)$.

Theorem 8.1 ([ABRR]). *The universal fusion matrix $J(\lambda)$ of $U_q(\mathfrak{g})$ is the unique solution of the form $1 + U'_q(\mathfrak{b}_-) \otimes U'_q(\mathfrak{b}_+)$ of the equation*

$$J(\lambda)(1 \otimes q^{2\theta(\lambda)}) = \mathcal{R}_0^{21}(1 \otimes q^{2\theta(\lambda)}) J(\lambda). \tag{8.1}$$

The universal fusion matrix $J(\lambda)$ of $U(\mathfrak{g})$ is the unique solution of the form $1 + U'(\mathfrak{b}_-) \otimes U'(\mathfrak{b}_+)$ of the equation

$$[J(\lambda), 1 \otimes \theta(\lambda)] = (\sum_{\alpha \in \Delta^+} e_{-\alpha} \otimes e_\alpha) J(\lambda) \tag{8.2}$$

We will call these equations the *ABRR equations* for $U_q(\mathfrak{g})$ and for $\mathfrak{g}$, respectively.

Proof. Let us first show the statement about uniqueness. Let $T(\lambda) \in 1 + U'_q(\mathfrak{b}_-) \otimes U'_q(\mathfrak{b}_+)$ be any solution of (8.1). Then

$$(\mathcal{R}_0^{21})^{-1} T(\lambda) = \text{Ad} \, (1 \otimes q^{2\theta(\lambda)}) T(\lambda)$$

$$\Leftrightarrow ((\mathcal{R}_0^{21})^{-1} - 1) T(\lambda) = (\text{Ad} \, (1 \otimes q^{2\theta(\lambda)}) - 1) T(\lambda)$$

$$\Leftrightarrow T(\lambda) = 1 + (\text{Ad} \, (1 \otimes q^{2\theta(\lambda)}) - 1)^{-1} ((\mathcal{R}_0^{21})^{-1} - 1) T(\lambda)$$

Now notice that $((\mathcal{R}_0^{21})^{-1} - 1) \in U'_q(\mathfrak{b}_-) \otimes U'_q(\mathfrak{b}_+)$. This implies that $T(\lambda)$ can be recusively constructed as follows. Set $T_0(\lambda) = 1$ and put

$$T_{n+1}(\lambda) = 1 + (\text{Ad} \, (1 \otimes q^{2\theta(\lambda)}) - 1)^{-1} ((\mathcal{R}_0^{21})^{-1} - 1) T_n(\lambda).$$

Then $\lim_{n \to \infty} T_n(\lambda) = T(\lambda)$ (the limit is in the sense of stabilization). In particular there exists a unique solution to (8.1) of the given form.

The proof in the rational case (i.e in the case of a simple Lie algebra $\mathfrak{g}$) is similar. In that case, the recursive construction is given by $T_0(\lambda) = 1$ and

$$T_{n+1}(\lambda) = 1 - \text{ad}(1 \otimes \theta(\lambda)^{-1})(\sum_{\alpha \in \Delta^+} e_{-\alpha} \otimes e_\alpha) T_n(\lambda).$$

We now give a proof that the fusion matrix $J(\lambda)$ actually satisfies the ABRR relation in the case of simple Lie algebras. The proof in the case of quantum groups is analogous but technically more challenging, and is given in Appendix B.

Let C be the quadratic Casimir operator in the center of the universal enveloping algebra $U\mathfrak{g}$:

$$C = \sum_i x_i^2 + 2\rho + 2\sum_{\alpha \in \Delta^+} e_{-\alpha} e_\alpha.$$

Then C acts on any highest weight representation of $\mathfrak{g}$ of highest weight λ by the scalar $(\lambda, \lambda + 2\rho)$. Now let V, W be two finite-dimensional $\mathfrak{g}$-modules and let $v \in V$, $w \in W$ be two homogeneous elements of weight $\text{wt}(v)$ and $\text{wt}(w)$. We compute the quantity

$$F(\lambda) = \langle v^*_{\lambda - \text{wt}(v) - \text{wt}(w)}, \Phi^w_{\lambda - \text{wt}(v)}(C \otimes 1)\Phi^v_\lambda v_\lambda \rangle$$

in two different ways. On one hand we have

$$F = (\lambda - \text{wt}(v), \lambda - \text{wt}(v) + 2\rho)J(\lambda)(w \otimes v). \tag{8.3}$$

On the other hand,

$$F = \langle v^*_{\lambda - \text{wt}(v) - \text{wt}(w)}, \{2((e_{-\alpha}e_\alpha)_1 + (e_{-\alpha}e_\alpha)_2 + (e_\alpha \otimes e_{-\alpha})_{12} + (e_{-\alpha} \otimes e_\alpha)_{12}$$
$$+ \rho_1 + \rho_2) + \sum_i (x_i^2)_1 + (x_i^2)_2 + 2(x_i \otimes x_i)_{12}\}\Phi^w_{\lambda - \text{wt}(v)}\Phi^v_\lambda v_\lambda \rangle.$$

Since $v^*_{\lambda - \text{wt}(v) - \text{wt}(w)}$ is a highest weight vector, it is clear that

$$\langle v^*_{\lambda - \text{wt}(v) - \text{wt}(w)}, (e_{-\alpha}e_\alpha)_1 \Phi^w_{\lambda - \text{wt}(v)}\Phi^v_\lambda v_\lambda \rangle$$
$$= \langle v^*_{\lambda - \text{wt}(v) - \text{wt}(w)}, (e_{-\alpha} \otimes e_\alpha)_{12}\Phi^w_{\lambda - \text{wt}(v)}\Phi^v_\lambda v_\lambda \rangle = 0.$$

Moreover, by the intertwining property again, we have

$$(e_\alpha \otimes e_{-\alpha})_{12}\Phi^w_{\lambda - \text{wt}(v)}\Phi^v_\lambda v_\lambda = -(e_{-\alpha}e_\alpha)_2 - (e_{-\alpha} \otimes e_\alpha)_{23}\Phi^w_{\lambda - \text{wt}(v)}\Phi^v_\lambda v_\lambda,$$
$$(\rho_1 + \rho_2)\Phi^w_{\lambda - \text{wt}(v)}\Phi^v_\lambda v_\lambda = -\rho_3 \Phi^w_{\lambda - \text{wt}(v)}\Phi^v_\lambda v_\lambda + \Phi^w_{\lambda - \text{wt}(v)}\Phi^v_\lambda \rho v_\lambda,$$
$$((x_i^2)_1 + (x_i^2)_2 + 2(x_i \otimes x_i)_{12})\Phi^w_{\lambda - \text{wt}(v)}\Phi^v_\lambda v_\lambda$$
$$= -(2(x_i \otimes x_i)_{13} + 2(x_i \otimes x_i)_{23} + (x_i^2)_3)\Phi^w_{\lambda - \text{wt}(v)}\Phi^v_\lambda v_\lambda + \Phi^w_{\lambda - \text{wt}(v)}\Phi^v_\lambda x_i^2 v_\lambda$$
$$= \Phi^w_{\lambda - \text{wt}(v)}\Phi^v_\lambda x_i^2 v_\lambda + ((x_i^2)_3 - 2\lambda_3)\Phi^w_{\lambda - \text{wt}(v)}\Phi^v_\lambda v_\lambda.$$

Summing up these equations, we finally obtain

$$F = (-2\rho_{|V} + 2(\rho, \lambda) + (\sum_i x_i^2)_{|V} - 2\lambda_{|V} + \lambda^2 - 2 \sum_{\alpha \in \Delta^+} e_{-\alpha} \otimes e_\alpha) J(\lambda) \quad (8.4)$$

Combining (8.3) and (8.4) yields

$$(\sum_i x_i^2 - 2(\lambda + \rho))_{|V} J(\lambda) - (\text{wt}(v)^2 - 2(\lambda + \rho, \text{wt}(v))) J(\lambda)$$

$$= -2(\sum_{\alpha \in \Delta^+} e_{-\alpha} \otimes e_\alpha) J(\lambda)$$

which is equivalent to (8.2). ∎

Example. Let us use the recursive procedure in the proof above to compute $J(\lambda)$ for $U(sl_2)$. Setting $J(\lambda) = 1 + \sum_{n \geq 1} J^{(n)}(\lambda)$ where $J^{(n)}(\lambda) \in U'(\mathfrak{b}_-)[-2n] \otimes U'(\mathfrak{b}_+)[2n]$, the ABRR equation reads

$$\frac{1}{2}[1 \otimes ((\lambda + 1)h - \frac{h^2}{2}), J - 1] = -(f \otimes e)J,$$

which gives the recurrence relation

$$1 \otimes ((\lambda + 1)n - nh + n^2) J^{(n)} = (-f \otimes e) J^{(n-1)}.$$

Hence

$$J^{(n)}(\lambda) = \frac{(-1)^n}{n!} f^n \otimes (\lambda - h + 2)^{-1} \dots (\lambda - h + n + 1)^{-1} e^n.$$

This formula and its quantum analogue were obtained in the pioneering paper [BBB], which was a motivation to the authors of [ABRR].

8.2. Classical limits of fusion and exchange matrices. The ABRR relations can also be used to derive the classical limits of the fusion (and thus of the exchange) matrices. Setting $q = e^{-\gamma/2}$, rescaling $\lambda \mapsto \frac{\lambda}{\gamma}$ and considering the limit $\gamma \to 0$ yields the following classical version of the ABRR equation:

$$\text{Ad}(1 \otimes e^{-\lambda}) j(\lambda) - j(\lambda) = \sum_{\alpha \in \Delta^+} e_{-\alpha} \otimes e_\alpha,$$

which admits the unique lower triangular solution

$$j(\lambda) = - \sum_{\alpha \in \Delta^+} \frac{e_{-\alpha} \otimes e_\alpha}{1 - e^{-(\alpha, \lambda)}}.$$

From this we deduce

$$r(\lambda) = r^{21} + j(\lambda) - j^{21}(\lambda) = \frac{1}{2}\Omega + \frac{1}{2} \sum_{\alpha > 0} \coth\left(\frac{1}{2}\langle \alpha, \lambda \rangle\right) (e_\alpha \otimes e_{-\alpha} - e_{-\alpha} \otimes e_\alpha)$$

which is consistent with Proposition 3.3.

The case of a simple Lie algebra $\mathfrak{g}$ is completely analogous; the classical version of the ABRR equation is

$$[j(\lambda), \lambda \otimes 1] = \sum_{\alpha \in \Delta^+} e_{-\alpha} \otimes e_\alpha,$$

which admits the unique solution

$$j(\lambda) = -\sum_{\alpha \in \Delta^+} \frac{e_{-\alpha} \otimes e_\alpha}{(\lambda, \alpha)}.$$

This yields Corollary 3.1.

9 Transfer matrices and Generalized Macdonald-Ruijsenaars equations

9.1. Transfer matrices. We first recall the well-known transfer matrix construction.

Let A be a Hopf algebra with a commutative Grothendieck ring, and let $\mathcal{R} \in A \otimes A$ be an element such that $(\Delta \otimes 1)(\mathcal{R}) = \mathcal{R}^{13}\mathcal{R}^{23}$. A basic example is: A is quasitriangular, $\mathcal{R}$ is its universal R-matrix.

For any finite-dimensional representation $\pi_V : A \to \operatorname{End} V$ of A, set

$$T_V = \operatorname{Tr}_{|V}(\pi_V \otimes 1)(\mathcal{R}) \in A.$$

These elements are called transfer matrices.

Lemma 9.1. *For any finite-dimensional A-modules V, W we have $T_V T_W = T_{V \otimes W} = T_W T_V$.*

Proof. By definition we have

$$(\pi_{V \otimes W} \otimes 1)\mathcal{R} = (\pi_V \otimes \pi_W \otimes 1)(\Delta \otimes 1)\mathcal{R} = (\pi_V \otimes 1)\mathcal{R}_{13}(\pi_W \otimes 1)\mathcal{R}_{23},$$

which implies the first equality. The second equality follows from the commutativity of the Grothendieck ring. ∎

The transfer matrix construction gives rise to interesting examples of quantum integrable systems which arise in statistical mechanics. For example, if A is the quantum affine algebra or the elliptic algebra, one gets transfer matrices of the 6-vertex and 8-vertex models, respectively.

We adapt the notion of transfer matrices in our dynamical setting in the following way. Let $\mathfrak{g}$ be a simple Lie algebra and let $U_q(\mathfrak{g})$ be the associated quantum group. For any two finite-dimensional $U_q(\mathfrak{g})$-modules V and W let $R_{VW}(\lambda)$ be the exchange matrix. It is more convenient to work with the shifted exchange matrix $\mathbb{R}(\lambda) = R(-\lambda - \rho)$.

Let $\mathcal{F}_V$ be the space of $V[0]$-valued meromorphic functions on $\mathfrak{h}^*$. For $\nu \in \mathfrak{h}^*$ let $T_\nu \in \mathrm{End}(\mathcal{F}_V)$ be the shift operator $(T_\nu f)(\lambda) = f(\lambda + \nu)$. As pointed out in [FV3], the role of the transfer matrix is played by the following difference operator

$$\mathcal{D}_W^V = \sum_\nu \mathrm{Tr}_{|W[\nu]}(\mathbb{R}_{WV}(\lambda))T_\nu.$$

It follows from the dynamical 2-cocycle condition for fusion matrices (see Proposition 2.3) that for any $U_q(\mathfrak{g})$-modules U, V, W we have

$$\mathcal{D}_{V \otimes W}^U = \mathcal{D}_V^U \mathcal{D}_W^U = \mathcal{D}_W^U \mathcal{D}_V^U.$$

Hence $\{\mathcal{D}_W^U\}$ span a commuting family of difference operators acting on $\mathcal{F}_U$.

9.2. Weighted trace functions.

Let V be a finite-dimensional $U_q(\mathfrak{g})$-module. Recall that, for any homogeneous vector $v \in V[\nu]$ and for generic $\mu \in \mathfrak{h}^*$ there exists a unique intertwiner $\Phi_\mu^v : M_\mu \to M_{\mu-\nu} \otimes V$ such that $\langle v_{\mu-\nu}^*, \Phi_\mu^v v_\mu \rangle = v$. Set

$$\Phi_\mu^V = \sum_{v \in \mathcal{B}} \Phi_\mu^v \otimes v^* \in \mathrm{Hom}_\mathbb{C}\left(M_\mu, \bigoplus_\nu M_{\mu-\nu} \otimes V[\nu] \otimes V^*[-\nu]\right),$$

where $\mathcal{B}$ is any homogeneous basis of V. Consider the weighted trace function

$$\Psi_V(\lambda, \mu) = \mathrm{Tr}\left(\Phi_\mu^V q^{2\lambda}\right) \in V[0] \otimes V^*[0]$$

where $q^{2\lambda}$ acts on any $\mathfrak{h}$-semisimple $U_q(\mathfrak{g})$-module U by $q_{|U[\nu]}^{2\lambda} = q^{2(\lambda,\nu)}Id$. It can be shown that $\Psi_V \in V[0] \otimes V^*[0] \otimes q^{2(\lambda,\mu)}\mathbb{C}(q^\lambda) \otimes \mathbb{C}(q^\mu)$. Let

$$\delta_q(\lambda) = \left(\mathrm{Tr}_{|M_{-\rho}}(q^{2\lambda})\right)^{-1} = q^{-2(\lambda,\rho)}\prod_{\alpha > 0}(1 - q^{-2(\lambda,\alpha)})$$

be the Weyl denominator, and set

$$Q(\lambda) = m^{op}(1 \otimes S^{-1})(J(-\lambda - \rho)),$$

where $m^{op} : U_q(\mathfrak{g}) \otimes U_q(\mathfrak{g}) \to U_q(\mathfrak{g}), a \otimes b \mapsto ba$ and where $J(\lambda)$ is the universal fusion matrix. It can be shown that $Q(\lambda)$ is invertible. Finally, set

$$F_V(\lambda, \mu) = Q^{-1}(\mu)_{|V^*}\Psi_V(\lambda, -\mu - \rho)\delta_q(\lambda).$$

Theorem 9.1 ([EV4], The Macdonald-Ruijsenaars equations). *For any two finite-dimensional $U_q(\mathfrak{g})$-modules, we have*

$$\mathcal{D}_W^{\lambda,V} F_V(\lambda,\mu) = \chi_W(q^{-2\mu}) F_V(\lambda,\mu)$$

where $\chi_W(q^x) = \sum \dim W[\nu] q^{\langle \nu, x \rangle}$ is the character of W.

Theorem 9.2 ([EV4], The dual Macdonald-Ruijsenaars equations). *For any two finite-dimensional $U_q(\mathfrak{g})$-modules, we have*

$$\mathcal{D}_W^{\mu,V^*} F_V(\lambda,\mu) = \chi_W(q^{-2\lambda}) F_V(\lambda,\mu)$$

In the above, we add a superscript to $\mathcal{D}$ to specify on which variable the difference operators act. Thus, in Theorem 9.1, $\mathcal{D}_W^V$ acts on functions of the variable λ in the component $V[0]$, and in Theorem 9.2, $\mathcal{D}_W^{V^*}$ acts on functions in the variable μ in the component $V^*[0]$.

From Theorems 9.1 and 9.2 it is not difficult to deduce the following result:

Theorem 9.3 ([EV4], The symmetry identity). *For any finite dimensional $U_q(\mathfrak{g})$-module we have*

$$F_V(\lambda,\mu) = F_{V^*}^*(\mu,\lambda),$$

where $* : V[0] \otimes V^*[0] \to V^*[0] \otimes V[0]$ is the permutation.

9.3. Relation to Macdonald theory.

Let us now restrict ourselves to the case of $\mathfrak{g} = sl_n$, and let V be the q-analogue of the representation $S^{mn}\mathbb{C}^n$. The zero-weight subspace of this representation is 1-dimensional, so the function Ψ_V can be regarded as a scalar function. We will denote this scalar function by $\Psi_m(q,\lambda,\mu)$.

Recall the definition of Macdonald operators [Ma, EK1]. They are operators on the space of functions $f(\lambda_1,...,\lambda_n)$ which are invariant under simultaneous shifting of the variables, $\lambda_i \to \lambda_i + c$, and have the form

$$M_r = \sum_{I \subset \{1,...,n\}: |I|=r} \left(\prod_{i \in I, j \notin I} \frac{tq^{2\lambda_i} - t^{-1}q^{2\lambda_j}}{q^{2\lambda_i} - q^{2\lambda_j}} \right) T_I,$$

where $T_I\lambda_j = \lambda_j$ if $j \notin I$ and $T_I\lambda_j = \lambda_j + 1$ if $j \in I$. Here q, t are parameters. We will assume that $t = q^{m+1}$, where m is a nonnegative integer.

It is known [Ma] that the operators M_r commute. From this it can be deduced that for a generic $\mu = (\mu_1,...,\mu_n)$, $\sum \mu_i = 0$, there exists a unique power series $f_{m0}(q,\lambda,\mu) \in \mathbb{C}[[q^{\lambda_2-\lambda_1}, ..., q^{\lambda_n-\lambda_{n-1}}]]$ such that the series $f_m(q,\lambda,\mu) := q^{2(\lambda,\mu-m\rho)} f_{m0}(q,\lambda,\mu)$ satisfies difference equations

$$M_r f_m(q,\lambda,\mu) = \Big(\sum_{I \subset \{1,...,n\}: |I|=r} q^{2\sum_{i \in I}(\mu+\rho)_i} \Big) f_m(q,\lambda,\mu).$$

Remark. The series f_{m0} is convergent to an analytic (in fact, a trigonometric) function.

The following theorem is contained in [EK1].

Theorem 9.4 ([EK1], Theorem 5). *One has*

$$f_m(q, \lambda, \mu) = \gamma_m(q, \lambda)^{-1} \Psi_m(q^{-1}, -\lambda, \mu),$$

where

$$\gamma_m(q, \lambda) := \prod_{i=1}^{m} \prod_{l<j} (q^{\lambda_l - \lambda_j} - q^{2i} q^{\lambda_j - \lambda_l}).$$

Let $\mathcal{D}_W(q^{-1}, -\lambda)$ denote the difference operator, obtained from the operator $\mathcal{D}_W$ defined in Section 1 by the transformation $q \to q^{-1}$ and the change of coordinates $\lambda \to -\lambda$. Let $\Lambda^r \mathbb{C}^n$ denote the q-analog of the r-th fundamental representation of sl_n.

Corollary 9.1.

$$\mathcal{D}_{\Lambda^r \mathbb{C}^n}(q^{-1}, -\lambda) = \delta_q(\lambda) \gamma_m(q, \lambda) \circ M_r \circ \gamma_m(q, \lambda)^{-1} \delta_q(\lambda)^{-1}.$$

Proof. This follows from Theorem 9.4 and Theorem 9.1.

Remark. In the theorems of this section, Verma modules M_μ can be replaced with finite dimensional irreducible modules L_μ with sufficiently large highest weight, and one can prove analogs of these theorems in this situation (in the same way as for Verma modules). In particular, one may set $\hat{\Psi}_m(q, \lambda, \mu) = \mathrm{Tr}(\hat{\Phi}_\mu^V q^{2\lambda})$, where $\hat{\Phi}_\mu^V : L_\mu \to L_\mu \otimes V \otimes V^*[0]$ is the intertwiner with highest coefficient 1 (Such an operator exists iff $\mu - m\rho \geq 0$, see [EK1]). Then one can show analogously to Theorem 9.1 (see [EK1]) that the function $\hat{f}_m(q, \lambda, \mu) := \gamma_m(q, \lambda)^{-1} \hat{\Psi}_m(q^{-1}, -\lambda, \mu + m\rho)$ is the Macdonald polynomial $P_\mu(q, t, q^{2\lambda})$ with highest weight μ (μ is a dominant integral weight). In this case, Theorem 9.1 says that Macdonald's polynomials are eigenfunctions of Macdonald's operators, Theorem 9.2 gives recursive relations for Macdonald's polynomials with respect to the weight (for $sl(2)$ – the usual 3-term relation for orthogonal polynomials), and Theorem 9.3 is the Macdonald symmetry identity (see [Ma]).

10 Appendix A: Classical dynamical r-matrices on a simple Lie algebra $\mathfrak{g}$ with respect to $\mathfrak{l} \subset \mathfrak{h}$.

A.1. Let $\mathfrak{g}$ be a simple complex Lie algebra and $\mathfrak{l}$ a commutative subalgebra of $\mathfrak{g}$ consisting of semisimple elements. Then $\mathfrak{l} \subset \mathfrak{h}$ for some Cartan subalgebra $\mathfrak{h}$. We keep the notations of Section 1. In this appendix, we give a classification of all classical dynamical r-matrices $\mathfrak{l}^* \to (\mathfrak{g} \otimes \mathfrak{g})^\mathfrak{l}$ with coupling constant 1. Note that we can suppose without loss of generality that the restriction of $\langle\,,\,\rangle$ to $\mathfrak{l}$ is nondegenerate. Indeed, given a dynamical r-matrix $r : \mathfrak{l}^* \to (\mathfrak{g} \otimes \mathfrak{g})^\mathfrak{l}$, we can always replace $\mathfrak{l}$ by the largest subalgebra of $\mathfrak{h}$ under which r is invariant, and this subalgebra is real.

A.2. Gauge transformations. Let $\Omega' \in \mathfrak{l}^\perp \otimes \mathfrak{l}^\perp$ (where the orthogonal complement is in $\mathfrak{g}$) be the inverse (Casimir) element to the form $\langle\,,\,\rangle$. If $r(\lambda) - \frac{1}{2}\Omega + (\varphi(\lambda) \otimes 1)\Omega'$, $\varphi : \mathfrak{l}^* \to \mathrm{End}_\mathfrak{l}(\mathfrak{l}^\perp)$ is a meromorphic function with values in $(\mathfrak{g} \otimes \mathfrak{g})^\mathfrak{l}$, and if $f : \mathfrak{l}^* \to \mathfrak{h}$ is any meromorphic function, set

$$r^f(\lambda) = \frac{1}{2}\Omega + (e^{-ad\, f(\lambda)}\varphi(\lambda)e^{ad\, f(\lambda)} \otimes 1)\Omega'.$$

Lemma 10.1. *The transformations $r(\lambda) \mapsto r^f(\lambda)$ preserve the set of classical dynamical r-matrices with coupling constant* 1.

Two r-matrices which can be obtained one from the other by such a transformation are called *gauge-equivalent*.

A.3. Classification of dynamical r-matrices. Let $\mathfrak{h}$ be a Cartan subalgebra of $\mathfrak{g}$, and let $\Pi \subset \mathfrak{h}^*$ be a system of simple roots in Δ. Let $\mathfrak{h}_0 \subset \mathfrak{h}$ be the orthogonal complement of $\mathfrak{l}$ in $\mathfrak{h}$.

Definition. A *generalized Belavin-Drinfeld triple* is a triple $(\Gamma_1, \Gamma_2, \tau)$ where $\Gamma_1, \Gamma_2 \subset \Pi$, and where $\tau : \Gamma_1 \xrightarrow{\sim} \Gamma_2$ is a norm-preserving isomorphism.

Given a generalized Belavin-Drinfeld triple $(\Gamma_1, \Gamma_2, \tau)$, we extend linearly the map τ to a norm-preserving bijection $\langle\Gamma_1\rangle \to \langle\Gamma_2\rangle$, where $\langle\Gamma_1\rangle$ (resp. $\langle\Gamma_2\rangle$) is the set of roots $\alpha \in \Delta$ which are linear combinations of simple roots from Γ_1 (resp. from Γ_2).

We say that a generalized Belavin-Drinfeld triple $(\Gamma_1, \Gamma_2, \tau)$ is $\mathfrak{l}$-admissible if $\tau(\alpha) - \alpha \in \mathfrak{l}^\perp$ for all $\alpha \in \Gamma_1$, if τ satisfies the following condition: for every cycle $\alpha \mapsto \tau(\alpha) \mapsto \ldots \mapsto \tau^r(\alpha) = \alpha$ we have $\alpha + \tau(\alpha) + \ldots + \tau^{r-1}\alpha \in \mathfrak{l}$.

If $(\Gamma_1, \Gamma_2, \tau)$ is a generalized Belavin-Drinfeld triple, let $\mathfrak{g}_{\Gamma_1}$, $\mathfrak{g}_{\Gamma_2}$ be the subalgebras generated by e_α, f_α $\alpha \in \Gamma_1$ (resp. generated by e_α, f_α $\alpha \in \Gamma_2$). The map $\tau : e_\alpha \mapsto e_{\tau(\alpha)}$, $\alpha \in \Gamma_1$ extends to an isomorphism $\tau : \mathfrak{g}_{\Gamma_1} \xrightarrow{\sim} \mathfrak{g}_{\Gamma_2}$. Finally, define an operator $K : \mathfrak{l}^* \to \mathrm{Hom}\,(\mathfrak{g}_{\Gamma_1}, \mathfrak{g})$ by

$$K(\lambda)e_\alpha = \sum_{n>0} e^{-n(\alpha,\lambda)}\tau^n(e_\alpha).$$

Notice that this sum is finite if τ acts nilpotently on α.

Theorem 10.1 ([S]). *Let $(\Gamma_1, \Gamma_2, \tau)$ be an $\mathfrak{l}$-admissible generalized Belavin-Drinfeld triple.*

(i) The equation

$$((\alpha - \tau(\alpha)) \otimes 1)r_0 = \frac{1}{2}((\tau(\alpha) + \alpha) \otimes 1)\Omega_{\mathfrak{h}_0},$$

where $\Omega_{\mathfrak{h}_0} \subset \mathfrak{h}_0 \otimes \mathfrak{h}_0$ is the inverse element to the form $\langle,\rangle$, has solutions $r_0 \in \Lambda^2 \mathfrak{h}_0$.

(ii) Let $r_0 \in \Lambda^2 \mathfrak{h}_0$ satisfy the equation from (i). Then

$$r(\lambda) = \frac{1}{2}\Omega + r_0 + \sum_{\substack{\alpha \in \Delta^+ \\ e_\alpha \in \mathfrak{g}_{\Gamma_1}}} K(\lambda)e_\alpha \wedge f_\alpha + \sum_{\alpha \in \Delta^+} \frac{1}{2}e_\alpha \wedge f_\alpha \qquad (10.1)$$

is a classical dynamical r-matrix. Conversely, any classical dynamical r-matrix $r : \mathfrak{l}^* \to (\mathfrak{g} \otimes \mathfrak{g})^{\mathfrak{l}}$ with coupling constant 1 is gauge equivalent to one of the above form, for suitable choices of Cartan subalgebra $\mathfrak{h}$ containing $\mathfrak{l}$, polarization of $\mathfrak{g}$ and $\mathfrak{l}$-admissible generalized Belavin-Drinfeld triple.

Proof. Let us prove statement (i); statement (ii) is proved in [S]. Let $\mathfrak{l}_{max}$ be the Lie algebra of all $x \in \mathfrak{h}$ such that $(\alpha - \tau(\alpha), x) = 0$, and let $\mathfrak{p}$ be the orthogonal complement of $\mathfrak{l}$ in $\mathfrak{l}_{max}$. Then we have an orthogonal direct sum decomposition $\mathfrak{h}_0 = \mathfrak{p} \oplus \mathfrak{l}_{max}^\perp$. Let us regard r_0 as a bilinear form on $\mathfrak{h}$ (via the standard inner product). The equation from (i) determines $r_0(x, y)$ where $x \in \mathfrak{l}_{max}^\perp$, and y is arbitrary. To check that r_0 can be extended to a skew symmetric form, it suffices to check that it is skew-symmetric on $\mathfrak{l}_{max}^\perp$. But using the equation from (i) we find that

$$r_0(\alpha - \tau(\alpha), \beta - \tau(\beta)) = 1/2(\alpha + \tau(\alpha), \beta - \tau(\beta)) = \frac{1}{2}((\beta, \tau(\alpha)) - (\alpha, \tau(\beta)))$$

(we use that τ preseves $(,)$), which is obviously skew symmetric. ∎

Remarks. 1. This classification is very similar in spirit to the classification of classical r-matrices $r \in \mathfrak{g} \otimes \mathfrak{g}$ satisfying $r + r^{21} = \Omega$ (quasitriangular structures) obtained by Belavin and Drinfeld, and reduces to it for $\mathfrak{l} = 0$ (see [BD]).

2. When $\mathfrak{l} = \mathfrak{h}$ one recovers the classification result Theorem 4.2: the only $\mathfrak{h}$-graded generalized Belavin-Drinfeld triples are of the form $\Gamma_1 = \Gamma_2, \tau = Id$, and in this case the r-matrix (10.1) corresponds to $r_X^1(\lambda)$, with $X = \Gamma_1$.

3. Theorem 10.1 is proved in [S] under the additional assumption that $\mathfrak{l}$ contains a regular semisimple element. However, the proof easily extends to the present situation.

11 Appendix B: Proof of the ABRR relation for $U_q(\mathfrak{g})$

We keep the notations of Section 8. Recall the Drinfeld construction of the quantum Casimir element of $U_q(\mathfrak{g})$. Let $\mathcal{R}$ be the universal R-matrix for $U_q(\mathfrak{g})$. Let us write $\mathcal{R} = \sum_i a_i \otimes b_i$ and set $u = \sum S(b_i)a_i$. Then $u = q^{2\rho}z$ where z is a central element in a completion of $U_q(\mathfrak{g})$, which is called the quantum Casimir element. Moreover, for any $\mu \in \mathfrak{h}^*$ we have

$$uv_\mu = q^{-\sum_i x_i^2}v_\mu = q^{-(\mu,\mu)}v_\mu$$

hence $zv_\mu = q^{-(\mu,\mu+2\rho)}v_\mu$ and $u_{|M_\mu} = q^{-(\mu,\mu+2\rho)}q^{2\rho}$.

Now let V and W be two finite-dimensional $U_q(\mathfrak{g})$-modules, $v \in V$ and $w \in W$ homogeneous vectors of weight μ_v and μ_w respectively, and consider the expectation value

$$X_{vw}(\mu) = \langle v^*_{\mu-\mu_v-\mu_w}, \Phi^w_{\mu-\mu_v}u_{|M_{\mu-\mu_v}}\Phi^v_\mu v_\mu\rangle.$$

We will compute $X_{vw}(\mu)$ in two different ways. On one hand,

$$\begin{aligned}
X_{vw}(\mu) &= \langle v^*_{\mu-\mu_v-\mu_w}, \Phi^w_{\mu-\mu_v}(q^{2\rho}z)_{|M_{\mu-\mu_v}}\Phi^v_\mu v_\mu\rangle\\
&= \langle v^*_{\mu-\mu_v-\mu_w}, \Phi^w_{\mu-\mu_v}z_{|M_{\mu-\mu_v}}q^{-2\rho}_{|V}\Phi^v_\mu q^{2\rho}v_\mu\rangle\\
&= q^{-(\mu-\mu_v,\mu-\mu_v+2\rho)}q^{2\rho}_{|V}q^{(2\rho,\mu)}J(\mu)(w\otimes v)\\
&= q^{-(\mu-\mu_v,\mu-\mu_v)+2(\rho,\mu_v)}q^{-2\rho}_{|V}J(\mu)(w\otimes v)
\end{aligned}$$

On the other hand, we have $(1\otimes\Delta^{op})\mathcal{R} = \mathcal{R}^{12}\mathcal{R}^{13}$ hence, by the intertwining property,

$$(1\otimes\Delta)(1\otimes S)\mathcal{R} = \sum_{ij} a_ia_j \otimes S(b_i) \otimes S(b_j).$$

Thus

$$X_{vw}(\mu) = \sum_{i,j} S(b_j)_{|W}\langle v^*_{\mu-\mu_v-\mu_w}, S(b_i)_{|M_{\mu-\mu_v-\mu_w}}\Phi^w_{\mu-\mu_v}a_ia_j{}_{|M_{\mu-\mu_v}}\Phi^v_\mu v_\mu\rangle. \tag{11.1}$$

Now, since $v^*_{\mu-\mu_v-\mu_w}$ is a lowest weight vector and since

$$\mathcal{R} \in (1 + U'_q(\mathfrak{b}_-)\otimes U'_q(\mathfrak{b}_+))q^{\sum_i x_i\otimes x_i}$$

equation (11.1) reduces to

$$\begin{aligned}
X_{vw}(\mu) &= \sum_j S(b_j)_{|W}\langle v^*_{\mu-\mu_v-\mu_w}, q^{-\sum x_i^2}_{|M_{\mu-\mu_v-\mu_w}}q^{-\sum x_i\otimes x_i}_{|M_{\mu-\mu_v-\mu_w}\otimes W}\Phi^w_{\mu-\mu_v}a_j{}_{|M_{\mu-\mu_v}}\Phi^v_\mu v_\mu\rangle\\
&= q^{-(\mu-\mu_v-\mu_w)^2}\sum_j \left(S(b_j)q^{-\mu+\mu_v+\mu_w}\right)_{|W}\langle v^*_{\mu-\mu_v-\mu_w}\Phi^w_{\mu-\mu_v}a_j{}_{|M_{\mu-\mu_v}}\Phi^v_\mu v_\mu\rangle
\end{aligned} \tag{11.2}$$

Using the relation

$$(\Delta \otimes S)\mathcal{R} = \sum a_l \otimes a_k \otimes S(b_k)S(b_l)$$

we have

$$\sum_j S(b_j)_{|W} \otimes \Phi^v_\mu a_{j|M_\mu} = \left(\sum_{k,l} (S(b_k)S(b_l))_{|W} \otimes a_{l|M_{\mu-\mu_v}} \otimes a_{k|V}\right)\Phi^v_\mu,$$

i.e

$$\left(\sum S(b_l)_{|W} \otimes a_{l|M_{\mu-\mu_v}}\right)\Phi^v_\mu = \left(\sum S(b_k)_{|W} \otimes a_{k|V}\right)^{-1}\sum S(b_j)_{|W} \otimes \Phi^v_\mu a_{j|M_\mu}.$$

Substitution of this in (11.2) gives

$$X_{vw}(\mu) = q^{-(\mu-\mu_v-\mu_w)^2}\left(\sum S(b_k)_{|W} \otimes a_{k|V}\right)^{-1} q_{|W}^{-2\mu+\mu_v+\mu_w} J(\mu)(w \otimes v) \quad (11.3)$$

Claim: we have $\left(\sum S(b_k) \otimes a_k\right)^{-1} = (1 \otimes q^{2\rho})\mathcal{R}(1 \otimes q^{-2\rho})$.
Proof. We have $S^2(a) = q^{2\rho}aq^{-2\rho}$ for any $a \in U_q(\mathfrak{g})$. Hence $\sum a_k \otimes S(b_k) = (1 \otimes q^{2\rho})(1 \otimes S^{-1})(1 \otimes q^{-2\rho})$. The claim now follows from the relation $(1 \otimes S^{-1})\mathcal{R} = \mathcal{R}^{-1}$.

Thus, combining (11.1) and (11.3), we get

$$q^{-(\mu-\mu_v,\mu-\mu_v+2\rho)}q_{|V}^{-2\rho}q^{2(\mu,\rho)} J(\mu)(w \otimes v)$$
$$= q^{-(\mu-\mu_v-\mu_w)^2} q_{|W}^{2\rho} \mathcal{R}^{21} q_{|W}^{\mu_v+\mu_w-2(\mu+\rho)} J(\mu)(w \otimes v).$$

This implies that

$$q^{-(\mu_v+\mu_w)^2}\mathcal{R}^{21} q_{|W}^{\mu_v+\mu_w-2(\mu+\rho)} J(\mu)(w \otimes v)$$
$$= q^{\mu^2-2(\mu,\mu_v+\mu_w)} q_{|V}^{-2\rho} q_{|W}^{-2\rho} q^{2(\mu,\rho)} q^{-(\mu-\mu_v,\mu-\mu_v+2\rho)} J(\mu)(w \otimes v) \quad (11.4)$$

Using the weight zero property of $\mathcal{R}$, we can rewrite the l.h.s of this last equation as

$$\mathcal{R}^{21} q_{|W\otimes V}^{-\sum x_i \otimes x_i} q_{|V}^{-\sum x_i^2} q_{|W}^{-2(\mu+\rho)} J(\mu)(w \otimes v) \quad (11.5)$$

Similarly, using the weight zero property of $J(\mu)$, it is easy to see that the right hand side is equal to

$$q^{-2(\mu+\rho,\mu_w)}q^{-\mu^2} J(\mu) = J(\mu)q_{|W}^{-2(\mu+\rho)} q_{|V}^{-\sum x_i^2}(w \otimes v) \quad (11.6)$$

The ABRR equation now follows from (11.4), (11.5), (11.6) and the weight zero property of $J(\mu)$.

12 Review of literature

In conclusion, we would like to give a brief review of the existing literature on the dynamical Yang-Baxter equations. We would like to make it clear that this list is by no means complete, and contains only some of the basic references which are relevant to this paper.

The physical paper in which the dynamical Yang-Baxter equation was first considered is [GN]; dynamical R-matrices are also discussed in [Fad1],[ΛF].

The classical dynamical Yang-Baxter equation and examples of its solutions were introduced in [F]. Its geometric interpretation in terms of Poisson groupoids of Weinstein [W] was introduced in [EV1]. Solutions of this equation were studied and classified in [EV1], [S]. The relationship of solutions of this equation to Poisson groupoids and Lie bialgebroids was further explored in [LX] and [BK-S]. The relationship of solutions of the classical dynamical Yang-Baxter equation (defined on noncommutative Lie algebras) to equivariant cohomology is discussed in [AM]. The relationship to integrable systems is discussed in [ABB].

Quantum groups associated to a dynamical R-matrix were first introduced in [F]. In the case when the R-matrix is elliptic, they are called elliptic quantum groups. These quantum groups and their representation theory (for the Lie algebras sl_n), as well as their relationship with integrable systems, were systematically studied in [FV1, FV2, FV3]. The papers [EV2, EV3] study the trigonometric versions of these quantum groups (for any simple Lie algebra).

Quantum groupoids were introduced by Maltsiniotis (in the case when base is classical), and by Lu in [Lu] in the general case. The interpretation of dynamical quantum groups as quantum groupoids was first discussed in [EV2, EV3], and further studied in [Xu1], [Xu2]. The interpretation of dynamical quantum groups as quasi-Hopf algebras is contained in [BBB], and was further developed in [Fr, JKOS], [ABRR],[EF]. The connection between these two interpretation is discussed in [Xu1].

Quantum KZB equations (which are not discussed in these notes) were introduced in [F], and studied in [FTV1, FTV2, MV, FV3, FV4, FV5, FV6, FV7]. Monodromy of quantum KZ equations [FR], which yields dynamical R-matrices of the elliptic quantum groups, is computed in [TV1, TV2].

The theory of traces of intertwining operators for Lie algebras and quantum groups and its applications to the theory of special functions (in particular, Macdonald theory) is developed in [B],[F],[E],[EK1, EK2, EK3, EK4],[K1, K2, K3], [EFK], [ES1, ES2]. The relationship of this theory with dynamical R-matrices is studied in [EV4].

References

[ABB] Avan J., Babelon O., Billey E., *The Gervais-Neveu-Felder equation and the quantum Calogero-Moser systems*, hep-th/9505091, Comm. Math. Phys., **178**, issue 2, (1996) 281-299.

[ABRR] D.Arnaudon, E.Buffenoir, E.Ragoucy, and Ph.Roche, *Universal Solutions of quantum dynamical Yang-Baxter equations*, Lett. Math. Phys. **44** (1998), no. 3, 201-214.

[AF] Alekseev A., Faddeev L., $(T^*G)_t$: *a toy model of conformal field theory*, Comm. Math. Phys., **141**, (1991) 413-422.

[AM] Alekseev A., Meinrenken M., *The noncommutative Weil algebra*, /math.DG 9903052, (1999).

[B] [B] Bernard, D., *On the Wess-Zumino-Witten models on the torus*, Nucl. Phys., B303, (1988) 77-93.

[BBB] Babelon, O., Bernard, D., Billey, E.,*A quasi-Hopf algebra interpretation of quantum 3-j and 6-j symbols and difference equations*, Phys. Lett. B, **375** (1996) 89-97.

[BK-S] Bangoura M., Kosmann-Schwarzbach Y., *Equations de Yang-Baxter dynamique classique et algébroides de Lie*, C.R.Acad Sci. Paris, **327** (1998), no. 6, 541-546.

[BD] Belavin A.A, Drinfeld V.G, *Triangle equations and simple Lie algebras*, Soviet Sci. reviews, Sect C **4**, 93-165. (1984).

[CP] Chari V., Pressley A., *A guide to quantum groups*, Cambridge University press, (1994).

[Dr] Drinfeld, V. *On some open problems in quantum group theory*, Lect. notes in Math, 1510, 1-8.

[E] Etingof, P.I., *Quantum integrable systems and representations of Lie algebras*, hep-th 9311132, J. Math. Phys. **36** (1995), no.6, 2636-2651.

[EF] Enriquez B., Felder G., *Elliptic quantum groups $E_{\tau,\eta}(sl_2)$ and quasi-Hopf algebras*, q-alg/9703018, Comm. Math. Phys., **195** (1998), no.3., 651-689.

[EFK] Etingof P., Frenkel I., Kirillov Jr. A., *Lectures on representation theory and Knizhnik-Zamolodchikov equations* AMS, (1998).

[EK] Etingof P., Kazhdan D., *Quantization of Lie bialgebras I*, Selecta Math. **2** (1996), 1-41.

[EK1] Etingof, P.I., Kirillov, A.A., Jr, *Macdonald's polynomials and representations of quantum groups*, Math. Res. Let., **1** (3), (1994) 279-296.

[EK2] Etingof, P.I., Kirillov, A.A., Jr, *Representation-theoretic proof of the inner product and symmetry identities for Macdonald's polynomials*, Comp.Math., **102**, (1996) 179-202.

[EK3] Etingof, P.I., Kirillov, A.A., Jr, *On Cherednik-Macdonald-Mehta identities*, math.QA 9712051, Electr. Res. Ann., (1998) 43-47.

[EK4] Etingof, P.I., Kirillov, A.A., Jr, On an affine analogue of of Jack and Macdonald polynomials, Duke Math.J., **78**, (2) (1995) 229 256.

[ES1] Etingof P., Styrkas K., *Algebraic integrability of Schrödinger operators and representations of Lie algebras*, hep-th 9403135, Compositio. Math., **98**, (1), (1995) 91-112.

[ES2] Etingof P., Styrkas K., *Algebraic integrability of Macdonald operators and representations of quantum groups*, q-alg 9603022, Compositio. Math. **114** (2) (1998), 125-152.

[EV1] Etingof P., Varchenko A., *Geometry and classification of solutions of the classical dynamical Yang-Baxter equation*, Commun. Math. Phys, **192** 77-120 (1998).

[EV2] Etingof P., Varchenko A., *Solutions of the quantum dynamical Yang-Baxter equation and dynamical quantum groups*, Commun. Math. Phys, **196** 591-640 (1998).

[EV3] Etingof P., Varchenko A., *Exchange dynamical quantum groups*, preprint math.QA/9801135.

[EV4] Etingof P., Varchenko A., *Traces of intertwiners for quantum groups and difference equations, I*, preprint math.QA/9907181.

[Fad1] Faddeev L., *On the exchange matrix of the WZNW model*, Comm. Math. Phys., **132** (1990), 131-138.

[F] Felder G., *Conformal field theory and integrable systems associated to elliptic curves*, Proceedings of the International Congress of Mathematicians, Zürich 1994, p.1247–1255, Birkhäuser, 1994; *Elliptic quantum groups*, preprint hep-th/9412207, to appear in the Proceedings of the ICMP, Paris 1994.

[FR] Frenkel I., Reshetikhin N., *Quantum affine algebras and holonomic difference equations*, Commun. Math. Phys. **146** (1992), 1-60.

[FTV1] Felder G., Tarasov V., Varchenko A., *Solutions of the elliptic QKZB equations and Bethe ansatz I*, q-alg/9606005, in Topics in Singularity Theory, V.I.Arnold's 60th Anniversary Collection, Advances in the Mathematical Sciences -34, AMS Translations, Series 2, **180** (1997), 45-76.

[FTV2] Felder G., Tarasov V., Varchenko A., *Monodromy of solutions of the elliptic quantum Knizhnik-Zamolodchikov-Bernard difference equations*, q-alg/9705017, (1997), 1-26.

[FV1] Felder G., Varchenko A., *On representations of the elliptic quantum group* $E_{\tau,\eta}(sl_2)$, Commun. Math. Phys. **181** (1996), 746–762.

[FV2] Felder G., Varchenko A., *Algebraic Bethe ansatz for the elliptic quantum group* $E_{\tau,\eta}(sl_2)$, Nuclear Physics B **480** (1996) 485-503.

[FV3] Felder G., Varchenko A., *Elliptic quantum groups and Ruijsenaars models*, J. Statist. Phys. **89** (1997), no. 5-6, 963-980.

[FV4] Felder G., Varchenko A., *Quantum KZB heat equation, modular transformations, and* $GL(3, Z)$, *I*, math.QA/9809139, (1998).

[FV5] Felder G., Varchenko A., *Resonance relations for solutions of the elliptic qKZB equations, fusion rules, and eigenvectors of transfer matrices of restricted interaction-round-a-face models*, math.QA/9901111, (1999).

[FV6] Felder G., Varchenko A., *Quantum KZB heat equation, modular transformations, and* $GL(3, Z)$, *II*, math.QA 9907061, (1999).

[FV7] Felder G., Varchenko A., *The elliptic gamma-function, and* $SL_3(\mathbb{Z}) \ltimes \mathbb{Z}^3$, math.QA 9907061, (1999).

[Fr] Fronsdal C., *Quasi-Hopf deformations of quantum groups*, Lett. Math. Phys, **40**, (1997) 117-134.

[GN] Gervais, J.-L., Neveu, A., *Novel triangle relation and absense of tachyons in Liouville string field theory*, Nucl. Phys. B, **238**, (1984), 125-141.

[JKOS] Jimbo M., Konno H., Odake S., Shiraishi J., *Quasi-Hopf twistors for elliptic quantum groups*, q-alg 9712029, (1997).

[K1] Kirillov, A. Jr., *Traces of intertwining operators and Macdonald polynomials*, PhD thesis, q-alg 9503012, (1995).

[K2] Kirillov, A.Jr., *On an inner product in modular tensor categories*, J. Amer. Math. Soc, **9**, (4), (1996) 1135-1169.

[K3] Kirillov, A. Jr, *On an inner product in modular tensor categories, II*, Adv.Theor.Math.Phys, **2**, (1998) 155-180.

[Lu] Lu J.H., *Hopf algebroids and quantum groupoids*, Inter. J. Math., **7** (1), (1996), 47-70.

[LX] Liu, Z.-J., Xu, P., *Dirac structures and dynamical r-matrices*, math.DG/9903119, (1999).

[M] Mackenzie K, *Lie groupoids and Lie algebroids in differential geometry*, Cambridge Univ. Press, 1997.

[Ma] Macdonald, I.G., *A new class of symmetric functions*, Publ. I.R.M.A. Strasbourg, 372/S-20, Actes 20 Seminaire Lotharingien (1988) 131-171.

[MV] Mukhin E., Varchenko A., *Solutions of the qKZB equations in tensor products of finite dimensional modules over the elliptic quantum group $E_{\tau,\eta}sl_2$*, 9712056, (1997).

[S] Schiffmann O, *On classification of dynamical r-matrices*, Math. Res. Letters, **5**, 13-30 (1998).

[TV1] Tarasov T., Varchenko A., *Geometry of q-Hypergeometric functions as a bridge between Yangians and Quantum Affine Algebras*, Inv.Math., **128**, (1997), 501-588.

[TV2] Tarasov V., Varchenko A., *Geometry of q-Hypergeometric Functions, Quantum Affine Algebras and Elliptic Quantum Groups*, q-alg/9703044, Asterisque, 246, (1997), pages 1-135.

[W] Weinstein A., *Coisotropic calculus and Poisson groupoids* J. Math. Soc. Japan **40**, 705-727, (1988).

[Xu1] Xu, P., *Quantum groupoids associated to universal dynamical R-matrices*, C.R.Acad. Sci. Paris, Serie I, **328**, (1999) 327-332.

[Xu2] Xu, P., *Quantum groupoids*, math.QA 9905192, (1999).

QUANTIZED PRIMITIVE IDEAL SPACES AS QUOTIENTS OF AFFINE ALGEBRAIC VARIETIES

K. R. GOODEARL

ABSTRACT. Given an affine algebraic variety V and a quantization $\mathcal{O}_q(V)$ of its coordinate ring, it is conjectured that the primitive ideal space of $\mathcal{O}_q(V)$ can be expressed as a topological quotient of V. Evidence in favor of this conjecture is discussed, and positive solutions for several types of varieties (obtained in joint work with E. S. Letzter) are described. In particular, explicit topological quotient maps are given in the case of quantum toric varieties.

INTRODUCTION

A major theme in the subject of quantum groups is the philosophy that in the passage from a classical coordinate ring to a quantized analog, the classical geometry is replaced by structures that should be treated as 'noncommutative geometry'. Indeed, much work has been invested into the development of theories of noncommutative differential geometry and noncommutative algebraic geometry. We would like to pose the question whether these theories are entirely noncommutative, or whether traces of classical geometry are to be found in the noncommutative geometry. This rather vague question can, of course, be focused in any number of different directions. We discuss one particular direction here, which was developed in joint work with E. S. Letzter [6]; it concerns situations in which quantized analogs of classical varieties contain certain quotients of these varieties.

To take an ideal-theoretic perspective on the question posed above, recall the way in which an affine algebraic variety V is captured in its coordinate ring $\mathcal{O}(V)$: the space of maximal ideals, $\max \mathcal{O}(V)$, is homeomorphic to V, under their respective Zariski topologies. In the passage from commutative to noncommutative algebras, the natural analog of a maximal ideal space is a primitive ideal space – this point of view is taken partly on practical grounds, because noncommutative rings often have too few maximal ideals for various purposes, but also in order to reflect the ideal theory connected with the study of irreducible representations. (Recall that the primitive

---

This research was partially supported by NATO Collaborative Research Grant 960250 and National Science Foundation research grants DMS-9622876, DMS-9970159.

ideals of an algebra A are precisely the annihilators of the irreducible A-modules.) Thus, given a quantized version of $\mathcal{O}(V)$, say $\mathcal{O}_q(V)$, the natural analog of V is the primitive ideal space $\operatorname{prim} \mathcal{O}_q(V)$. The aspect of the general problem which we wish to discuss here, then, is that of finding classical geometric structure in $\operatorname{prim} \mathcal{O}_q(V)$, and relating it to the structure of V.

We report below on some success in expressing $\operatorname{prim} \mathcal{O}_q(V)$ as a quotient of V with respect to the respective Zariski topologies. Modulo a minor technical assumption, this is done in three situations – when V is an algebraic torus, a full affine space, or (generalizing both of those cases) an affine toric variety. We discuss these cases in Sections 2, 3, and 4, respectively; we sketch some of the methods and indicate relationships among the three cases. In particular, in the last section we establish precise formulas for topological quotient maps in the case of quantum toric varieties, maps were only given an existence proof in [**6**].

Recall that a map $\phi : X \rightarrow Y$ between topological spaces is a *topological quotient map* provided ϕ is surjective and the topology on Y coincides with the quotient topology induced by ϕ, that is, a subset $C \subseteq Y$ is closed in Y precisely when $\phi^{-1}(C)$ is closed in X. In that case, Y is completely determined (as a topological space) by the topology on X together with the partition of X into fibres of ϕ.

Throughout the paper, we fix an algebraically closed base field k. All the varieties we discuss will be affine algebraic varieties over k.

1. QUANTUM SEMISIMPLE GROUPS

We follow the usual practice in writing 'quantum semisimple groups' as an abbreviation for 'quantized coordinate rings of semisimple groups'. Thus, 'quantum SL_n' refers to any quantization of the coordinate ring $\mathcal{O}(SL_n(k))$. We begin by displaying the primitive ideal space of the most basic example, the standard single parameter quantization of SL_2.

1.1. Example. Consider the algebra $\mathcal{O}_q(SL_2(k))$, where $q \in k^\times$ is a non-root of unity. This is the k-algebra with generators X_{11}, X_{12}, X_{21}, X_{22} satisfying the following relations:

$$X_{11}X_{12} = qX_{12}X_{11} \qquad X_{11}X_{21} = qX_{21}X_{11} \qquad X_{12}X_{21} = X_{21}X_{12}$$
$$X_{21}X_{22} = qX_{22}X_{21} \qquad X_{12}X_{22} = qX_{22}X_{12}$$
$$X_{11}X_{22} - X_{22}X_{11} = (q - q^{-1})X_{12}X_{21} \qquad X_{11}X_{22} - qX_{12}X_{21} = 1$$

Since q is not a root of unity, the primitive ideal space of $\mathcal{O}_q(SL_2(k))$ is not large, and can be completely described as follows:

$$\text{.................} \langle X_{11} - \alpha,\ X_{12},\ X_{21},\ X_{22} - \alpha^{-1} \rangle \text{.................}$$

$$(\alpha \in k^\times)$$

$$\langle X_{12} \rangle \hspace{8cm} \langle X_{21} \rangle$$

$$\text{.................} \langle X_{12} - \beta X_{21} \rangle \text{.................}$$

$$(\beta \in k^\times)$$

With the Zariski topology, this space seems more like a scheme than an affine variety, since it has many non-closed points. When taken apart as pictured, however, it can be viewed as a disjoint union of four classical varieties: two single points, and two punctured affine lines. Taken as a whole, prim $\mathcal{O}_q(SL_2(k))$ can be related to the variety $SL_2(k)$ via the following map:

$$\begin{pmatrix} \alpha & 0 \\ 0 & \alpha^{-1} \end{pmatrix} \mapsto \langle X_{11} - \alpha,\ X_{12},\ X_{21},\ X_{22} - \alpha^{-1} \rangle \qquad (\alpha \in k^\times)$$

$$\begin{pmatrix} \alpha & 0 \\ \gamma & \alpha^{-1} \end{pmatrix} \mapsto \langle X_{12} \rangle \qquad (\alpha, \gamma \in k^\times)$$

$$\begin{pmatrix} \alpha & \beta \\ 0 & \alpha^{-1} \end{pmatrix} \mapsto \langle X_{21} \rangle \qquad (\alpha, \beta \in k^\times)$$

$$\begin{pmatrix} \alpha & \beta \\ \gamma & \delta \end{pmatrix} \mapsto \langle X_{12} - \beta\gamma^{-1} X_{21} \rangle \qquad (\alpha, \delta \in k;\ \beta, \gamma \in k^\times)$$

$$(\alpha\delta - \beta\gamma = 1)$$

We leave as an exercise for the reader to show that the map $SL_2(k) \to$ prim $\mathcal{O}_q(SL_2(k))$ given above is Zariski continuous; in fact, it is a topological quotient map.

1.2. Let G be a connected semisimple algebraic group over $\mathbb{C}$, and let $q \in \mathbb{C}^\times$ be a non-root of unity. Quantized coordinate rings of G have been defined in both single parameter and multiparameter versions, which we denote $\mathcal{O}_q(G)$ and $\mathcal{O}_{q,p}(G)$ respectively. We shall not recall the definitions here, but refer the reader to [8], [9], [10], [12], [13], or [4]. The first classification results for primitive ideals in this context were proved in the single parameter case:

Theorem. [Hodges-Levasseur, Joseph] *There exists a bijection*

$$\text{prim}\, \mathcal{O}_q(G) \longleftrightarrow \{\, \text{symplectic leaves in } G \,\},$$

where the symplectic leaves are computed relative to the associated Poisson structure on G arising from the quantization.

Proof. This was proved for the cases $G = SL_3(\mathbb{C})$ and $G = SL_n(\mathbb{C})$ by Hodges and Levasseur in [8, Theorem 4.4.1] and [9, Theorem 4.2], and then for arbitrary G by Joseph [**12**, Theorem 9.2], [**13**, Theorems 10.3.7, 10.3.8]. □

Let us rewrite this result to bring in the points of G more directly: There exists a surjection $G \twoheadrightarrow \text{prim}\, \mathcal{O}_q(G)$ whose fibres are precisely the symplectic leaves in G. In the multiparameter case, the situation is slightly more complicated, as follows.

1.3. Theorem. [Hodges-Levasseur-Toro] *There exists a surjection* $G \twoheadrightarrow$ *prim* $\mathcal{O}_{q,p}(G)$*, but the fibres are symplectic leaves only for certain choices of p.*

Proof. The first statement follows from the results in [**10**, Section 4]; for the second, see [**10**, Theorems 1.8, 4.18]. □

1.4. We can fill in a bit more detail about these surjections by bringing in a group of symmetries. Let H be a maximal torus in G, acting on $\mathcal{O}_{q,p}(G)$ by 'winding automorphisms' (cf. [**13**, (1.3.4)]). For example, if $G = SL_2(\mathbb{C})$ and $H = \left\{ \left(\begin{smallmatrix} \alpha & 0 \\ 0 & \alpha^{-1} \end{smallmatrix} \right) \mid \alpha \in \mathbb{C}^\times \right\}$, then H acts on $\mathcal{O}_q(G)$ so that $\left(\begin{smallmatrix} \alpha & 0 \\ 0 & \alpha^{-1} \end{smallmatrix} \right) \cdot \left(\begin{smallmatrix} X_{11} & X_{12} \\ X_{21} & X_{22} \end{smallmatrix} \right) = \left(\begin{smallmatrix} \alpha X_{11} & \alpha X_{12} \\ \alpha^{-1} X_{21} & \alpha^{-1} X_{22} \end{smallmatrix} \right)$. In this case, the induced action of H on $\text{prim}\, \mathcal{O}_q(G)$ has four orbits, which are the sets displayed in Example 1.1.

It follows from the analyses in [**8**, **9**, **12**, **13**, **10**] that
- The number of H-orbits in $\text{prim}\, \mathcal{O}_{q,p}(G)$ is finite.
- Each of these H-orbits is Zariski homeomorphic to H modulo the relevant stabilizer.
- The preimage of each H-orbit, under the surjection in Theorem 1.3, is a locally closed subset of G.

These properties hint strongly that there is some topological connection between G and $\text{prim}\, \mathcal{O}_{q,p}(G)$.

1.5. Conjecture. There exists an H-equivariant surjection

$$G \to \text{prim}\, \mathcal{O}_{q,p}(G)$$

which is a topological quotient map (with respect to the Zariski topologies).

If this conjecture were verified, then it together with knowledge of the fibres of the map would completely determine $\text{prim}\, \mathcal{O}_{q,p}(G)$ as a topological space.

In the case of $\mathcal{O}_q(SL_2(\mathbb{C}))$, the conjecture can be readily verified 'by hand', as already suggested above. There is a further piece of evidence in the SL_3 case: Brown and the author have constructed an H-equivariant topological quotient map $B^+ \twoheadrightarrow \text{prim}\, \mathcal{O}_q(B^+)$ where B^+ is the upper triangular Borel subgroup of $SL_3(\mathbb{C})$ [Work in progress].

The major difficulty in the way of the conjecture is that no good global description of the Zariski topology on prim $\mathcal{O}_{q,p}(G)$ is known, only the Zariski topologies on the separate H-orbits.

In order to try to obtain a better feel for the problem, let us widen the focus:

1.6. Conjecture. If $\mathcal{O}_{\mathbf{q}}(V)$ is a quantized coordinate ring of an affine algebraic variety (or group) V, there exists a topological quotient map $V \twoheadrightarrow \text{prim}\,\mathcal{O}_{\mathbf{q}}(V)$, equivariant with respect to an appropriate group of automorphisms.

Here $\mathcal{O}_{\mathbf{q}}(V)$ is meant to stand for any quantization of $\mathcal{O}(V)$; a phrase which is only slightly less vague than the symbolism, since no definition of a "quantization" of a coordinate ring exists. For descriptions of many commonly studied classes of examples, see [4].

In the remainder of the paper, we discuss some cases where the extended conjecture has been verified.

2. QUANTUM TORI

The simplest case in which to examine our conjecture is that of a quantization of the coordinate ring of an algebraic torus. The parameters required can be arranged as a matrix $\mathbf{q} = (q_{ij}) \in M_n(k^\times)$ which is *multiplicatively antisymmetric*, that is, $q_{ii} = 1$ and $q_{ji} = q_{ij}^{-1}$ for all i, j. (These conditions are assumed in order to avoid degeneracies in the resulting algebra.) The *quantum torus* over k with respect to these parameters is the algebra

$$\mathcal{O}_{\mathbf{q}}((k^\times)^n) := k\langle x_1^{\pm 1}, \ldots, x_n^{\pm 1} \mid x_i x_j = q_{ij} x_j x_i \text{ for all } i, j\rangle.$$

The structure of this algebra has been analyzed in various cases by many people, among them McConnell and Pettit [15], De Concini, Kac, and Procesi [3], Hodges [7], Vancliff [16], Brown and the author [2], Letzter and the author [5], and Ingalls [11]. In particular, it is known that prim $\mathcal{O}_{\mathbf{q}}((k^\times)^n)$ and spec $\mathcal{O}_{\mathbf{q}}((k^\times)^n)$ are homeomorphic to the maximal and prime spectra, respectively, of the center $Z(\mathcal{O}_{\mathbf{q}}((k^\times)^n))$, via contraction and extension, and that the center is a Laurent polynomial ring over k (e.g., [5, Lemma 1.2, Corollary 1.5]).

2.1. Theorem. [6, Theorem 3.11] *For any multiplicatively antisymmetric matrix* $\mathbf{q} \in M_n(k^\times)$, *there exist topological quotient maps*

$$(k^\times)^n \twoheadrightarrow \text{prim}\,\mathcal{O}_{\mathbf{q}}((k^\times)^n)$$
$$\text{spec}\,k[y_1^{\pm 1}, \ldots, y_n^{\pm 1}] \twoheadrightarrow \text{spec}\,\mathcal{O}_{\mathbf{q}}((k^\times)^n),$$

equivariant with respect to the natural actions of the torus $(k^\times)^n$. □

This case, in many ways, works out too smoothly to really put Conjecture 1.6 to the test. In that regard, the case of quantum affine spaces is much more interesting. Before moving on to that case, however, we describe the form of the maps in Theorem 2.1, since this will preview essential aspects of the quantum affine space case. We proceed by rewriting quantum tori as twisted group algebras, as follows.

2.2. Fix a quantum torus $A = \mathcal{O}_q((k^\times)^n)$. Set $\Gamma = \mathbb{Z}^n$, and define $\sigma : \Gamma \times \Gamma \to k^\times$ by the rule

$$\sigma(\alpha, \beta) = \prod_{i,j=1}^n q_{ij}^{\alpha_i \beta_j}.$$

Then σ is an alternating bicharacter on Γ, and it determines the commutation rules in A, since $x^\alpha x^\beta = \sigma(\alpha, \beta) x^\beta x^\alpha$ for all $\alpha, \beta \in \Gamma$, where we have used the standard multi-index notation $x^\alpha = x_1^{\alpha_1} x_2^{\alpha_2} \cdots x_n^{\alpha_n}$. Recall that the *radical* of σ is the subgroup

$$\mathrm{rad}(\sigma) = \{\alpha \in \Gamma \mid \sigma(\alpha, -) \equiv 1\}$$

of Γ. It plays the following role: $Z(A) = k[x^\alpha \mid \alpha \in \mathrm{rad}(\sigma)]$ (cf. [**5**, Lemma 1.2] or [**11**, Proposition 5.2]).

Now choose a 2-cocycle $c : \Gamma \times \Gamma \to k^\times$ such that $c(\alpha, \beta) c(\beta, \alpha)^{-1} = \sigma(\alpha, \beta)$ for $\alpha, \beta \in \Gamma$. (This can be done in many ways – see [**1**, Proposition 1, p. 888] or [**6**, (3.2)], for instance.) We may identify A with the twisted group algebra $k^c\Gamma$, that is, the k-algebra with a basis $\{x_\alpha \mid \alpha \in \Gamma\}$ such that

$$x_\alpha x_\beta = c(\alpha, \beta) x_{\alpha+\beta}$$

for $\alpha, \beta \in \Gamma$.

Let us write the group algebra $k\Gamma$ in terms of a basis $\{y_\alpha \mid \alpha \in \Gamma\}$, where $y_\alpha y_\beta = y_{\alpha+\beta}$ for $\alpha, \beta \in \Gamma$. There is a k-linear (vector space) isomorphism

$$\Phi_c : A \to k\Gamma$$

such that $\Phi_c(x_\alpha) = y_\alpha$ for $\alpha \in \Gamma$. While this map is not multiplicative, it satisfies $\Phi_c(x_\alpha x_\beta) = c(\alpha, \beta) \Phi_c(x_\alpha) \Phi_c(x_\beta)$ for $\alpha, \beta \in \Gamma$ (cf. [**6**, (3.5)]).

The final ingredient needed to describe the topological quotient maps in the current situation is the torus $H = (k^\times)^n$ and its natural actions on A and $k\Gamma$ by k-algebra automorphisms. The induced actions of H on the prime and primitive spectra of A and $k\Gamma$ will also be required. We identify H with $\mathrm{Hom}(\Gamma, k^\times)$, which allows us to write the above actions as

$$h.x_\alpha = h(\alpha) x_\alpha \quad \text{and} \quad h.y_\alpha = h(\alpha) y_\alpha$$

for $h \in H$ and $\alpha \in \Gamma$. Observe that the map Φ_c is H-equivariant. The identification $H = \text{Hom}(\Gamma, k^\times)$ provides a pairing $H \times \Gamma \to k^\times$, and we use $\perp$ to denote orthogonals with respect to this pairing. In particular,

$$S^\perp = \{ h \in H \mid \ker h \supseteq S \}$$

for $S \subseteq \Gamma$.

2.3. It is convenient to have a compact notation for the intersection of the orbit of an ideal under a group of automorphisms. If P is an ideal of a ring B, and T is a group acting on B by ring automorphisms, we write

$$(P : T) = \bigcap_{t \in T} t(P).$$

This ideal can also be described as the largest T-invariant ideal of B contained in P. The same notation can also be applied to any subset of B.

2.4. Theorem. [**6**, Theorem 3.11] *Let $A = \mathcal{O}_q((k^\times)^n)$, and fix $\Gamma, \sigma, c, \Phi_c, H$ as above. Assume that $c \equiv 1$ on $\text{rad}(\sigma) \times \Gamma$. (Such a choice for c is always possible; cf.* [**6**, Lemma 3.12]*.) Then the rule $P \mapsto \Phi_c^{-1}(P : \text{rad}(\sigma)^\perp)$, for prime ideals P of $k\Gamma$, defines H-equivariant topological quotient maps*

$$\text{spec}\, k\Gamma \twoheadrightarrow \text{spec}\, A \qquad and \qquad \max k\Gamma \twoheadrightarrow \text{prim}\, A.$$

The fibres of the second map are exactly the $\text{rad}(\sigma)^\perp$-orbits in $\max k\Gamma$. $\square$

2.5. As the formula in Theorem 2.4 indicates, the given maps are compositions of two maps, which themselves have nice properties. To express this in more detail, write $S = \text{rad}(\sigma)$, and recall that an $S^\perp$-*prime ideal* of $k\Gamma$ is any proper $S^\perp$-invariant ideal Q such that whenever Q contains a product of $S^\perp$-invariant ideals, it must contain one of the factors. Denote by $S^\perp$- $\text{spec}\, k\Gamma$ the set of $S^\perp$-prime ideals of $k\Gamma$, and observe that this set supports a Zariski topology (defined in the obvious way). The first map in Theorem 2.4 can be factored in the form

$$\text{spec}\, k\Gamma \twoheadrightarrow S^\perp\text{-}\,\text{spec}\, k\Gamma \xrightarrow{\approx} \text{spec}\, A,$$

where the map $P \mapsto (P : S^\perp)$ from $\text{spec}\, k\Gamma$ to $S^\perp$- $\text{spec}\, k\Gamma$ is a topological quotient map, and the map $Q \mapsto \Phi_c^{-1}(Q)$ from $S^\perp$- $\text{spec}\, k\Gamma$ to $\text{spec}\, A$ is a homeomorphism. That $P \mapsto (P : S^\perp)$ is a topological quotient map is an easy exercise (cf. [**6**, Proposition 1.7] for a generalization); the bulk of the work in proving Theorem 2.4 goes into establishing the homeomorphism $S^\perp$- $\text{spec}\, k\Gamma \approx \text{spec}\, A$ (cf. [**6**, Proposition 3.10]). The key point is that the hypothesis on c ensures that Φ_c is 'central-semilinear', that is, $\Phi_c(za) = \Phi_c(z)\Phi_c(a)$ for $z \in Z(A)$ and $a \in A$ [**6**, Lemma 3.5].

The factorization above respects maximal and primitive ideals in the following way. Let $S^\perp$-$\max k\Gamma$ denote the set of maximal proper $S^\perp$-invariant ideals of $k\Gamma$; this is a subset of $S^\perp$-$\operatorname{spec} k\Gamma$, which we equip with the relative topology. Then we have the factorization

$$\max k\Gamma \twoheadrightarrow S^\perp\text{-}\max k\Gamma \xrightarrow{\approx} \operatorname{prim} A,$$

where $P \mapsto (P : S^\perp)$ gives a topological quotient map from $\max k\Gamma$ to $S^\perp$-$\max k\Gamma$, and $Q \mapsto \Phi_c^{-1}(Q)$ gives a homeomorphism from $S^\perp$-$\max k\Gamma$ to $\operatorname{prim} A$ (cf. [6, Propositions 3.9, 3.10]).

3. Quantum Affine Spaces

Let $\mathbf{q} \in M_n(k^\times)$ again be a multiplicatively antisymmetric matrix. The corresponding *quantum affine space* over k is the algebra

$$\mathcal{O}_{\mathbf{q}}(k^\times) := k\langle x_1, \dots, x_n \mid x_i x_j = q_{ij} x_j x_i \text{ for all } i, j\rangle.$$

3.1. Denote the algebra $\mathcal{O}_{\mathbf{q}}(k^\times)$ by A for purposes of discussion. The torus $H = (k^\times)^n$ again acts naturally as k-algebra automorphisms on A. The primitive spectrum of A consists of 2^n H-orbits, which we denote $\operatorname{prim}_w A$, indexed by the subsets $w \subseteq \{1, \dots, n\}$ [5, Theorem 2.3]. Here

$$\operatorname{prim}_w A = \{P \in \operatorname{prim} A \mid x_i \in P \Longleftrightarrow i \in w\}.$$

Similarly, $\operatorname{spec} A$ is a disjoint union of 2^n subsets

$$\operatorname{spec}_w A = \{P \in \operatorname{spec} A \mid x_i \in P \Longleftrightarrow i \in w\}.$$

Each $\operatorname{spec}_w A$ is homeomorphic, via localization, to the prime spectrum of the quantum torus

$$A_w = \left(A/\langle x_i \mid i \in w\rangle\right)[x_j^{-1} \mid j \notin w],$$

and likewise $\operatorname{prim}_w A$ is homeomorphic to $\operatorname{prim} A_w$ (cf. [5, Theorem 2.3]).

The partition $\operatorname{prim} A = \bigsqcup_w \operatorname{prim}_w A$ is a quantum analog of the standard stratification of affine n-space by its H-orbits. To emphasize the parallel, we label the H-orbits in k^n as

$$(k^n)_w = \{(\alpha_1, \dots, \alpha_n) \in k^n \mid \alpha_i = 0 \Longleftrightarrow i \in w\},$$

for $w \subseteq \{1, \dots, n\}$. It follows from Theorem 2.1 that there are H-equivariant topological quotient maps

$$(k^n)_w \twoheadrightarrow \operatorname{prim} A_w \approx \operatorname{prim}_w A$$

for all w. The problem is then to patch these maps together coherently. Modulo a minor technical assumption, that can be done:

3.2. Theorem. [6, Theorem 4.11] *Let* $\mathbf{q} = (q_{ij}) \in M_n(k^\times)$ *be a multiplicatively antisymmetric matrix. Assume that either* $-1 \notin \langle q_{ij} \rangle$ *or* char $k = 2$. *Then there exist topological quotient maps*

$$k^n \twoheadrightarrow \text{prim}\, \mathcal{O}_{\mathbf{q}}(k^n)$$

$$\text{spec}\, k[y_1, \ldots, y_n] \twoheadrightarrow \text{spec}\, \mathcal{O}_{\mathbf{q}}(k^n),$$

equivariant with respect to the natural actions of the torus $(k^\times)^n$. $\square$

Here $\langle q_{ij} \rangle$ denotes the subgroup of $k^\times$ generated by the q_{ij}.

To describe the maps in the above theorem, and identify the fibres of the first, we use a setup analogous to that in the quantum torus case. In particular, we write $\mathcal{O}_{\mathbf{q}}(k^n)$ as a twisted semigroup algebra with respect to a suitable cocycle.

3.3. Fix a quantum affine space $A = \mathcal{O}_{\mathbf{q}}(k^n)$. Set $\Gamma = \mathbb{Z}^n$ and $\Gamma^+ = (\mathbb{Z}^+)^n$, and define $\sigma : \Gamma \times \Gamma \to k^\times$ as in (2.2). As in the previous case, σ determines the commutation rules in A.

Next, choose a 2-cocycle $c : \Gamma \times \Gamma \to k^\times$ such that $c(\alpha, \beta)c(\beta, \alpha)^{-1} = \sigma(\alpha, \beta)$ for $\alpha, \beta \in \Gamma$. We identify A with the twisted semigroup algebra $k^c\Gamma^+$ with a basis $\{x_\alpha \mid \alpha \in \Gamma^+\}$, and we write the corresponding semigroup algebra $R = k\Gamma^+$ in terms of a basis $\{y_\alpha \mid \alpha \in \Gamma^+\}$. There are partitions $\text{spec}\, R = \bigsqcup_w \text{spec}_w R$ and $\max R = \bigsqcup_w \max_w R$ analogous to those for $\text{spec}\, A$ and $\text{prim}\, A$.

Again as before, identify the torus $H = (k^\times)^n$ with $\text{Hom}(\Gamma, k^\times)$, so that the natural actions of H on A and $k\Gamma^+$ are given by

$$h.x_\alpha = h(\alpha)x_\alpha \qquad \text{and} \qquad h.y_\alpha = h(\alpha)y_\alpha.$$

3.4. It is convenient to identify the localizations A_w with subalgebras of the twisted group algebra $k^c\Gamma$, and to identify the corresponding localizations R_w of R with subalgebras of the group algebra $k\Gamma$. This is done as follows:

$$A_w = \sum_{\alpha \in \Gamma_w} kx_\alpha \qquad \text{and} \qquad R_w = \sum_{\alpha \in \Gamma_w} ky_\alpha = k\Gamma_w,$$

where $\Gamma_w = \{\alpha \in \Gamma \mid \alpha_i = 0 \text{ for } i \in w\}$. (Here we have extended the x_α and the y_α to Γ-indexed bases for $k^c\Gamma$ and $k\Gamma$.) As in (2.2), there is an H-equivariant k-linear isomorphism $\Phi = \Phi_c : k^c\Gamma \to k\Gamma$ such that $\Phi(x_\alpha) = y_\alpha$ for $\alpha \in \Gamma$. This map restricts to H-equivariant k-linear isomorphisms from A onto R and from A_w onto R_w for each w.

Next, let σ_w and c_w denote the restrictions of σ and c to Γ_w. Then the identification above can be restated as $A_w = k^{c_w}\Gamma_w$. We shall need the radical of σ_w, and we emphasize that this is a subgroup of Γ_w. There is a natural pairing between Γ_w and a quotient group of H, but it is more useful to take orthogonals with respect to the pairing of H and Γ; thus, we write

$$\text{rad}(\sigma_w)^\perp = \{h \in H \mid \ker h \supseteq \text{rad}(\sigma_w)\}.$$

3.5. Theorem. [6, Theorem 4.11] *Let $A = \mathcal{O}_q(k^n)$, and assume that either $-1 \notin \langle q_{ij} \rangle$ or $\operatorname{char} k = 2$. Fix $\Gamma, \sigma, c, \Phi, H, \sigma_w$ as above. Assume that c is an alternating bicharacter on Γ with $c^2 = \sigma$ and such that $c(\alpha, \beta) = 1$ whenever $\sigma(\alpha, \beta) = 1$. (Such a choice for c is possible by [6, Lemma 4.2].) Then there are H-equivariant topological quotient maps*

$$\operatorname{spec} k\Gamma^+ \twoheadrightarrow \operatorname{spec} A \qquad and \qquad \max k\Gamma^+ \twoheadrightarrow \operatorname{prim} A$$

such that $P \mapsto \Phi^{-1}(P : \operatorname{rad}(\sigma_w)^\perp)$ for $P \in \operatorname{spec}_w k\Gamma^+$. The fibres of the second map over points in $\operatorname{prim}_w A$ are exactly the $\operatorname{rad}(\sigma_w)^\perp$-orbits within $\max_w k\Gamma^+$. $\square$

While the precise formula given for the maps in Theorem 3.5 is particularly useful for computations, it may round out the picture to give a global formula (independent of w), as follows.

3.6. Lemma. *Let $\phi : \operatorname{spec} k\Gamma^+ \twoheadrightarrow \operatorname{spec} A$ be the topological quotient map given in Theorem 3.5. For all $P \in \operatorname{spec} k\Gamma^+$, the prime ideal $\phi(P) \in \operatorname{spec} A$ equals the largest ideal of A contained in the set $\Phi^{-1}(P)$.*

Proof. We shall need the generators $y_i = y_{\epsilon_i} \in k\Gamma^+$ and $x_i = x_{\epsilon_i} \in A$, where $\epsilon_1, \ldots, \epsilon_n$ is the standard basis for Γ.

Let $P \in \operatorname{spec} k\Gamma^+$; then $P \in \operatorname{spec}_w k\Gamma^+$ for some w. First suppose that $P \in \max_w k\Gamma^+$. Then $\phi(P) \in \operatorname{prim}_w A$, and so $\phi(P)$ is a maximal element of $\operatorname{spec}_w A$ [5, Theorem 2.3]. Now $\phi(P)$ contains the ideal $J_w := \langle x_i \mid i \in w \rangle$. On the other hand, the multiplicative set generated by $\{y_j \mid j \notin w\}$ in $k\Gamma^+$ is disjoint from P, and Φ^{-1} sends elements of this set to scalar multiples of elements in the multiplicative set X_w generated by $\{x_j \mid j \notin w\}$ in A. Hence, $\Phi^{-1}(P)$ is disjoint from X_w.

Let I be the largest ideal of A which is contained in $\Phi^{-1}(P)$. Then $J_w \subseteq \phi(P) \subseteq I$ and I is disjoint from X_w. Hence, I/J_w induces a proper ideal, call it IA_w, in the localization $A_w = (A/J_w)[X_w^{-1}]$. Let M be a maximal ideal of A_w containing IA_w, and let Q be the inverse image of M under the localization map $A \to A/J_w \to A_w$. Then $Q \in \operatorname{spec}_w A$ and $\phi(P) \subseteq I \subseteq Q$. Since $\phi(P)$ is maximal in $\operatorname{spec}_w A$, we obtain $\phi(P) = Q$ and thus $\phi(P) = I$. Therefore the lemma holds for $P \in \max_w k\Gamma^+$.

Now consider an arbitrary $P \in \operatorname{spec}_w k\Gamma^+$. Since $k\Gamma^+$ is a commutative affine k-algebra, P is an intersection of maximal ideals. It follows easily, as in [2, Proposition 1.3(a)], that $P = \bigcap_r P_r$ for some maximal ideals $P_r \in \max_w k\Gamma^+$. (Namely, $P = \bigcap_{v \subseteq \{1,\ldots,n\}} Q_v$ where each Q_v is an intersection of maximal ideals from $\max_v k\Gamma^+$. Observe that $Q_v = k\Gamma^+$ when $v \not\supseteq w$, and that $Q_v \supsetneq P$ when $v \supsetneq w$. Consequently, $P = Q_w$.) Since the set functions $(- : \operatorname{rad}(\sigma_w)^\perp)$ and Φ^{-1} preserve intersections, we see that $\phi(P) = \bigcap_r \phi(P_r)$ and $\Phi^{-1}(P) = \bigcap_r \Phi^{-1}(P_r)$. By the previous paragraph, each $\phi(P_r)$ is the largest ideal of A contained in $\Phi^{-1}(P_r)$. Therefore $\phi(P)$ is the largest ideal of A contained in $\Phi^{-1}(P)$. $\square$

3.7. The maps in Theorem 3.5 have analogous factorizations to those in (2.5), which we write

$$\text{spec}\, k\Gamma^+ \twoheadrightarrow \mathcal{G}\text{-spec}\, k\Gamma^+ \xrightarrow{\approx} \text{spec}\, A$$

$$\max k\Gamma^+ \twoheadrightarrow \mathcal{G}\text{-max}\, k\Gamma^+ \xrightarrow{\approx} \text{prim}\, A.$$

Here $\mathcal{G}$ stands for the indexed family of groups $\{S_w^\perp \mid w \subseteq \{1, \ldots, n\}\}$ where $S_w = \text{rad}(\sigma_w)$, while

$$\mathcal{G}\text{-spec}\, k\Gamma^+ = \bigsqcup_{w \subseteq \{1,\ldots,n\}} \{(P : S_w^\perp) \mid P \in \text{spec}_w\, k\Gamma^+\}$$

$$\mathcal{G}\text{-max}\, k\Gamma^+ = \bigsqcup_{w \subseteq \{1,\ldots,n\}} \{(P : S_w^\perp) \mid P \in \max_w\, k\Gamma^+\}.$$

Closed sets for Zariski topologies on $\mathcal{G}$-spec $k\Gamma^+$ and $\mathcal{G}$-max $k\Gamma^+$ are defined in the usual way; a compatibility condition on the groups in $\mathcal{G}$ ensures that the result actually is a topology [**6**, (2.4), (4.9)]. A fairly general piece of commutative algebra, developed in [**6**, Section 2], provides the topological quotient maps from spec $k\Gamma^+$ and max $k\Gamma^+$ onto $\mathcal{G}$-spec $k\Gamma^+$ and $\mathcal{G}$-max $k\Gamma^+$. The homeomorphisms of $\mathcal{G}$-spec $k\Gamma^+$ and $\mathcal{G}$-max $k\Gamma^+$ onto spec A and prim A are constructed by patching together homeomorphisms of $S_w^\perp$-spec $k\Gamma_w$ and $S_w^\perp$-max $k\Gamma_w$ onto spec A_w and prim A_w obtained as in (2.5).

3.8. Example. To illustrate the form of these topological quotients, we give the basic example in dimension 3, where various differences between the classical and quantum cases appear. Recall that the single parameter quantum affine spaces, relative to a scalar $q \in k^\times$, are the algebras

$$\mathcal{O}_q(k^n) = k\langle x_1, \ldots, x_n \mid x_i x_j = q x_j x_i \text{ for } i < j \rangle.$$

(a) Assume first that q is not a root of unity, and let p be a square root of q in k. In this case, the topological quotient map

$$k^3 \approx \max k(\mathbb{Z}^+)^3 \twoheadrightarrow \text{prim}\, \mathcal{O}_q(k^3)$$

described in Theorem 3.5 can be computed as follows [**6**, (5.2)], where the scalars λ_i are all assumed to be nonzero:

$$(0,0,0) \mapsto \langle x_1, x_2, x_3 \rangle \qquad\qquad (\lambda_1, \lambda_2, 0) \mapsto \langle x_3 \rangle$$
$$(\lambda_1, 0, 0) \mapsto \langle x_1 - \lambda_1, x_2, x_3 \rangle \qquad (\lambda_1, 0, \lambda_3) \mapsto \langle x_2 \rangle$$
$$(0, \lambda_2, 0) \mapsto \langle x_1, x_2 - \lambda_2, x_3 \rangle \qquad (0, \lambda_2, \lambda_3) \mapsto \langle x_1 \rangle$$
$$(0, 0, \lambda_3) \mapsto \langle x_1, x_2, x_3 - \lambda_3 \rangle \qquad (\lambda_1, \lambda_2, \lambda_3) \mapsto \langle \lambda_2 x_1 x_3 - p\lambda_1 \lambda_3 x_2 \rangle$$

The right column demonstrates both the compression effect of the quotient process and the fact that $\mathcal{O}_q(k^3)$ has 'many fewer' primitive ideals than $\mathcal{O}(k^3)$. Note the factor p in the last term. Without this factor, we obtain only a surjective map, which is not Zariski continuous.

(b) For contrast, consider the case where q is a primitive t-th root of unity for some odd $t > 1$, and take $p = q^{(t+1)/2}$. Then the topological quotient map $k^3 \twoheadrightarrow \operatorname{prim} \mathcal{O}_q(k^3)$ has the form below [**6**, (5.3)]:

$$(0,0,0) \mapsto \langle x_1, x_2, x_3 \rangle \qquad (\lambda_1, \lambda_2, 0) \mapsto \langle x_1^t - \lambda_1^t, x_2^t - \lambda_2^t, x_3 \rangle$$

$$(\lambda_1, 0, 0) \mapsto \langle x_1 - \lambda_1, x_2, x_3 \rangle \qquad (\lambda_1, 0, \lambda_3) \mapsto \langle x_1^t - \lambda_1^t, x_2, x_3^t - \lambda_3^t \rangle$$

$$(0, \lambda_2, 0) \mapsto \langle x_1, x_2 - \lambda_2, x_3 \rangle \qquad (0, \lambda_2, \lambda_3) \mapsto \langle x_1, x_2^t - \lambda_2^t, x_3^t - \lambda_3^t \rangle$$

$$(0, 0, \lambda_3) \mapsto \langle x_1, x_2, x_3 - \lambda_3 \rangle$$

$$(\lambda_1, \lambda_2, \lambda_3) \mapsto \langle x_1^t - \lambda_1^t, x_2^t - \lambda_2^t, x_3^t - \lambda_3^t, \lambda_2 x_1 x_3 - p \lambda_1 \lambda_3 x_2 \rangle$$

In this case, $\operatorname{prim} \mathcal{O}_q(k^3)$ is more 'classical', in that all its points are closed.

4. QUANTUM TORIC VARIETIES AND COCYCLE TWISTS

In this section, we discuss quantizations of toric varieties [**11**] and extend Theorem 3.5 to that setting. There are several ways in which (affine) toric varieties may be defined. For our purposes, the simplest definition is a variety on which a torus acts (morphically) with a dense orbit. Some equivalent conditions are given in the following theorem.

4.1. Theorem. [**14**, Satz 5, p. 105] *Let H be an algebraic torus, acting morphically on an irreducible affine variety V. The following conditions are equivalent:*

(a) *There are only finitely many H-orbits in V.*

(b) *Some H-orbit is dense in V.*

(c) *The fixed field $k(V)^H = k$.*

(d) *All H-eigenspaces in $k(V)$ are 1-dimensional.* $\square$

For comparison, we state a ring-theoretic version of this theorem. Since it will not be needed below, however, we leave the proof to the reader.

4.2. Theorem. *Let R be a commutative affine domain over k, and let H be an algebraic torus acting rationally on R by k-algebra automorphisms. Then the following conditions are equivalent:*

(a) *There are only finitely many H-orbits in $\max R$.*

(b) *Some H-orbit is dense in $\max R$.*

(c) $(\operatorname{Fract} R)^H = k$, *where $\operatorname{Fract} R$ is the quotient field of R.*

(d) *All H-eigenspaces in R are 1-dimensional.*

(e) *There are only finitely many H-invariant prime ideals in R.*   □

In the noncommutative setting, consider a prime, noetherian, affine k-algebra R, let Fract R be the Goldie quotient ring of R, and replace the maximal ideal space by prim R. The conditions of Theorem 4.2 are no longer equivalent, since prim R may well have a dense point (if 0 is a primitive ideal). It is thus natural to focus on the stronger conditions, (c) and (d).

Following Ingalls [11], we define a *quantum (affine) toric variety* to be an affine domain over k equipped with a rational action of an algebraic torus H by k-algebra automorphisms, such that the H-eigenspaces are 1-dimensional. If we fix H and require the action to be faithful, such algebras can be classified by elements of $\wedge^2 H$ together with finitely generated sub-monoids that generate the character group $\widehat{H}$ [11, Theorem 2.6].

Quantum toric varieties may also be presented as 'cocycle twists' (see below) of coordinate rings of classical toric varieties, analogous to the way quantum tori and quantum affine spaces are twisted versions of (Laurent) polynomial rings. Furthermore, quantum toric varieties can be written as factor algebras of quantum affine spaces, which shows, in particular, that they are noetherian. It also allows us to apply Theorem 3.2 and express their primitive spectra as topological quotients of classical varieties. For this purpose, a presentation as a cocycle twist of a commutative affine algebra suffices; one-dimensionality of eigenspaces is not needed. Hence, we can work with a somewhat larger class of algebras than quantum toric varieties.

4.3. Suppose that R is a k-algebra graded by an abelian group G, and that $c : G \times G \to k^\times$ is a 2-cocycle. The underlying G-graded vector space of R can be equipped with a new multiplication $*$, where $r * s = c(\alpha, \beta)rs$ for all homogeneous elements $r \in R_\alpha$ and $s \in R_\beta$. This new multiplication is associative, and gives R a new structure as G-graded k-algebra, called the *twist of R by c* (see [1, Section 3] for details). The following alternative presentation is helpful in keeping computations straight. Let R' be an isomorphic graded vector space copy of R, equipped with a G-graded k-linear isomorphism $R \to R'$ denoted $r \mapsto r'$. Then R' becomes a G-graded k-algebra such that $r's' = c(\alpha, \beta)(rs)'$ for $r \in R_\alpha$ and $s \in R_\beta$. The *twist map* $r \mapsto r'$ then gives an isomorphism of the twist of R by c onto R'.

Now suppose that R is a commutative affine k-algebra, that G is torsionfree, and that either -1 is not in the subgroup of $k^\times$ generated by the image of c or char $k = 2$. Then there exist topological quotient maps

$$\text{spec } R \twoheadrightarrow \text{spec } R' \qquad \text{and} \qquad \text{max } R \twoheadrightarrow \text{prim } R'$$

[6, Theorem 6.3]. We shall show that these maps can be given by formulas analogous to those in Theorem 3.5.

It is convenient to immediately make a reduction to a special choice of c. First, we can replace G by any finitely generated subgroup containing the support of R. Then, as in the proof of [**6**, Theorem 6.3], there exists an alternating bicharacter d on G, satisfying the same hypotheses as c, such that R' is isomorphic to the twist of R by d. Thus, there is no loss of generality in assuming that c is an alternating bicharacter, and we shall describe the topological quotient maps above under this assumption.

The torsionfreeness hypothesis on G is only needed for the reduction just indicated. Hence, we drop this assumption in the discussion to follow.

4.4. To recap, we are now assuming that R is a commutative affine k-algebra graded by an abelian group G, that $c : G \times G \to k^\times$ is an alternating bicharacter, and that either $-1 \notin \langle \operatorname{im} c \rangle$ or $\operatorname{char} k = 2$. Let $A = R'$ be the twist of R by c.

Choose homogeneous k-algebra generators $r_1, \ldots, r_n$ for R, and set $a_i = r_i' \in A$ for $i = 1, \ldots, n$. We use these elements to define sets $\operatorname{spec}_w R$, $\operatorname{spec}_w A$, $\operatorname{max}_w R$, $\operatorname{prim}_w A$ for $w \subseteq \{1, \ldots, n\}$ along the same lines as in (3.1). Thus

$$\operatorname{spec}_w R = \{P \in \operatorname{spec} R \mid r_i \in P \Longleftrightarrow i \in w\}$$
$$\operatorname{spec}_w A = \{P \in \operatorname{spec} A \mid a_i \in P \Longleftrightarrow i \in w\},$$

while $\operatorname{max}_w R = (\operatorname{max} R) \cap (\operatorname{spec}_w R)$ and $\operatorname{prim}_w A = (\operatorname{prim} A) \cap (\operatorname{spec}_w A)$. Of course, some of these sets may be empty.

Set $H = \operatorname{Hom}(G, k^\times)$, an abelian group under pointwise multiplication. Because R and A are G-graded k-algebras, H acts on them by k-algebra automorphisms such that $h.r = h(x)r$ and $h.a = h(x)a$ for $h \in H$, $r \in R_x$, $a \in A_x$. Note that the sets $\operatorname{spec}_w R$ and $\operatorname{spec}_w A$ are invariant under the induced H-actions, since the elements r_i and a_i are H-eigenvectors.

For $i = 1, \ldots, n$, let $\delta_i \in G$ denote the degree of r_i and a_i. For $w \subseteq \{1, \ldots, n\}$, set $G_w = \sum_{i \notin w} \mathbb{Z}\delta_i$ and let c_w denote the restriction of c to G_w. By definition of H, we have a pairing $H \times G \to k^\times$, and we use $\perp$ to denote orthogonals with respect to this pairing. In particular, for each w we can define $\operatorname{rad}(c_w)^\perp$, which is a subgroup of H.

We can now state the following more precise version of [**6**, Theorem 6.3]:

4.5. Theorem. *Let R be a commutative affine k-algebra, graded by an abelian group G. Let $c : G \times G \to k^\times$ be an alternating bicharacter, and let $A = R'$ be the twist of R by c. Assume that either $-1 \notin \langle \operatorname{im} c \rangle$ or $\operatorname{char} k = 2$.*

With notation as in (4.4), there exist H-equivariant topological quotient maps

$$\phi_s : \operatorname{spec} R \twoheadrightarrow \operatorname{spec} A \qquad \text{and} \qquad \phi_m : \operatorname{max} R \twoheadrightarrow \operatorname{prim} A$$

such that $P \mapsto (P : \operatorname{rad}(c_w)^\perp)'$ for $P \in \operatorname{spec}_w R$. Alternatively, $\phi_s(P)$ is the largest ideal of A contained in P'. Moreover, the fibres of ϕ_m over points in $\operatorname{prim}_w A$ are precisely the $\operatorname{rad}(c_w)^\perp$-orbits in $\operatorname{max}_w R$.

Note that the second description of ϕ_s shows that this map depends only on R, G, and c, that is, it is independent of the choice of homogeneous generators $r_i \in R$ as in (4.4).

The proof of Theorem 4.5 will be distributed over the following three subsections.

4.6. Let $\Phi : A \to R$ be the inverse of the twist map; this is a G-graded k-linear isomorphism. In particular, $X' = \Phi^{-1}(X)$ for subsets $X \subseteq R$.

Set $\Gamma = \mathbb{Z}^n$ and $\Gamma^+ = (\mathbb{Z}^+)^n$, and let $\rho : \Gamma \to G$ be the group homomorphism given by the rule $\rho(\alpha_1, \ldots, \alpha_n) = \alpha_1 \delta_1 + \cdots + \alpha_n \delta_n$. Set $\tilde{c} = c \circ (\rho \times \rho)$, which is an alternating bicharacter on Γ, as is $\sigma = \tilde{c}^2$. For $\alpha = (\alpha_1, \ldots, \alpha_n) \in \Gamma^+$, define

$$r^\alpha = r_1^{\alpha_1} r_2^{\alpha_2} \cdots r_n^{\alpha_n} \in R_{\rho(\alpha)} \qquad \text{and} \qquad a_\alpha = (r^\alpha)' \in A_{\rho(\alpha)}.$$

Then $a_\alpha a_\beta = \tilde{c}(\alpha, \beta) a_{\alpha+\beta}$ for $\alpha, \beta \in \Gamma^+$.

Next, set $\widetilde{R} = k\Gamma^+$ and $\widetilde{A} = k^{\tilde{c}}\Gamma^+$, expressed with k-bases $\{y_\alpha\}$ and $\{x_\alpha\}$ such that $y_\alpha y_\beta = y_{\alpha+\beta}$ and $x_\alpha x_\beta = \tilde{c}(\alpha, \beta) x_{\alpha+\beta}$ for $\alpha, \beta \in \Gamma^+$. Let $\widetilde{\Phi} : \widetilde{A} \to \widetilde{R}$ be the inverse of the twist map; this is a Γ-graded k-linear isomorphism such that $\widetilde{\Phi}(x_\alpha) = y_\alpha$ for $\alpha \in \Gamma^+$. There exist surjective k-algebra maps $\tau : \widetilde{R} \to R$ and $\pi : \widetilde{A} \to A$ such that $\tau(y_\alpha) = r^\alpha$ and $\pi(x_\alpha) = a_\alpha$ for $\alpha \in \Gamma^+$. Thus, we obtain a commutative diagram

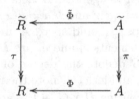

Let $\widetilde{H} = \mathrm{Hom}(\Gamma, k^\times)$, acting via k-algebra automorphisms on $\widetilde{A}$ and $\widetilde{R}$ in the usual way. For $w \subseteq \{1, \ldots, n\}$, define $\mathrm{spec}_w \widetilde{A}$, $\mathrm{spec}_w \widetilde{R}$, Γ_w, $\widetilde{A}_w$, $\widetilde{R}_w$, σ_w, $\tilde{c}_w$ as in (3.1), (3.3), (3.4). (To define $\mathrm{spec}_w \widetilde{A}$ and $\mathrm{spec}_w \widetilde{R}$, for instance, use the elements $x_{\epsilon_i} \in \widetilde{A}$ and $y_{\epsilon_i} \in \widetilde{R}$, where $\epsilon_1, \ldots, \epsilon_n$ denotes the standard basis for Γ.) Because of our hypotheses on c, the group $\langle \mathrm{im}\,\tilde{c} \rangle$ contains no elements of order 2, and hence $\mathrm{rad}(\sigma_w) = \mathrm{rad}(\tilde{c}_w)$ for all w. We set $\widetilde{S}_w = \mathrm{rad}(\tilde{c}_w)$.

Now Theorem 3.5 gives us $\widetilde{H}$-equivariant topological quotient maps

$$\widetilde{\phi}_s : \mathrm{spec}\,\widetilde{R} \twoheadrightarrow \mathrm{spec}\,\widetilde{A} \qquad \text{and} \qquad \widetilde{\phi}_m : \mathrm{max}\,\widetilde{R} \twoheadrightarrow \mathrm{prim}\,\widetilde{A}$$

such that $P \mapsto \widetilde{\Phi}^{-1}(P : \widetilde{S}_w^\perp)$ for $P \in \mathrm{spec}_w \widetilde{R}$. Moreover, the fibres of $\widetilde{\phi}_m$ over points in $\mathrm{prim}_w \widetilde{A}$ are the $\widetilde{S}_w^\perp$-orbits in $\mathrm{max}_w \widetilde{R}$.

Set $V = \{P \in \operatorname{spec} \widetilde{R} \mid P \supseteq \ker \tau\}$ and $W = \{P \in \operatorname{spec} \widetilde{A} \mid P \supseteq \ker \pi\}$. The proof of [6, Theorem 6.3] shows that $\widetilde{\phi}_s$ and $\widetilde{\phi}_m$ restrict to topological quotient maps $V \twoheadrightarrow W$ and $V \cap \max \widetilde{R} \twoheadrightarrow W \cap \operatorname{prim} \widetilde{A}$. Therefore there are topological quotient maps ϕ_s and ϕ_m such that the following diagrams commute:

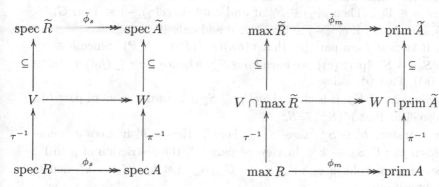

It remains to show that ϕ_s and ϕ_m have the properties announced in the statement of Theorem 4.5.

Given any $P \in \operatorname{spec} R$, we have $\pi^{-1}(\phi_s(P)) = \widetilde{\phi}_s(\tau^{-1}(P))$, which by Lemma 3.6 equals the largest ideal of $\widetilde{A}$ contained in $\widetilde{\Phi}^{-1}\tau^{-1}(P)$. If I is an ideal of A contained in $P' = \Phi^{-1}(P)$, then $\pi^{-1}(I)$ is an ideal of $\widetilde{A}$ contained in $\pi^{-1}\Phi^{-1}(P) = \widetilde{\Phi}^{-1}\tau^{-1}(P)$. Hence, $\pi^{-1}(I) \subseteq \pi^{-1}(\phi_s(P))$, and so $I \subseteq \phi_s(P)$. This shows that $\phi_s(P)$ is the largest ideal of A contained in P'.

The following lemma records some facts and conditions that we shall need to finish the proof.

4.7. Lemma. Let $w \subseteq \{1, \ldots, n\}$, and set $S_w = \operatorname{rad}(c_w)$.

(a) $\ker \tau \subseteq (\tau^{-1}(P) : \widetilde{S}_w^\perp) \subseteq h(\tau^{-1}(P))$ for $P \in \operatorname{spec}_w R$ and $h \in \widetilde{S}_w^\perp$.

(b) $(h.(-)) \circ \tau = \tau \circ ((h\rho).(-))$ for $h \in H$.

(c) $\rho^{-1}(S_w) \cap \Gamma_w = \widetilde{S}_w$ and so $\rho(\widetilde{S}_w) = S_w$.

(d) $(h\rho)(\tau^{-1}(P)) = \tau^{-1}(h(P))$ for $P \in \operatorname{spec}_w R$ and $h \in S_w^\perp$.

(e) $\widetilde{S}_w^\perp = \rho^*(S_w^\perp)\Gamma_w^\perp$, where $\rho^* : H \to \widetilde{H}$ denotes the homomorphism given by composition with ρ.

(f) For $P \in \operatorname{spec}_w R$, the set $\{\tau^{-1}(h(P)) \mid h \in S_w^\perp\}$ equals the $\widetilde{S}_w^\perp$-orbit of $\tau^{-1}(P)$.

Proof. (a) Note that $\tau^{-1}(P) \in \operatorname{spec}_w \widetilde{R}$ and $\widetilde{\phi}_s(\tau^{-1}(P)) \in W$. Thus $\widetilde{\Phi}^{-1}(\tau^{-1}(P) : \widetilde{S}_w^\perp) \supseteq \ker \pi$, and so $(\tau^{-1}(P) : \widetilde{S}_w^\perp) \supseteq \ker \tau$. The remaining inclusion is clear.

(b) This follows from the observation that

$$h.\tau(y_\alpha) = h.r^\alpha = h\rho(\alpha)r^\alpha = \tau(h\rho(\alpha)y_\alpha) = \tau((h\rho).y_\alpha)$$

for $h \in H$ and $\alpha \in \Gamma^+$.

(c) Observe that $\rho(\Gamma_w) = G_w$. Since $S_w \subseteq G_w$ and $\widetilde{S}_w \subseteq \Gamma_w$, the second part of the claim will follow from the first. Now consider an arbitrary element $\gamma \in \Gamma_w$. Then $\rho(\gamma) \in S_w$ if and only if $c(\rho(\gamma), -) \equiv 1$ on $G_w = \rho(\Gamma_w)$, if and only if $\tilde{c}(\gamma, -) \equiv 1$ on Γ_w, if and only if $\gamma \in \widetilde{S}_w$.

(d) It is clear from part (b) that $\tau(h\rho)(\tau^{-1}(P)) = h(P)$. Since $h \in S_w^\perp$ and $\rho(\widetilde{S}_w) = S_w$ (part (c)), we have $h\rho \in \widetilde{S}_w^\perp$, whence $\ker \tau \subseteq (h\rho)(\tau^{-1}(P))$ (part (a)). Part (d) follows.

(e) Since $\widetilde{S}_w \subseteq \Gamma_w$, it is clear that $\Gamma_w^\perp \subseteq \widetilde{S}_w^\perp$. As in the proof of part (d), it is also clear that $\rho^*(S_w^\perp) \subseteq \widetilde{S}_w^\perp$.

Now consider $h' \in \widetilde{S}_w^\perp$; thus $\widetilde{S}_w \subseteq \ker h'$. Hence, h' induces a homomorphism $h_1 : \Gamma/\widetilde{S}_w \to k^\times$. In view of part (c), the restriction of ρ to Γ_w induces an isomorphism $\rho_w : \Gamma_w/\widetilde{S}_w \to G_w/S_w$. Let $h_2 : G_w \to k^\times$ denote the composition

$$G_w \xrightarrow{\text{quo}} G_w/S_w \xrightarrow{\rho_w^{-1}} \Gamma_w/\widetilde{S}_w \xrightarrow{\subseteq} \Gamma/\widetilde{S}_w \xrightarrow{h_1} k^\times,$$

and observe that the restrictions of $h_2\rho$ and h' to Γ_w coincide. Since k is algebraically closed, $k^\times$ is a divisible group, and hence $k^\times$ is injective in the category of abelian groups. Consequently, h_2 can be extended to a homomorphism $h_3 \in H$. On one hand, $h_3\rho$ and h' agree on Γ_w, whence $h'(h_3\rho)^{-1} \in \Gamma_w^\perp$. On the other hand, $S_w \subseteq \ker h_2 \subseteq \ker h_3$, whence $h_3 \in S_w^\perp$. Therefore $h' \in \rho^*(S_w^\perp)\Gamma_w^\perp$, and part (e) is proved.

(f) If $h \in S_w^\perp$, then $h\rho \in \widetilde{S}_w^\perp$ and we see from part (d) that $\tau^{-1}(h(P))$ lies in the $\widetilde{S}_w^\perp$-orbit of $\tau^{-1}(P)$.

Conversely, consider $h' \in \widetilde{S}_w^\perp$. By part (e), $h' = (h\rho)h_0$ for some $h \in S_w^\perp$ and $h_0 \in \Gamma_w^\perp$. We claim that h_0 leaves $\tau^{-1}(P)$ invariant; it will then follow from part (d) that $h'(\tau^{-1}(P)) = \tau^{-1}(h(P))$.

Since $P \in \operatorname{spec}_w R$, we see that $\tau^{-1}(P)$ contains the ideal $J_w = \langle y_{\epsilon_i} \mid i \in w \rangle$, where $\epsilon_1, \ldots, \epsilon_n$ denotes the standard basis for Γ. Note that J_w is invariant under H. Since $\widetilde{R}/J_w$ is spanned by the cosets $y_\alpha + J_w$ for $\alpha \in \Gamma_w \cap \Gamma^+$, we see that the induced action of $\Gamma_w^\perp$ on $\widetilde{R}/J_w$ is trivial. Thus $\tau^{-1}(P)$ is indeed invariant under h_0, which establishes part (f). $\square$

4.8. (Proof of Theorem 4.5) Let $w \subseteq \{1, \ldots, n\}$, and consider $P \in \operatorname{spec}_w R$. In view of Lemma 4.7(a) and the commutative diagrams in (4.6), we see that

$$\phi_s(P) = \pi\widetilde{\phi}_s(\tau^{-1}(P)) = \pi\widetilde{\Phi}^{-1}(\tau^{-1}(P) : \widetilde{S}_w^\perp)$$
$$= \Phi^{-1}\tau(\tau^{-1}(P) : \widetilde{S}_w^\perp) = \left(\tau(\tau^{-1}(P) : \widetilde{S}_w^\perp)\right)'.$$

On the other hand, Lemma 4.7(f) implies that

$$(\tau^{-1}(P) : \widetilde{S}_w^{\perp}) = \bigcap_{h \in S_w^{\perp}} \tau^{-1}(h(P)) = \tau^{-1}(P : S_w^{\perp}),$$

whence $\tau(\tau^{-1}(P) : \widetilde{S}_w^{\perp}) = (P : S_w^{\perp})$. Therefore $\phi_s(P) = (P : S_w^{\perp})'$.

Since H is abelian, the set functions $(- : S_w^{\perp})$ are H-equivariant, as is the twist map $(-)'$. Hence, it follows from the formula just proved that ϕ_s is H-equivariant.

Finally, note that the fibre of ϕ_m over any point of $\mathrm{prim}_w A$ is contained in $\mathrm{max}_w R$. Consider $P, P' \in \mathrm{max}_w R$. Taking account of (4.6), we see that $\phi_m(P) = \phi_m(P')$ if and only if $\widetilde{\phi}_m(\tau^{-1}(P)) = \widetilde{\phi}_m(\tau^{-1}(P'))$, if and only if $\tau^{-1}(P)$ and $\tau^{-1}(P')$ lie in the same $\widetilde{S}_w^{\perp}$-orbit. By Lemma 4.7(f), this occurs if and only if $\tau^{-1}(P') = \tau^{-1}(h(P))$ for some $h \in S_w^{\perp}$, and thus if and only if P and P' lie in the same $S_w^{\perp}$-orbit. $\square$

Once the precise form of the maps in Theorem 4.5 is given, we can easily construct topological quotient maps between different cocycle twists of R, as follows.

4.9. Corollary. *Let R be a commutative affine k-algebra, graded by an abelian group G. Let $c_1, c_2 : G \times G \to k^{\times}$ be alternating bicharacters, and let A_i be the twist of R by c_i. Assume that either $-1 \notin \langle \mathrm{im}\, c_1 \rangle \cup \langle \mathrm{im}\, c_2 \rangle$ or char $k = 2$. Set $H = \mathrm{Hom}(G, k^{\times})$, and let $\phi_i : \mathrm{spec}\, R \to \mathrm{spec}\, A_i$ be the H-equivariant topological quotient map given in Theorem 4.5.*

Assume that $c_2(x, y) = 1 \implies c_1(x, y) = 1$, for any $x, y \in G$. Then there exists an H-equivariant topological quotient map $\phi : \mathrm{spec}\, A_1 \twoheadrightarrow \mathrm{spec}\, A_2$ such that the following diagram commutes:

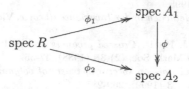

Proof. Set up notation for A_i and ϕ_i as in (4.4), and abbreviate $(c_i)_w$ by c_{iw}. Let $\Phi_i : A_i \to R$ be the inverse of the twist map. Then ϕ_i is given by the rule

$$\phi_i(P) = \Phi_i^{-1}(P : \mathrm{rad}(c_{iw})^{\perp}) \qquad \text{for } P \in \mathrm{spec}_w R.$$

Our hypotheses on c_1, c_2 imply that $\mathrm{rad}(c_{2w}) \subseteq \mathrm{rad}(c_{1w})$ for all w, whence $\mathrm{rad}(c_{2w})^{\perp} \supseteq \mathrm{rad}(c_{1w})^{\perp}$ and so

$$(P : \mathrm{rad}(c_{2w})^{\perp}) = ((P : \mathrm{rad}(c_{1w})^{\perp}) : \mathrm{rad}(c_{2w})^{\perp})$$

for all $P \in \operatorname{spec}_w R$. Hence, we can define a map $\phi : \operatorname{spec} A_1 \to \operatorname{spec} A_2$ such that

$$\phi(Q) = \Phi_2^{-1}(\Phi_1(Q) : \operatorname{rad}(c_{2w})^\perp)$$

for $Q \in \operatorname{spec}_w A_1$. It is clear that $\phi\phi_1 = \phi_2$. Since ϕ_1 and ϕ_2 are H-equivariant topological quotient maps, so is ϕ. $\square$

REFERENCES

1. M. Artin, W. Schelter, and J. Tate, *Quantum deformations of GL_n*, Communic. Pure Appl. Math. **44** (1991), 879-895.
2. K. A. Brown and K. R. Goodearl, *Prime spectra of quantum semisimple groups*, Trans. Amer. Math. Soc. **348** (1996), 2465-2502.
3. C. De Concini, V. Kac, and C. Procesi, *Some remarkable degenerations of quantum groups*, Comm. Math. Phys. **157** (1993), 405-427.
4. K. R. Goodearl, *Prime spectra of quantized coordinate rings*, in Proc. Euroconference on Interactions between Ring Theory and Representations of Algebras (Murcia 1998) (F. Van Oystaeyen, ed.), (to appear); xxx.lanl.gov/abs/math.QA/9903091.
5. K. R. Goodearl and E. S. Letzter, *Prime and primitive spectra of multiparameter quantum affine spaces*, in Trends in Ring Theory. Proc. Miskolc Conf. 1996 (V. Dlab and L. Márki, eds.), Canad. Math. Soc. Conf. Proc. Series **22** (1998), 39-58.
6. _____, *Quantum n-space as a quotient of classical n-space*, Trans. Amer. Math. Soc., (to appear); xxx.lanl.gov/abs/math.RA/9905055.
7. T. J. Hodges, *Quantum tori and Poisson tori*, Unpublished Notes, 1994.
8. T. J. Hodges and T. Levasseur, *Primitive ideals of $\mathbf{C}_q[SL(3)]$*, Commun. Math. Phys. **156** (1993), 581-605.
9. _____, *Primitive ideals of $\mathbf{C}_q[SL(n)]$*, J. Algebra **168** (1994), 455-468.
10. T. J. Hodges, T. Levasseur, and M. Toro, *Algebraic structure of multi-parameter quantum groups*, Advances in Math. **126** (1997), 52-92.
11. C. Ingalls, *Quantum toric varieties*, Preprint, 1999.
12. A. Joseph, *On the prime and primitive spectra of the algebra of functions on a quantum group*, J. Algebra **169** (1994), 441-511.
13. _____, *Quantum Groups and Their Primitive Ideals*, Ergebnisse der Math. (3) 29, Springer-Verlag, Berlin, 1995.
14. H. Kraft, *Geometrische Methoden in der Invariantentheorie*, Vieweg, Braunschweig, 1984.
15. J. C. McConnell and J. J. Pettit, *Crossed products and multiplicative analogs of Weyl algebras*, J. London Math. Soc. (2) **38** (1988), 47-55.
16. M. Vancliff, *Primitive and Poisson spectra of twists of polynomial rings*, Algebras and Representation Theory **3** (1999), 269-285.

DEPARTMENT OF MATHEMATICS, UNIVERSITY OF CALIFORNIA, SANTA BARBARA, CA 93106

E-mail address: goodearl@math.ucsb.edu

REPRESENTATIONS OF SEMISIMPLE LIE ALGEBRAS IN POSITIVE CHARACTERISTIC AND QUANTUM GROUPS AT ROOTS OF UNITY

IAIN GORDON

1. INTRODUCTION

1.1.   If A is a finite dimensional algebra then its blocks are in one-to-one correspondence with its primitive central idempotents. The aim of this paper is to study this interaction for a class of noetherian algebras arising naturally in representation theory. This class includes the universal enveloping algebra of a reductive Lie algebra in positive characteristic and its quantised counterpart, the quantised enveloping algebra of a Borel subalgebra and the quantised function algebra of a semisimple algebraic group at roots of unity.

1.2.   More generally this paper is concerned with the role the centre of these algebras plays in their representation theory. The techniques used fall into two categories: local and global. The local approach is concerned principally with the behaviour of certain finite dimensional factors of these noetherian algebras whilst the global approach focuses on general properties of these algebras. The aim in both cases is to understand the structure of these finite dimensional factor algebras. In the first case we use a little deformation theory to piece things together whilst in the second case we can use some geometric tools before passing to the factors.

1.3.   In Section 2 we introduce the class of algebras we wish to study and present some general properties these have in common. In the following three sections we apply this theory to the study of enveloping algebras and quantised enveloping algebras of Lie algebras and to quantised function algebras. We end with an appendix on the structure of the centre of a quantised Borel algebra. Most of this paper surveys results from the articles [5] and [6]. The approach to Theorem 3.6 using deformation theory is new whilst several results in Section 5 tie up loose ends from [6].

I am grateful to the organisers of the Durham Symposium on Quantum Groups for the opportunity to talk to the conference and to a submit a paper to the proceedings. I have, as always, benefitted from conversations with Ken Brown. I also thank Gerhard Röhrle for useful discussions. Financial support was provided by TMR grant ERB FMRX-CT97-0100 at the University of Bielefeld.

2. Generalities

2.1. Throughout K denotes an algebraically closed field. We consider a triple of K-algebras

$$R \subseteq Z \subseteq H$$

where H is a prime Hopf algebra with centre Z and R is an affine sub-Hopf algebra of H over which H (and hence Z) are finitely generated modules. We have four examples in mind.

(A) Let K have positive characteristic p and let $\mathfrak{g}$ be a finite dimensional restricted Lie algebra over K. Then H is $U(\mathfrak{g})$, the enveloping algebra of $\mathfrak{g}$, and R the p-centre of H.

(B) Let $K = \mathbb{C}$, let $\mathfrak{g}$ be a finite dimensional semisimple Lie algebra over $\mathbb{C}$ and let $\epsilon \in \mathbb{C}$ be a primitive ℓ^{th} root of unity, for ℓ an odd integer greater than 1. Then H is the quantised enveloping algebra $U_\epsilon(\mathfrak{g})$ and R the ℓ-centre of H.

(C) Let K, $\mathfrak{g}$ and ℓ be as above. Then H is $U_\epsilon^{\leq 0}$, the subalgebra of $U_\epsilon(\mathfrak{g})$ corresponding to a Borel subalgebra of $\mathfrak{g}$, and R the ℓ-centre of H.

(D) Let $K = \mathbb{C}$, let G be a simply-connected, semisimple algebraic group over $\mathbb{C}$ and let ℓ be as above. Then H is $\mathcal{O}_\epsilon[G]$, the quantised function algebra of G, and R the ℓ-centre of H.

2.2. It is straightforward to show that there is an upper bound on the dimension of the simple H-modules, namely the PI degree of H, [4, Proposition 3.1]. In particular, each simple H-module is annihilated by a maximal ideal of R. As a result the family of finite dimensional algebras

$$\left\{ \frac{H}{\mathfrak{m}H} : \mathfrak{m} \text{ a maximal ideal in } R \right\}$$

captures an important slice of the representation theory of H: each simple H-module is a simple module for exactly one algebra in this family.

2.3. To firm up this notion of a "family of finite dimensional algebras" recall the following definition of the *variety of n-dimensional algebras* over K, [19]. Let

$$\mathrm{Bil}(n) = \{ \text{ bilinear maps } m : K^n \times K^n \longrightarrow K^n \} \cong \mathbb{A}_K^{n^3}$$

and

$$\mathrm{Alg}(n) = \{ \text{ associative, bilinear } m \text{ which have an identity} \} \subseteq \mathrm{Bil}(n).$$

It can be shown that $\mathrm{Alg}(n)$ is an affine variety, locally closed in $\mathrm{Bil}(n)$.

Let $Q(R)$ be the quotient field of R and let $Q(H) = H \otimes_R Q(R)$. Since H is a finitely generated R-module there is an integer n such that $Q(H)$ is an n-dimensional $Q(R)$-module.

Lemma. *Let n be as above. There is a morphism of varieties*

$$\alpha : \mathrm{Maxspec}(R) \longrightarrow \mathrm{Alg}(n),$$

sending $\mathfrak{m}$ to $H/\mathfrak{m}H$.

Proof. By [35] our hypotheses in 2.1 ensure that H is a free R-module of rank n. Let $\{x_1, \ldots, x_n\}$ be a basis for H over R and define $c_{ij}^k \in R$ for $1 \le i, j, k \le n$ by the following equations,

$$x_i x_j = \sum_k c_{ij}^k x_k.$$

For any maximal ideal, $\mathfrak{m}$, of R the structure constants of $H/\mathfrak{m}H$ with respect to the basis $\{x_i + \mathfrak{m}H\}$ are given by the scalars $(c_{ij}^k + \mathfrak{m}) \in R/\mathfrak{m}R \cong K$. It follows that α is a morphism of varieties, as required. $\square$

In 2.2 we could have equally considered the family of algebras $\{H/MH : M$ a maximal ideal of $Z\}$. This family, however, does not behave very well in general since the extension $Z \subseteq H$ need not be flat. For instance in examples (A), (B), (D) and often in (C) the presence of singular points in $\mathrm{Maxspec}(Z)$ prevents flatness since H has finite global dimension, [4].

2.4. Müller's Theorem. The first result on the block structure of the algebras $H/\mathfrak{m}H$ is a striking analogue of the finite dimensional case.

Theorem. *The blocks of $H/\mathfrak{m}H$ are in one-to-one correspondence with the maximal ideals of Z lying over $\mathfrak{m}$.*

¿From now on we write $\mathcal{B}_M$ to denote the block of $H/(M \cap R)H$ corresponding to M.

Remark. This result first appears in a different and more general context in [28, Theorem 7]; the interpretation here is discussed in [5, 2.10].

2.5. There is also a result on the local level about the number of blocks of $H/\mathfrak{m}H$.

Proposition. [19, Proposition 2.7] *Let $s \in \mathbb{N}$. The set*

$$X_s = \{\mathfrak{m} \in \mathrm{Maxspec}(R) : H/\mathfrak{m}H \text{ has no more than } s \text{ blocks}\}$$

is closed in $\mathrm{Maxspec}(R)$.

2.6. There is one type of block that is controlled by Z, a block corresponding to a point on the *Azumaya locus* of H:

$$\mathcal{A}_H = \{M \in \mathrm{Maxspec}(Z) : \mathcal{B}_M \text{ has a simple module of maximal dimension}\}.$$

This definition is not standard, but under the hypotheses of 2.1 is equivalent to the usual notion, see [5, 2.5].

Proposition. [5, Proposition 2.5] *Let M be a maximal ideal of Z belonging to $\mathcal{A}_H$ and let $\mathfrak{m} = M \cap R$. There is an algebra isomorphism*

$$\mathcal{B}_M \cong \mathrm{Mat}_n \left(\frac{Z_M}{\mathfrak{m}Z_M} \right)$$

where $Z_M/\mathfrak{m}Z_M$ is the primary component of $Z/\mathfrak{m}Z$ associated to M.

Recall a block is *primary* if it has a unique simple module. The proposition shows that blocks corresponding to Azumaya points are primary.

2.7.   Being Azumaya is a generic condition, that is $\mathcal{A}_H$ is a dense open set in Maxspec(Z), [34, Section 1.9]. Under the hypotheses of 2.1 $\mathcal{A}_H$ is contained in the smooth locus of Maxspec(Z), [4, Lemma 3.3]. In sufficiently well-behaved situations the converse holds.

Theorem. [4, Theorem 3.8] *Suppose H has finite global dimension. If*

$$\operatorname{codim}(\operatorname{Maxspec}(Z) \setminus \mathcal{A}_H) \geq 2$$

then $\mathcal{A}_H$ equals the smooth locus of Maxspec(Z).

It is not true in general that Maxspec(Z) is Azumaya in codimension one, [6, Proposition 2.6].

2.8.   We finish this section with a comparison of $Z/\mathfrak{m}Z$ and $Z(H/\mathfrak{m}H)$ for $\mathfrak{m} \in$ Maxspec(R). Quite generally there is an homomorphism

$$\iota : \frac{Z}{\mathfrak{m}Z} \longrightarrow Z\left(\frac{H}{\mathfrak{m}H}\right).$$

Lemma. *The map ι is generically an isomorphism. Moreover, if K has characteristic zero then it is always injective.*

Proof. The morphism $\pi : \operatorname{Maxspec}(Z) \longrightarrow \operatorname{Maxspec}(R)$ is finite since Z is a finitely generated R-module. Since finite morphisms are closed the set $\mathcal{F}_H = \operatorname{Maxspec}(R) \setminus \pi(\operatorname{Maxspec}(Z) \setminus \mathcal{A}_H)$ is a dense open set containing precisely those maximal ideals, $\mathfrak{m}$, of R whose fibre $\pi^{-1}(\mathfrak{m})$ is contained in $\mathcal{A}_H$. It follows from Proposition 2.6 that for any $\mathfrak{m} \in \mathcal{F}_H$ we have an isomorphism

$$(1) \qquad\qquad \frac{H}{\mathfrak{m}H} \cong \operatorname{Mat}_n\left(\frac{Z}{\mathfrak{m}Z}\right),$$

proving the first claim.

Now suppose that K has characteristic zero. Then there is a Z-module map, the reduced trace, $Tr : H \longrightarrow Z$ splitting the inclusion, [33, 9.8 and Theorem 10.1]. Thus Z is a direct summand of H and so $\mathfrak{m}H \cap Z = \mathfrak{m}Z$ for all $\mathfrak{m} \in$ Maxspec(R), as required.   □

2.9.   Under the hypotheses of 2.1 the algebras $H/\mathfrak{m}H$ are all Frobenius, [20, Theorem 3.4].

Lemma. [13, cf. I.3.9] *Suppose that $H/\mathfrak{m}H$ is a symmetric algebra. Then*

$$\dim(\operatorname{soc}(Z(H/\mathfrak{m}H))) \geq \text{the number of simple } H/\mathfrak{m}H\text{-modules}.$$

In particular, if $\iota(Z/\mathfrak{m}Z)$ is self-injective then ι is surjective only if all blocks of $H/\mathfrak{m}H$ are primary.

Proof. Write A for $H/\mathfrak{m}H$. Let $\{S_1, \ldots, S_t\}$ be a complete set of representatives for the isomorphism classes of simple A-modules and let P_i be the projective cover of S_i for $1 \leq i \leq t$. Since Morita equivalence preserves symmetry and the centre of an algebra, [2, Volume I, Proposition 2.2.7], without loss of generality we may assume that A is basic.

The homomorphism

$$\theta_i : P_i \longrightarrow S_i \longrightarrow P_i$$

defines an element in $A \cong \operatorname{End}_A(\oplus P_i)^{\mathrm{op}}$. Write an arbitrary element of A as $f = \sum_i \lambda_{f,i} id_{P_i} + f'$ where f' is a radical morphism. Then, by construction, $\theta_i f = \lambda_{f,i} \theta_i = f\theta_i$. In particular θ_i is central and lies in the socle of $Z(H/\mathfrak{m}H)$.

Let $Z' = \iota(Z/\mathfrak{m}Z)$ and decompose Z' into primary components. By hypothesis each of these is self-injective and so Frobenius, [1, Example IV.3]. In particular each component has a simple socle. Suppose that $Z' = Z(A)$. If S_i and S_j belong to the same block of A then θ_i and θ_j belong to the same primary component of Z', contradicting the simplicity of the socle. □

Remark. Let K have positive characteristic p and let $\mathfrak{l}$ be the Heisenberg Lie algebra over K, that is the Lie algebra with basis $\{x, y, z\}$ such that $[x, y] = z$ and z is central. Let $H = U(\mathfrak{l})$ and $R = Z = K[z, x^p, y^p]$. Let $\mathfrak{m} = (z, x^p, y^p)Z$, a maximal ideal of Z. Then $H/\mathfrak{m}H$ is a truncated polynomial ring, showing that the converse of the second claim is false in general. This also shows that in general not all primary blocks are Azumaya.

2.10. The following proposition provides a partial adjunct to Lemma 2.9.

Proposition. [19, Proposition 2.7] *Let* $s \in \mathbb{N}$. *The set*

$$Y_s = \{\mathfrak{m} \in \operatorname{Maxspec}(R) : \dim Z(H/\mathfrak{m}H) \geq s\}$$

is closed in $\operatorname{Maxspec}(R)$.

3. ENVELOPING ALGEBRAS

3.1. We follow the notation used in [25]. Let G be a connected, reductive algebraic group over K, an algebraically closed field of positive characteristic p, and let $\mathfrak{g} = \operatorname{Lie}(G)$. We assume that G satisfies the following hypotheses:

(1) The derived group $\mathcal{D}G$ of G is simply-connected;
(2) The prime p is good for $\mathfrak{g}$;
(3) There exists a G-invariant non-degenerate bilinear form on $\mathfrak{g}$.

More details can be found in [25, Section 6]. Let $T \subseteq B = U.T$ be a maximal torus contained in a Borel subgroup of G and let $\mathfrak{h} = \operatorname{Lie}(T)$, $\mathfrak{n} = \operatorname{Lie}(U)$ and $\mathfrak{b} = \operatorname{Lie}(B)$. Let X be the character group of T and let $\Lambda = X/pX$. Let Φ be the set of roots associated with $\mathfrak{g}$ and let Φ^+ be set of positive roots corresponding to the choice of B. Let W be the Weyl group of G. We will be interested only in the "dot action" of W on X (and hence on Λ). By definition this is just an affine translation of the natural action of W on X. Given a K-vector space V, let $V^{(1)}$ denote the twist of V along the automorphism of K which sends λ to λ^p.

3.2. Being the Lie algebra of G, $\mathfrak{g}$ has a restriction map $x \longmapsto x^{[p]}$. We have a triple

$$Z_0 \subseteq Z \subseteq U(\mathfrak{g})$$

where $U(\mathfrak{g})$ is the enveloping algebra of $\mathfrak{g}$ and $Z_0 = K[x^p - x^{[p]} : x \in \mathfrak{g}]$ is the p-centre of $\mathfrak{g}$, a central sub-Hopf algebra of $U(\mathfrak{g})$. Standard arguments with the PBW theorem imply that $Z_0 \cong \mathcal{O}(\mathfrak{g}^{*(1)})$, the ring of regular functions on $\mathfrak{g}^{*(1)}$ and that $U(\mathfrak{g})$ is a free Z_0-module of rank $p^{\dim \mathfrak{g}}$, [25, Proposition 2.3].

Thus $U(\mathfrak{g})$ satisfies the hypotheses of 2.1. By Lemma 2.3 we have a morphism of varieties

$$\alpha : \mathfrak{g}^{*(1)} \longrightarrow \mathrm{Alg}(p^{\dim \mathfrak{g}})$$

sending χ to the algebra $U_\chi = U(\mathfrak{g})/(x^p - x^{[p]} - \chi(x))$, a *reduced enveloping algebra*. Note that if $\chi = 0$ then U_0 is the restricted enveloping algebra of $\mathfrak{g}$. It is straightforward to check that $U_{g \cdot \chi} \cong U_\chi$ for $g \in G$ acting on $\mathfrak{g}^{*(1)}$ by the coadjoint action, [25, 2.9].

3.3. Hypothesis 3.1.3 yields a G-equivariant isomorphism $\theta : \mathfrak{g}^{*(1)} \longrightarrow \mathfrak{g}^{(1)}$. In particular, given $\chi \in \mathfrak{g}^{*(1)}$ let $y = \theta(\chi)$ and write $y = y_s + y_n$, the Jordan decomposition of y in $\mathfrak{g}$. Then $\chi = \chi_s + \chi_n$ where $\chi_s = \theta^{-1}(y_s)$ and $\chi_n = \theta^{-1}(y_n)$. We call $\chi = \chi_s + \chi_n$ the *Jordan decomposition* of χ.

Let $\mathfrak{z}_\mathfrak{g}(\chi) = \{x \in \mathfrak{g} : \chi([x, \mathfrak{g}]) = 0\}$ and $Z_G(\chi) = \{g \in G : g \cdot \chi = \chi\}$. Under the hypotheses in 3.1 we have that $Z_G(\chi_s)$ is a connected, reductive algebraic group such that $\mathfrak{z}_\mathfrak{g}(\chi_s) = \mathrm{Lie}(Z_G(\chi_s))$ and $Z_G(\chi_s)$ satisfies 3.1.1,3.1.2, and 3.1.3, [25, 6.5 and 7.4]. Note that χ can be considered as an element of $\mathfrak{z}_\mathfrak{g}(\chi_s)^{*(1)}$.

3.4. **Reduction Theorem.** It is reasonable to be concerned mostly with almost simple G and simple $\mathfrak{g}$. The following reduction theorem in conjunction with 3.3, however, justifies the general hypotheses of 3.1.

Theorem. [37, Theorem 2], [18, Theorem 3.2] *Let* $\chi = \chi_s + \chi_n \in \mathfrak{g}^{*(1)}$ *be the Jordan decomposition. Let* $d = \frac{1}{2}(\dim G \cdot \chi_s)$. *Then there is an algebra isomorphism*

$$U_\chi(\mathfrak{g}) \cong \mathrm{Mat}_{p^d}\left(U_{\chi_n}(\mathfrak{z}_\mathfrak{g}(\chi_s))\right).$$

3.5. Thanks to Theorem 3.4, without loss of generality we can work under the hypothesis $\chi = \chi_s + \chi_n$ where $\mathfrak{z}_\mathfrak{g}(\chi_s) = \mathfrak{g}$. Since there is a finite number of nilpotent orbits in $\mathfrak{g}$, [21, Chapter 3], the classification of simple $\mathfrak{g}$-modules essentially becomes a finite problem.

3.6. **Blocks of** $U(\mathfrak{g})$. Recall that we consider the dot action of the Weyl group W on $\Lambda = X/pX$.

Theorem. *Let* χ *be as in 3.5. Then* U_χ *has* $|\Lambda/W|$ *blocks.*

Proof. Let $\mathcal{N} = \{\eta \in \mathfrak{g}^{*(1)} : \theta(\eta)$ nilpotent$\} \subseteq \mathfrak{g}^{*(1)}$ be the nilpotent cone in $\mathfrak{g}^{*(1)}$. By 3.5 affine translation $\mathcal{N} \longrightarrow \chi_s + \mathcal{N}$ is a G-equivariant isomorphism of varieties. In particular $\chi_s + \mathcal{N}$ is irreducible and has a unique dense orbit consisting of regular elements, that is of elements whose centraliser has minimal dimension, [21, Chapter 4]. Moreover, every G-orbit in $\chi_s + \mathcal{N}$ contains χ_s in its closure, [31, Theorem 2.5].

Let

$$\mathcal{O} = \{\eta \in \chi_s + \mathcal{N} : U_\eta \text{ has } |\Lambda/W| \text{ blocks}\}.$$

Clearly $\mathcal{O}$ is G-stable and by Proposition 2.5 $\mathcal{O}$ is locally closed in $\chi_s + \mathcal{N}$. By [25, Section 10] $\mathcal{O}$ contains both χ_s and the regular orbit. The result follows. $\qquad \square$

This theorem first appeared in [5], confirming a conjecture of Humphreys in [22]. The proof given in [5], however, was based on Müller's Theorem, 2.4, and less representation theoretic than the above. Moreover the case $p = 2$ was omitted.

Henceforth we write

$$U_\chi = \bigoplus_{\lambda \in \Lambda/W} \mathcal{B}_{\chi,\lambda}.$$

We often abuse notation by writing $\mathcal{B}_{\chi,\lambda}$ for $\lambda \in \Lambda$ or even $\lambda \in X$.

3.7. **Baby Verma modules.** Let $\chi = \chi_s + \chi_n \in \mathfrak{g}^{*(1)}$ be as in 3.5. We can assume without loss of generality that $\chi(\mathfrak{n}) = 0$, [25, Lemma 6.6]. Then any element $\lambda \in \Lambda$ gives rise to K_λ, a one dimensional representation of $U_\chi(\mathfrak{b})$, a reduced enveloping algebra of the Lie algebra of $\mathfrak{b}$. Indeed, by [25, 11.1] and [5, 3.19] there is a W-equivariant isomorphism

$$\Lambda \cong \{\lambda \in \mathfrak{h}^* : \lambda(h)^p - \lambda(h^{[p]}) = \chi(h) \text{ for all } h \in \mathfrak{h}\}.$$

The induced module $V_\chi(\lambda) = U_\chi \otimes_{U_\chi(\mathfrak{b})} K_\lambda$, a *baby Verma module*, plays an important role in the representation theory of U_χ. For instance it follows from [25, 10.11] and Theorem 3.6 that $V_\chi(\lambda)$ belongs to a block of U_χ and further that we can choose the labelling of blocks such that $V_\chi(\lambda)$ belongs to $\mathcal{B}_{\chi,\lambda}$. In particular $V_\chi(\lambda)$ and $V_\chi(w \bullet \lambda)$ belong to the same block for all $w \in W$.

3.8. **Primary Blocks.** We can describe when a block $\mathcal{B}_{\chi,\lambda}$ is primary.

Proposition. *Assume that $p \neq 5$ if R is of type E_7. Then the block $\mathcal{B}_{\chi,\lambda}$ is primary if and only if it corresponds to an Azumaya point of* $\mathrm{Maxspec}(Z)$.

Proof. Sufficiency follows from 2.6. For the converse, let $\lambda \in X$ and suppose that L is the unique simple module in the block $\mathcal{B}_{\chi,\lambda}$ and let $P(L)$ be its projective cover. By [23, B.12(2)] we have

$$\dim P(L) = p^N |W \bullet (\lambda + pX)| [V_\chi(\lambda) : L],$$

where $N = |\Phi^+|$, the number of positive roots. On the other hand, by [23, C.2], there is a projective $\mathcal{B}_{\chi,\lambda}$-module P such that

$$\dim P = p^N |W \bullet \lambda|.$$

We deduce that for $\mu \in W \bullet \lambda + pX$

$$(2) \qquad [V_\chi(\mu) : L] \text{ divides } \frac{|W \bullet \mu|}{|W \bullet (\mu + pX)|}.$$

Moreover $[V_\chi(\mu) : L] = p^i$ since $V_\chi(\mu)$ belongs to $\mathcal{B}_{\chi,\lambda}$. So we must find an element $\mu \in W \bullet \lambda + pX$ which forces i to be zero.

Arguing as in [23, H.1 Remarks] we can assume without loss of generality that G is almost simple.

Let

$$C_0 = \{\lambda \in X : 0 \leq \langle \lambda + \rho, \beta^\vee \rangle \leq p \text{ for all } \beta \in \Phi^+\}$$

and

$$C_0' = \{\lambda \in X : 0 \leq \langle \lambda + \rho, \beta^\vee \rangle < p \text{ for all } \beta \in \Phi^+\}.$$

Since C_0 is a fundamental domain for the dot action of the affine Weyl group $W \ltimes pX$ on X, we can assume without loss of generality that $\mu \in C_0$. By [23, C.1 Lemma] if $\mu \in C_0'$ then $|W\cdot\mu| = |W\cdot(\mu + pX)|$ and so, by (2), $i = 0$ as required. By [23, H.1 Proposition], if Φ is not exceptional then we can choose $\mu \in C_0'$ finishing the proof. Hence we need only consider the case $\mu \in C_0 \setminus C_0'$ and Φ exceptional. In particular if p is prime to the order of W then (2) forces $i = 0$. The only remaining cases are $p = 5$ and Φ of type E_6, E_7 or, E_8 and $p = 7$ and Φ of type E_7 or E_8. A case-by-case analysis shows that the only possible exception to $i = 0$ occurs when Φ has type E_7, $p = 5$ and $\mu = \varpi_2 + \varpi_5$ (where we've followed the numbering of [3]). $\square$

3.9. The previous proposition has a consequence for the structure of the centre of U_χ. This phenomenon was observed in the case $\chi = 0$ by Premet and noted in [5, 3.17].

Corollary. *Suppose that $p \neq 2$ and that $p \neq 5$ if Φ has a component of type E_7. Then the centre of U_χ is isomorphic to $Z/\mathfrak{m}_\chi Z$ if and only if χ is regular.*

Proof. By [36] and [18, 1.2] the algebras U_χ are symmetric. By [5, Theorem 3.5(4)] Z is a complete intersection ring (this is where we require $p \neq 2$), so in particular Gorenstein. By [5, Theorem 3.5(6)] Z is a free Z_0-module so, since Z_0 is smooth, standard commutative algebra implies that each factor $Z/\mathfrak{m}_\chi Z$ is Gorenstein, or, in other words, self-injective. Premet has proved the following, [32],

the natural map $\iota : Z/\mathfrak{m}_\chi Z \longrightarrow Z(U_\chi)$ is injective.

The result follows from Lemma 2.9, Proposition 3.8 and [5, Proposition 3.15].
    $\square$

3.10. Example. Let $G = SL_2(K)$ where K has odd characteristic. If χ satisfies the condition in 3.5 then either χ is regular nilpotent or $\chi = 0$. The algebra U_χ has $(p+1)/2$ blocks, labelled by the set of integers $\{-1, 0, \ldots, (p-3)/2\}$. In all cases the block corresponding to -1 is associated with a simple projective module, the *Steinberg module*, and is isomorphic to $\mathrm{Mat}_p(K)$. If χ is regular then $\mathcal{B}_{\chi,i} \cong \mathrm{Mat}_p(K[X]/(X^2))$ for $0 \leq i \leq (p-3)/2$. This follows, for instance, from [5, Proposition 3.16]. If $\chi = 0$ then it is shown in [16] that for $0 \leq i \leq (p-3)/2$ the block $\mathcal{B}_{0,i}$ is Morita equivalent to the path algebra of the quiver

with relations $\alpha\delta = \beta\gamma = 0$; $\delta\alpha = \gamma\beta = 0$; $\alpha\gamma = \beta\delta$; $\gamma\alpha = \delta\beta$, an eight-dimensional algebra. In this case the centre of the block is spanned by the linearly independent elements $1, \alpha\gamma$ and $\gamma\alpha$. In particular we have

$$\dim Z(U_0) = 3 \left(\frac{p-1}{2} \right) + 1 = \frac{3p-1}{2},$$

whilst $\dim Z/\mathfrak{m}_0 Z = p$.

3.11. Let's conclude with a number of remarks.
1) The structure of $Z/\mathfrak{m}_\chi Z$ is known, [27]. To the best of my knowledge, however, the structure of $Z(U_\chi)$ is, in general, unknown.
2) Proposition 3.8 completes the classification of blocks of finite representation type begun in [29]. Indeed, thanks to [15, Theorem 3.2], any block of finite representation type is uniserial so in particular primary. Then Proposition 3.8 tells us that the block is Azumaya. Consequently the results in [29] · and [5] can be applied.
3) The structure of Z has a number of further consequences for the representation theory of U thanks to Theorem 2.7 which is valid in this situation, [4, Theorem 4.10] and [5, Theorem 2.9.1]. In particular see [5, 3.11 and 3.13].

4. QUANTISED ENVELOPING ALGEBRAS

4.1. We follow the notation of [11]. Let G be a simply-connected, semisimple algebraic group of rank r over $\mathbb{C}$ and let $\mathfrak{g} = \mathrm{Lie}(G)$. Let B^+ be a Borel subgroup of G and let B^- be the Borel subgroup opposite to B^+ in G. Let $T = B^+ \cap B^-$, a maximal torus of G, so that $B^\pm = T.U^\pm$. Let $W = N_G(T)/T$ be the Weyl group of G.

Let $\epsilon \in \mathbb{C}$ be a primitive ℓ^{th} root of unity where ℓ is an odd integer greater than 1 and prime to 3 if G has a component of type G_2.

4.2. Let $U_\epsilon(\mathfrak{g})$ be the simply-connected quantised enveloping algebra of $\mathfrak{g}$ at a root of unity ϵ, as defined in [8]. In particular $U_\epsilon(\mathfrak{g})$ is a Hopf algebra. We have a triple

$$Z_0 \subseteq Z \subseteq U_\epsilon(\mathfrak{g}),$$

where Z_0 is ℓ-centre of $U_\epsilon(\mathfrak{g})$, a central sub-Hopf algebra generated by the ℓth powers of certain generators of $U_\epsilon(\mathfrak{g})$, [8]. Then $U_\epsilon(\mathfrak{g})$ is a free Z_0-module of rank $\ell^{\dim \mathfrak{g}}$. Thanks to Lemma 2.3 we have a map

$$\mathrm{Maxspec}(Z_0) \longrightarrow \mathrm{Alg}(\ell^{\dim \mathfrak{g}})$$

where $\chi \in \mathrm{Maxspec}(Z_0)$ is sent to $U_{\epsilon,\chi}$.

Remark. The algebra $U_\epsilon(\mathfrak{g})$ is not the quantised enveloping algebra considered by Lusztig in [26]. Lusztig's algebra is a quantum analogue of the hyperalgebra of $\mathfrak{g}$ whilst $U_\epsilon(\mathfrak{g})$ is a quantum analogue of the enveloping algebra of $\mathfrak{g}$. However the reduced quantum group of Lusztig, a finite dimensional sub-Hopf algebra of Lusztig's quantised enveloping algebra, is isomorphic to a skew group extension of the algebra $U_{\epsilon,1}$, where $1 \in \mathrm{Maxspec}(Z_0)$ is the augmentation ideal of Z_0.

4.3. The structure of Z_0 is well-understood, [8]. In particular there is an unramified covering of degree 2^r

$$\pi : \mathrm{Maxspec}(Z_0) \cong (U^- \times U^+) \rtimes T \longrightarrow B^- B^+ \subseteq G.$$

Moreover, given $\tilde{\chi} \in (U^- \times U^+) \rtimes T$ there exists $\chi = \chi_u \chi_s \in U^- \rtimes T = B^-$ such that $\pi(\chi_u)\pi(\chi_s)$ is the Jordan decomposition of $\pi(\chi)$ in G and $U_{\epsilon,\chi} \cong U_{\epsilon,\tilde{\chi}}$.

4.4. The algebra $U_\epsilon(\mathfrak{g})$ can be defined as the specialisation of an integral form of $U_q(\mathfrak{g})$, the quantised enveloping algebra of $\mathfrak{g}$ at a transcendental parameter q. As a result we can find a central subalgebra $Z_1 \subseteq U_\epsilon(\mathfrak{g})$, the specialisation of the centre of $U_q(\mathfrak{g})$.

Theorem. [11, Section 21] *The algebra Z is a complete intersection ring and there is an isomorphism $Z \cong Z_0 \otimes_{Z_0 \cap Z_1} Z_1$. Moreover, Z is a free Z_0-module of rank ℓ^r.*

4.5. Blocks of $U_\epsilon(\mathfrak{g})$. Theorem 4.4 allows us to determine the structure of the algebra $Z/\mathfrak{m}_\chi Z$ for $\chi = \chi_u \chi_s \in B^-$, [5, Theorem 4.5]. In particular we find that the primary components of $Z/\mathfrak{m}_\chi Z$ are in natural bijection with the elements of $R_\chi = \{t \in T : t^\ell \in W\chi_s^2\}/W$. Theorem 2.4 gives the following result.

Theorem. [5, Theorem 4.8] *Let $\chi = \chi_u \chi_s \in B^-$ be as above. Then the blocks of $U_{\epsilon,\chi}$ naturally correspond to elements of R_χ.*

If we have $\chi_s = 1$ then $R_\chi = \{t \in T : t^\ell = 1\}/W$ and we have the quantum analogue of Theorem 3.6.

4.6. We conclude with a number of remarks.
1) There is an analogue of Theorem 3.4 for the algebras $U_{\epsilon,\chi}$ which shows that the classification of the simple $U_\epsilon(\mathfrak{g})$-modules is a finite problem, [7]. The reduction theorem, however, is difficult to use in practise: there is no guarantee that we remain in the class of simply-connected quantised enveloping algebras and it is not enough in general to consider only unipotent central characters.
2) The algebra $U_{\epsilon,1}$ is a Hopf algebra whose antipode squared is inner, [24, 4.9(1)]. Since $U_{\epsilon,1}$ has a unique one dimensional module it follows from [15, Lemma 3.1] and [17, Lemma 1.5] that $U_{\epsilon,1}$ is symmetric. Since the block containing this module is never primary we deduce from Lemma 2.9 and Theorem 4.4 that $Z/\mathfrak{m}_1 Z$ is a proper subalgebra of $Z(U_{\epsilon,1})$. Note that $Z/\mathfrak{m}_\chi Z$ is always a subalgebra of $Z(U_{\epsilon,\chi})$ thanks to Lemma 2.8.
3) More details on the influence of Z on the representation theory of $U_\epsilon(\mathfrak{g})$ can be found in [5, Section 4].

5. QUANTISED FUNCTION ALGEBRAS AND QUANTUM BORELS

5.1. Let G, B^+, B^-, T and W be as in 4.1. Let X be the character group of T and let Φ be the root system of G. Let Φ^+ be the positive roots of Φ with respect to B^+ and let $\{\alpha_i : 1 \leq i \leq r\}$ be the simple roots in Φ^+ and $\{\varpi_i : 1 \leq i \leq r\}$ be the fundamental weights in X. Let $(\,,\,)$ be the natural pairing between the root lattice, Q, and the weight lattice, P.

There is a stratification of G

$$G = \coprod_{w_1, w_2 \in W} X_{w_1, w_2}$$

where $X_{w_1, w_2} = B^+ w_1 B^+ \cap B^- w_2 B^-$. This restricts to a stratification of B^+

$$B^+ = \coprod_{w \in W} X_{e, w}.$$

For $w \in W$ let $\ell(w)$, respectively $s(w)$, equal the minimal length of an expression for w as a product of simple, respectively arbitrary, reflections.

Let w_0 be the longest word in W and let $N = \ell(w_0)$. Since w_0 sends Φ^+ to $-\Phi^+$ there is an involution σ on the set of integers $[1, r]$ defined by $w_0 \alpha_i = -\alpha_{\sigma(i)}$. It can be shown that the number of fixed points of σ equals $2s(w_0) - r$.

5.2. Let $\epsilon \in \mathbb{C}$ be a primitive ℓ^{th} root of unity where ℓ is an odd integer greater than one and good, that is ℓ is prime to the bad primes of Φ.

Let $U_\epsilon^{\leq 0}$ be the non-positive subalgebra of $U_\epsilon(\mathfrak{g})$ as defined in [9], the quantised enveloping algebra of a Borel subalgebra of $\mathfrak{g}$. We have a triple

$$Z_0^{\leq 0} \subseteq Z(U_\epsilon^{\leq 0}) \subseteq U_\epsilon^{\leq 0}$$

where $Z_0^{\leq 0}$ is the ℓ-centre of $U_\epsilon^{\leq 0}$, that is the intersection of the ℓ-centre of $U_\epsilon(\mathfrak{g})$ with $U_\epsilon^{\leq 0}$. By [10] $U_\epsilon^{\leq 0}$ is a free $Z_0^{\leq 0}$-module of rank $\ell^{\dim B^+}$. By [10] $Z_0^{\leq 0}$ is isomorphic as a Hopf algebra to $\mathcal{O}[B^+]$, the ring of regular functions on B^+. Given $b \in B^+$ we let $\mathfrak{m}_b$ denote the corresponding maximal ideal of $Z_0^{\leq 0}$. By Theorem 2.3 we have a morphism of varieties

$$\alpha : \text{Maxspec}(Z_0^{\leq 0}) \longrightarrow \text{Alg}(\ell^{\dim B^+}).$$

We denote $\alpha(\mathfrak{m}_b)$ by $U_\epsilon^{\leq 0}(b)$.

Let $\mathcal{O}_\epsilon[G]$ be the quantised function algebra of G at a root of unity ϵ, as defined in [10]. We have a triple

$$Z_0 \subseteq Z(\mathcal{O}_\epsilon[G]) \subseteq \mathcal{O}_\epsilon[G],$$

where Z_0 is the ℓ-centre of $\mathcal{O}_\epsilon[G]$, [10]. By [10] and [35] $\mathcal{O}_\epsilon[G]$ is a free Z_0-module of rank $\ell^{\dim G}$. By [10] Z_0 is isomorphic as a Hopf algebra to $\mathcal{O}[G]$. Given $g \in G$ we let $\mathfrak{m}_g$ denote the corresponding maximal ideal of Z_0. By Theorem 2.3 we have a morphism of varieties

$$\beta : \text{Maxspec}(Z_0) \longrightarrow \text{Alg}(\ell^{\dim G}).$$

We denote $\beta(\mathfrak{m}_g)$ by $\mathcal{O}_\epsilon[G](g)$.

5.3. The following theorem shows us that there is only a finite number of isomorphism classes of algebras $U_\epsilon^{\leq 0}(b)$ and $\mathcal{O}_\epsilon[G](g)$. It also demonstrates that the PI degree of $U_\epsilon^{\leq 0}$, respectively of $\mathcal{O}_\epsilon[G]$, is $\ell^{\frac{1}{2}(N + s(w_0))}$, respectively ℓ^N.

Theorem. [9, Theorem 4.4], [10, Section 9], [12, Theorem 4.4 and Proposition 4.10] *Let $w, w_1, w_2 \in W$ and let $b, b' \in X_{e,w}$ and $g, g' \in X_{w_1, w_2}$.*
(a)(i) There is an algebra isomorphism $U_\epsilon^{\leq 0}(b) \cong U_\epsilon^{\leq 0}(b')$.
(ii) There are precisely $\ell^{r-s(w)}$ simple $U_\epsilon^{\leq 0}(b)$-modules, each having dimension $\ell^{\frac{1}{2}(\ell(w)+s(w))}$.
(b)(i) There is an algebra isomorphism $\mathcal{O}_\epsilon[G](g) \cong \mathcal{O}_\epsilon[G](g')$.
(ii) There are precisely $\ell^{r-s(w_2^{-1}w_1)}$ simple $\mathcal{O}_\epsilon[G](g)$-modules, each having dimension $\ell^{\frac{1}{2}(\ell(w_1)+\ell(w_2)+s(w_2^{-1}w_1))}$.

5.4. Centres. We begin by introducing some distinguished elements in $\mathcal{O}[B^+]$ and $\mathcal{O}[G]$. Let $V(\varpi_i)$ be the simple G-module with highest weight ϖ_i and let $V(\varpi_i)^*$ be its dual. For each i choose a highest weight vector v_i, respectively a lowest weight vector v_i', of $V(\varpi_i)$. Let f_i, respectively f_i', be the unique weight vector in $V(\varpi_i)^*$ dual to v_i, respectively v_i'. For each i we have elements a_i, b_i and c_i of $\mathcal{O}[G]$ defined by

$$a_i(g) = f_{\sigma(i)}'(gv_{\sigma(i)}'), \quad b_i(g) = f_i'(gv_i), \quad c_i(g) = f_{\sigma(i)}(gv_{\sigma(i)}').$$

These elements restrict to elements of $\mathcal{O}[B^+]$ which we will denote by a_i, b_i and c_i too.

Given a subset $I \subseteq [1, r]$ let $S(I)$ be the subalgebra of $\mathbb{C}[X_1, Y_1, \ldots, X_r, Y_r]$ generated by $X_i^k Y_i^{\ell-k}$ for $i \in I$ and $0 \leq k \leq \ell$. Then $S(I)$ is a free module of rank $\ell^{|I|}$ over $S_0(I) = \mathbb{C}[X_i^\ell, Y_i^\ell : i \in I]$.

Theorem. *a) Let $\tilde{I} = \{1 \leq i \leq r : \sigma(i) \neq i\}$ and let I be a set of orbit representatives for the σ-action on $\tilde{I}$. Then the centre of $U_\epsilon^{\leq 0}$ is isomorphic to the algebra*

$$\mathcal{O}[B^+] \otimes_{S_0(I)} S(I),$$

where $S_0(I) \longrightarrow \mathcal{O}[B^+]$ sends X_i^ℓ to $a_i c_i$ and Y_i^ℓ to $a_{\sigma(i)} c_{\sigma(i)}$. In particular $Z(U_\epsilon^{\leq 0})$ is a free $Z_0^{\leq 0}$-module of rank $\ell^{r-s(w_0)}$.
b) The centre of $\mathcal{O}_\epsilon[G]$ is isomorphic to the algebra

$$\mathcal{O}[G] \otimes_{S_0([1,r])} S([1,r]),$$

where $S_0([1,r]) \longrightarrow \mathcal{O}[G]$ is obtained by sending X_i^ℓ to b_i and Y_i^ℓ to c_i. In particular $Z(\mathcal{O}_\epsilon[G])$ is a free Z_0-module of rank ℓ^r.

Proof. Part (a) can be found in the appendix: it follows the method of proof of (b) in [14]. □

Remark. If G only has components of type B_r, C_r, D_r (r even),E_7, E_8, F_4 or G_2 then $s(w_0) = r$ so $\tilde{I} = \emptyset$ and consequently $Z(U_\epsilon^{\leq 0}) \cong \mathcal{O}[B^+]$. It can be shown that this is the only case when one of the algebras above is Gorenstein.

5.5. In contrast to Section 3 the centre of $U_\epsilon^{\leq 0}(b)$ or $\mathcal{O}_\epsilon[G](g)$ is easy to describe.

Lemma. *a) Let $b \in B^+$. Then $Z(U_\epsilon^{\leq 0}(b)) \cong Z/\mathfrak{m}_b Z$.*
b) Let $g \in G$. Then $Z(\mathcal{O}_\epsilon[G](g)) \cong Z/\mathfrak{m}_g Z$.

Proof. (a) By Lemma 2.8 the natural map $\iota : Z/\mathfrak{m}_b Z \longrightarrow Z(U_\epsilon^{\leq 0}(b))$ is injective, so by Theorem 5.4(a), it is enough to show that $\dim Z(U_\epsilon^{\leq 0}(b)) = \ell^{r-s(w_0)}$ for all $b \in B^+$. Since $\overline{X_{e,w}} = \coprod_{w' \preceq w} X_{e,w'}$, where $\preceq$ denotes the Bruhat-Chevalley order on W, the identity of B^+ is contained in the closure of any cell $X_{e,w}$. By Proposition 2.10 $\dim Z(U_\epsilon^{\leq 0}(b)) \leq \dim Z(U_\epsilon^{\leq 0}(1))$ for all $b \in B^+$. We will prove that $\dim Z(U_\epsilon^{\leq 0}(1)) = \ell^{r-s(w_0)}$.

Let $\overline{U}_1 = U_\epsilon^{\leq 0}(1)$. Fix a reduced expression $w_0 = s_{i_1} \ldots s_{i_N}$ and hence an ordering $\beta_1 < \ldots < \beta_N$ in R^+. The algebra $\overline{U}_1$ is generated by the elements F_{β_j} $(1 \leq j \leq N)$ and K_i $(1 \leq i \leq r)$ subject to the relations

(3) $$K_i K_j = K_j K_i, \quad K_i^\ell = 1$$

(4) $$K_i F_{\beta_j} = \epsilon^{-(\beta_j, \varpi_i)} F_{\beta_j} K_i, \quad F_{\beta_j}^\ell = 0$$

(5) $$F_{\beta_j} F_{\beta_k} = \epsilon^{-(\beta_j, \beta_k)} F_{\beta_k} F_{\beta_j} + p_{jk}^1 \quad j < k$$

where p_{jk}^1 is a polynomial in the variables F_{β_h} for $j < h < k$, [11, Theorem 9.3].

Suppose we have defined inductively an $\ell^{\dim B^+}$-dimensional algebra $\overline{U}_m$ with generators F_{β_j} and K_i satisfying (3), (4) and

(6) $$F_{\beta_j} F_{\beta_k} = \epsilon^{-(\beta_j, \beta_k)} F_{\beta_k} F_{\beta_j} + p_{jk}^m$$

where p_{jk}^m is a polynomial in the variables F_{β_h} for $j < h < \min\{k, N+1-m\}$. We have a morphism

$$\mathbb{C} \longrightarrow \mathrm{Alg}(\ell^{\dim B^+})$$

where $t \in \mathbb{C}$ is sent to the algebra $\overline{U}_m(t)$ with generators F_{β_j} and K_i satisfying relations (3), (4) and (6') obtained from (6) by replacing $F_{\beta_{N+1-m}}$ with $t F_{\beta_{N+1-m}}$. Then $\overline{U}_m(t) \cong \overline{U}_m$ for $t \in \mathbb{C}^*$ and, by definition, $\overline{U}_{m+1} = \overline{U}_m(0)$.

The algebra $\overline{U}_N$ is generated by F_{β_j} and K_i subject to relations (3), (4) and

$$F_{\beta_j} F_{\beta_k} = \epsilon^{-(\beta_j, \beta_k)} F_{\beta_k} F_{\beta_j}.$$

By a repeated application of Proposition 2.10 we find that $\dim Z(U_\epsilon^{\leq 0}(1)) \leq \dim Z(\overline{U}_N)$. Using the techniques of [11, Chapter 2 and Section 10] we can see that the dimension of $Z(\overline{U}_N)$ is $\ell^{r-s(w_0)}$ as required.

(b) This is proved in a similar manner using [20, Section 2.9]. $\square$

5.6. Blocks. Given $w, w_1, w_2 \in W$ let

$$\mathcal{S}(w) = \{1 \leq i \leq r : \sigma(i) \neq i \text{ and } w_0 w, w w_0 \in \mathrm{Stab}_W(\varpi_i)\}$$

and

$$\mathcal{T}(w_1, w_2) = \{1 \leq i \leq r : w_0 w_1, w_0 w_2 \in \mathrm{Stab}_W(\varpi_i)\}.$$

Note that $\mathcal{S}(w)$ is σ-stable.

Theorem. (a) Let $b \in X_{e,w}$. Then $U_\epsilon^{\leq 0}(b)$ has $\ell^{\frac{1}{2}|\mathcal{S}(w)|}$ blocks.
(b) Let $g \in X_{w_1, w_2}$. Then $\mathcal{O}_\epsilon[G](g)$ has $\ell^{|\mathcal{T}(w_1, w_2)|}$ blocks.

Proof. (a) Recall that I is a set of representatives for the σ-action on $\tilde{I} = \{1 \leq i \leq r : \sigma(i) \neq i\}$. The number of blocks of $U_\epsilon^{\leq 0}(b)$ equals the number of maximal ideals of $Z/\mathfrak{m}_b Z$, either by Theorem 2.4 or by Lemma 5.5. The description of $Z(U_\epsilon^{\leq 0})$ shows that $Z/\mathfrak{m}_b Z$ is the tensor product of a discrete family of ℓ-dimensional algebras indexed by elements of I, where

(i) if $a_i c_i(b) \neq 0 \neq a_{\sigma(i)} c_{\sigma(i)}(b)$ then $Z_i \cong \mathbb{C}^\ell$;

(ii) if $a_i c_i(b) \neq 0$ and $a_{\sigma(i)} c_{\sigma(i)}(b) = 0$ (or vice-versa) then $Z_i \cong \mathbb{C}[X]/(X^\ell)$;

(iii) if $a_i c_i(b) = 0 = a_{\sigma(i)} c_{\sigma(i)}(b)$ then $Z_i \cong \mathbb{C}[X_1, \ldots X_{\ell-1}]/(X_j X_k : 1 \leq j, k \leq \ell - 1)$.

It follows that $U_\epsilon^{\geq 0}(b)$ has ℓ^d blocks where $d = |\{i \in I : a_i c_i(b) \neq 0 \neq a_{\sigma(i)} c_{\sigma(i)}(b)\}|$.

By definition a_i is a co-ordinate function on the maximal torus T so is non-vanishing on B^+. By [6, Lemma 7.2] c_i does not vanish on $X_{e,w}$ if and only if $w_0 w \in \text{Stab}_W(\varpi_i)$, or equivalently $w w_0 \in \text{Stab}_W(\varpi_{\sigma(i)})$. The first part follows.

(b) This is proved similarly. Details can be found in [6, Section 7]. $\qquad\square$

5.7. Azumaya locus.

The precise structure of $Z(U_\epsilon^{\leq 0}(b))$ and $Z(\mathcal{O}_\epsilon[G](g))$ can be determined using Lemma 5.5 and the methods in the proof of Theorem 5.6. The result, however, is a little awkward to present in general, but in the Azumaya case things are simpler.

Theorem. *(a) Let $w \in W$ be such that $\ell(w) + s(w) = N + s(w_0)$. Let $n = r - s(w)$ and $d = \frac{1}{2}(N + s(w_0))$. Then for $b \in X_{e,w}$ we have*

$$U_\epsilon^{\leq 0}(b) \cong \bigoplus^{\ell^n} \text{Mat}_{\ell^d} \left(\frac{\mathbb{C}[X_1, \ldots, X_{s(w)-s(w_0)}]}{(X_1^\ell, \ldots, X_{s(w)-s(w_0)}^\ell)} \right).$$

(b) Let $w_1, w_2 \in W$ be such that $\ell(w_1) + \ell(w_2) + s(w_2^{-1} w_1) = 2N$. Let $n = r - s(w_2^{-1} w_1)$. Then for $g \in X_{w_1, w_2}$ we have

$$\mathcal{O}_\epsilon[G](g) \cong \bigoplus^{\ell^n} \text{Mat}_{\ell^N} \left(\frac{\mathbb{C}[X_1, \ldots, X_{s(w_2^{-1} w_1)}]}{(X_1^\ell, \ldots, X_{s(w_2^{-1} w_1)}^\ell)} \right).$$

Proof. (a) By Theorem 5.3 every simple $U_\epsilon^{\leq 0}(b)$-module has maximal dimension so it follows from (1) that $U_\epsilon^{\leq 0}(b) \cong \text{Mat}(Z/\mathfrak{m}_b Z)$. The argument of [6, Proposition 3.2] shows that a point $(b, x) \in \text{Maxspec}(Z(U_\epsilon^{\leq 0}))$ such that $a_i c_i(b) = a_{\sigma(i)} c_{\sigma(i)}(b) = 0$ for some $i \in I$ is singular. Since the Azumaya locus is contained in the smooth locus of $\text{Maxspec}(Z(U_\epsilon^{\leq 0}))$ it follows that only cases (i) and (ii) in the proof of Theorem 5.6 can occur. The result follows from 5.3 by counting the number of simple $Z/\mathfrak{m}_b Z$-modules.

(b) This is proved similarly. Details can be found in [6, Section 3]. $\qquad\square$

Remark. This theorem is the crucial step in the determination of the representation type of the algebras $U_\epsilon^{\leq 0}(b)$ and $\mathcal{O}_\epsilon[G](g)$, see [6, Section 4].

APPENDIX A

A.1. We would like to describe the centre of $U_\epsilon^{\leq 0}$. We follow the ideas of [14] and use the notation of [12] without further ado. We recall that there is an isomorphism of Hopf algebras

$$U_\epsilon^{\leq 0} \cong \mathcal{O}_\epsilon[B^+]$$

where $\mathcal{O}_\epsilon[B^+]$ is the factor algebra of $\mathcal{O}_\epsilon[G]$ essentially obtained by restricting functions from $U_q(\mathfrak{g})$ to $U_q^{\geq n}$, [10]. Let

$$\pi : \mathcal{O}_\epsilon[G] \longrightarrow \mathcal{O}_\epsilon[B^+]$$

be the canonical map. We often abuse notation by continuing to write x for $\pi(x)$.

A.2. For $1 \leq i \leq r$ the following elements of $\mathcal{O}_\epsilon[B^+]$ are defined as matrix coefficients (compare 5.4)

$$x_i = c_{\phi^{w_0\varpi_i}, v-\varpi_i}^{V(-w_0\varpi_i)}, \qquad y_i = c_{\phi^{\varpi_i}, v-\varpi_i}^{V(-w_0\varpi_i)}.$$

We have a general commutation rule, [12, 1.2]

$$(7) \qquad c_{\phi,v} c_{\psi,w} = \epsilon^{-(\mu_1,\mu_2)+(\nu_1,\nu_2)} c_{\psi,w} c_{\phi,v} + \sum_j c_{\psi_j,w_j} c_{\phi_j,v_j},$$

where $\psi_j \otimes \phi_j = p_j(M_j(E) \otimes M_j(F))\psi \otimes \psi$, $w_j \otimes v_j = p'_j(M'_j(E) \otimes M'_j(F))w \otimes v$ and p_j, p'_j are scalars and M_j, M'_j are monomials, at least one of which is non-constant.

A.3. We define

$$z_{i,k} = (x_i y_i)^k (x_{\sigma(i)} y_{\sigma(i)})^{\ell - k}.$$

Lemma. *For $1 \leq i \leq r$ and $0 \leq k \leq \ell$ the element $z_{i,k}$ is central in $U_\epsilon^{\leq 0}$.*

Proof. It can be checked, using (7), that $z_{i,k}$ commutes with all matrix coefficients of the form $c_{\psi,v}^{V(\lambda)}$ where v is a lowest weight of $V(\lambda)$. By [12, Lemma 2.3(2)] $\mathcal{O}_\epsilon[B^+]$ is generated by such matrix coefficients, together with the inverse of $c_{\phi^\rho, v-\rho}^{V(\rho)}$. $\qquad\qquad \square$

A.4. Let $\tilde{I} = \{1 \leq i \leq r : \sigma(i) \neq i\}$ and let $n = |\tilde{I}|$. Let R be the algebra with generators X_j and Y_j for $j \in I$ satisfying the following relations

$$(8) \qquad X_j X_{j'} = X_{j'} X_j, \quad Y_j Y_{j'} = Y_{j'} Y_j, \quad X_j Y_{j'} = \epsilon^{(\varpi_{j'}, w_0 \varpi_j - \varpi_j)} Y_{j'} X_j.$$

Let R_0 be the subalgebra generated by X_j^ℓ and Y_j^ℓ, a polynomial ring in $2n$ variables. It is immediate that R is a free R_0-module of rank ℓ^{2n} with basis $\{X_{j_1}^{a_1} \dots X_{j_n}^{a_n} Y_{j_1}^{b_n} \dots Y_{j_n}^{b_n} : 0 \leq a_i, b_i < \ell\}$.

By (7) we have an algebra map

$$\psi : R \longrightarrow \mathcal{O}_\epsilon[B^+]$$

which sends X_j to x_j and Y_j to y_j. There is a commutative diagram,

$$
\begin{array}{ccc}
U_q^{\geq 0}(\mathfrak{sl}_2, i) & \longrightarrow & U_q^{\geq 0}(\mathfrak{g}) \\
\downarrow & & \downarrow \\
U_q(\mathfrak{sl}_2, i) & \longrightarrow & U_q(\mathfrak{g})
\end{array}
$$

which, in particular, induces maps

$$
\eta_i : \mathcal{O}_\epsilon[B^+] \longrightarrow \mathcal{O}_\epsilon[B^+(SL(2)), i]
$$

and

$$
\tau_i : \mathcal{O}_\epsilon[G] \longrightarrow \mathcal{O}_\epsilon[SL(2), i],
$$

see [12, 2.4] for example. Let $w_0 = s_{i_1} \ldots s_{i_N}$ be a fixed reduced expression for the longest word of the Weyl group. We define two algebra maps

$$
\sigma : \mathcal{O}_\epsilon[B^+] \xrightarrow{\Delta^{(N-1)}} \mathcal{O}_\epsilon[B^+]^{\otimes N} \xrightarrow{\eta_{i_1} \otimes \cdots \otimes \eta_{i_N}} \bigotimes_{j=1}^{N} \mathcal{O}_\epsilon[B^+(SL(2)), i_j],
$$

and

$$
\tau : \mathcal{O}_\epsilon[G] \xrightarrow{\Delta^{(N-1)}} \mathcal{O}_\epsilon[G]^{\otimes N} \xrightarrow{\tau_{i_1} \otimes \cdots \otimes \tau_{i_N}} \bigotimes_{j=1}^{N} \mathcal{O}_\epsilon[SL(2), i_j].
$$

The map σ is injective by [12, Theorem 3.2]. By construction we have another commutative diagram

(9)
$$
\begin{array}{ccc}
\mathcal{O}_\epsilon[B^+] & \xrightarrow{\eta} & \bigotimes_{j=1}^{N} \mathcal{O}_\epsilon[B^+(SL(2)), i_j] \\
\pi \uparrow & & \uparrow \pi_{i_1} \otimes \cdots \otimes \pi_{i_N} \\
\mathcal{O}_\epsilon[G] & \xrightarrow{\tau} & \bigotimes_{j=1}^{N} \mathcal{O}_\epsilon[SL(2), i_j]
\end{array}
$$

A.5. For any algebra $\mathcal{O}_\epsilon[B^+(SL(2)), i]$ we can construct elements analogous to x_j and y_j, which we denote by $x(i)$ and $y(i)$.

Lemma. *Let* $j \in J$. *Then* $\eta(x_j) = \otimes_i x(i)^{m_j(i)}$ *and* $\eta(y_j) = \otimes_i y(i)^{m'_j(i)}$.

Proof. This first equality follows from [14, Proposition 3.1] together with (9). For the second part we note that in $\mathcal{O}_\epsilon[G]$ we have

$$
\Delta(y_i) = \Delta_{\phi^{\varpi_i}, v_{-\varpi_i}} = \sum_s c_{\phi^{\varpi_i}, v_s} \otimes c_{\phi^s, v_{-\varpi_i}},
$$

where $\{\phi^s\}$ and $\{v_s\}$ are dual bases for $V(-w_0\varpi_i)^*$ and $V(-w_0\varpi_i)$ respectively, chosen so that $\phi^1 = \phi^{\varpi_i}$ and $v_1 = v_{-\varpi_i}$. In $\mathcal{O}_\epsilon[B^+]$ we have $c_{\phi^{\varpi_i}, v_s} = 0$ unless $s = 1$. So we deduce that in $\mathcal{O}_\epsilon[B^+]$

$$
\Delta^{(N-1)}(y_i) = y_i \otimes \ldots \otimes y_i.
$$

Hence we need only describe $\tau_j(y_i)$ to complete the lemma. A simple calculation yields the following

$$
\tau_i(y_j) = \begin{cases} y_j & \text{if } i = j \\ 1 & \text{otherwise.} \end{cases}
$$

$\square$

A.6. The above lemma provides us with a pair of linear maps

$$m : \mathbb{N}^n \longrightarrow \mathbb{N}^N, \quad m' : \mathbb{N}^n \longrightarrow \mathbb{N}^N,$$

which are determined on the components by m_j and m'_j respectively. These induce linear maps

$$\overline{m} : \left(\frac{\mathbb{N}}{\ell\mathbb{N}}\right)^n \longrightarrow \left(\frac{\mathbb{N}}{\ell\mathbb{N}}\right)^N, \quad \overline{m}' : \left(\frac{\mathbb{N}}{\ell\mathbb{N}}\right)^n \longrightarrow \left(\frac{\mathbb{N}}{\ell\mathbb{N}}\right)^N.$$

Lemma. *The maps $\overline{m}$ and $\overline{m}'$ are injective.*

Proof. For $\overline{m}$ this is proved in [14, Proposition 3.2 and Proposition 6.1]. That $\overline{m}'$ is injective follows from the explicit description given in the proof of Lemma A.5. $\qquad\square$

A.7. We now have two algebra maps

$$\theta_1 : \mathcal{O}[B^+] \longrightarrow \bigotimes_{j=1}^{N} \mathcal{O}[B^+(SL(2)), i_j],$$

and

$$\theta_2 : R \longrightarrow \bigotimes O_\epsilon[B^+(SL(2)), i_j]$$

where θ_1 is the restriction of σ to $\mathcal{O}[B^+]$ and θ_2 is the composition $\sigma \circ \psi$.

Let I be a set of orbit representatives for the σ-action on $\tilde{I}$. In particular $|I| = r - s(w_0)$. Let $Z_{j,k} = (X_j Y_j)^k (X_{\sigma(j)} Y_{\sigma(j)})^{\ell-k}$ for $j \in I$ and $0 \le k \le \ell$ and let $R' \subseteq R$ be the subalgebra generated by the elements $Z_{j,k}$. It follows from (8) that R' is commutative and it is straightforward to check that R' is free over the subring R'_0 generated by $Z_{j,0}$ and $Z_{j,\ell}$ with $j \in I$, a polynomial ring in n variables.

Lemma. *There exists an algebra map*

$$(10) \qquad \theta : \mathcal{O}[B^+] \otimes_{R'_0} R' \longrightarrow \bigotimes_{j=1}^{N} \mathcal{O}_\epsilon[B^+(SL(2)), i_j],$$

such that θ is injective. In particular $\mathcal{O}[B^+] \otimes_{R'_0} R'$ is an integral domain of Krull dimension $\dim B^+$.

Proof. The map θ is obtained by combining θ_1 and θ_2. That this is well-defined follows immediately from construction. Let $t = r - s(w_0)$, $\{j_1, \ldots, j_t\} = I$. The left hand side of (10) is a free $\mathcal{O}[B^+]$-module with basis $\{Z_{j_1,k_1} \ldots Z_{j_t,k_t} : 0 \le k_i < \ell\}$ whilst the right hand side is a free $\otimes_{j=1}^{N} \mathcal{O}[B^+(SL(2)), i_j]$-module with basis $\{\otimes_{j=1}^{N} x_j^{b_j} y_j^{c_j} : 0 \le b_j, c_j < \ell\}$. By Lemma A.6 θ_2 is injective with respect to these bases, proving the first claim. The second follows from the fact that the right hand side of (10) is a domain and that the left hand side is a finite extension of $\mathcal{O}[B^+]$. $\qquad\square$

A.8. Now we have an algebra map

$$\phi : \mathcal{O}[B^+] \otimes_{R'_0} R' \longrightarrow \mathcal{O}_\epsilon[B^+],$$

whose image is generated by $\mathcal{O}[B^+]$ and the elements $z_{i,k}$.

Theorem. *The centre of* $U_\epsilon^{\leq 0}$ *is isomorphic to* $\mathcal{O}[B^+] \otimes_{R'_0} R'$ *under the map* ϕ.

Proof. Let Z' be the image of ϕ. By Lemma A.3 Z' is central in $\mathcal{O}_\epsilon[B^+]$. Since Z' is an integral domain of Krull dimension $\dim B^+$ it follows from the second claim of Lemma A.7 that ϕ is an injection.

Let $P^{w_0} = \{\lambda \in P : w_0\lambda = \lambda\}$. Write $\lambda = \sum a_i\varpi_i$. It is clear that $w_0\lambda = \lambda$ if and only if $\sum_{i=1}^r (a_i + a_{\sigma(i)})\varpi_i = 0$. It follows that $P^{w_0} = \mathbb{Z}[\varpi_i - \varpi_{\sigma(i)} : i \in I]$. Now the quotient ring of Z' must equal the quotient ring of $Z(\mathcal{O}_\epsilon[B^+])$ thanks to the description given in [12, Theorem 4.5]. Since $Z(\mathcal{O}_\epsilon[B^+])$ is a finite extension of Z' it is therefore enough to show that Z' is integrally closed. But the arguments of [14, Section 7] can be applied verbatim, confirming this. $\qquad\square$

References

[1] M. Auslander, I. Reiten, and S.O. Smalø. *Representation Theory of Artin Algebras.* Number 36 in Cambridge studies in advanced mathematics. Cambridge University Press, first paperback edition, 1995.

[2] D.J. Benson. *Representations and Cohomology.* Number 30 (1) in Cambridge studies in advanced mathematics. Cambridge University Press, 1991.

[3] N. Bourbaki. *Groupes et algèbres de Lie, Chapitres 4,5 et 6.* Éléments de Mathématique. Hermann, 1968.

[4] K.A. Brown and K.R. Goodearl. Homological aspects of noetherian PI Hopf algebras and irreducible modules of maximal dimension. *J. Alg.*, 198(1):240–265, 1997.

[5] K.A. Brown and I. Gordon. The ramification of centres: Lie algebras in positive characteristic and quantised enveloping algebras. University of Glasgow preprint no. 99/16.

[6] K.A. Brown and I. Gordon. The ramification of centres: quantised function algebras at roots of unity. University of Glasgow preprint no. 99/46.

[7] C. De Concini and V.G. Kac. Representations of quantum groups at roots of 1: reduction to the exceptional case. *Adv. Ser. Math. Phys.*, 16:141–149, 1992.

[8] C. De Concini, V.G. Kac, and C. Procesi. Quantum coadjoint action. *J. Amer. Math. Soc.*, 5(1):151–189, 1992.

[9] C. De Concini, V.G. Kac, and C. Procesi. Some quantum analogues of solvable Lie groups. In *Geometry and Analysis*, pages 41–65. Tata Inst. Fund. Res., Bombay, 1992.

[10] C. De Concini and V. Lyubashenko. Quantum function algebras at roots of 1. *Adv. Math.*, 108:205–262, 1994.

[11] C. De Concini and C. Procesi. Quantum groups. Springer Lecture Notes in Mathematics 1565. 31-140.

[12] C. De Concini and C. Procesi. Quantum Schubert cells and representations at roots of 1. In G.I. Lehrer, editor, *Algebraic groups and Lie groups*, number 9 in Australian Math. Soc. Lecture Series. Cambridge University Press, Cambridge, 1997.

[13] K. Erdmann. *Blocks of Tame Representation Type and Related Algebras*, number 1428 in Springer Lecture Notes in Mathematics, 1990.

[14] B. Enriquez. Le centre des algèbres de coordonnées des groupes quantiques aux racines p^α-ièmes de l'unité. *Bull. Soc. Math. France*, 122(4), 1994.

[15] R. Farnsteiner. Periodicity and representation type of modular Lie algebras. *J. reine angew. Math.*, 464:47–65, 1995.

[16] G. Fischer. *Darstellungtheorie des ersten Frobeniuskerns der SL_2*, PhD thesis, Universität Bielefeld, 1982.

[17] D. Fischman, S. Montgomery, and H. J. Schneider. Frobenius extensions of subalgebras of Hopf algebras. *Trans. Amer. Math. Soc.*, 349(12):4857–4895, 1997.

[18] E.M. Friedlander and B.J. Parshall. Modular representation theory of Lie algebras. *Amer. J. Math.*, 110(6):1055–1093, 1988.

[19] P. Gabriel. Finite representation type is open. In V. Dlab and P. Gabriel, editors, *Representations of Algebras*, number 488 in Springer Lecture Notes in Mathematics, pages 132–155, 1974.

[20] I. Gordon. Complexity of representations of quantised function algebras and representation type. Preprint, University of Glasgow, 1998.

[21] J.E. Humphreys. *Conjugacy Classes in Semisimple Algebraic Groups*, volume 43 of *Math. Surveys Monographs*. Amer. Math. Soc., Providence, RI, 1995.

[22] J.E. Humphreys. Modular representations of simple Lie algebras. *Bull. Amer. Math. Soc.*, 35(2):105–122, 1998.

[23] J. C. Jantzen. Subregular nilpotent representations of Lie algebras in prime characteristic. *Represent. Theory*, 3:139–152, 1999.

[24] J.C. Jantzen. *Lectures on Quantum Groups*. Number 6 in Graduate Studies in Mathematics. Amer. Math. Soc., Providence, R.I., 1996.

[25] J.C. Jantzen. Representations of Lie algebras in prime characteristic. In A. Broer, editor, *Representation Theories and Algebraic Geometry*, Proceedings Montréal 1997 (NATO ASI series C 514), pages 185–235. Dordrecht etc, Kluwer, 1998.

[26] G. Lusztig. Quantum groups at roots of 1. *Geom. Dedicata*, 35:89–114, 1990.

[27] I. Mirković and D. Rumynin. Centers of reduced enveloping algebras. *Math. Zeit.*, 231:123–132, 1999.

[28] B. J. Müller. Localization in non-commutative Noetherian rings. *Canad. J. Math.*, 28:600–610, 1976.

[29] D. K. Nakano and R. D. Pollack. Blocks of finite type in reduced enveloping algebras for classical Lie algebras. Preprint 1998.

[30] R. Pollack. Restricted Lie algebras of bounded type. *Bull. Amer. Math. Soc.*, 74:326–331, 1968.

[31] A. Premet. An analogue of the Jacobson-Morozov theorem for Lie algebras of reductive groups of good characteristics. *Trans. Amer. Math. Soc.*, 347:2961-2988, 1995.

[32] A. Premet. Private communication.

[33] I. Reiner. *Maximal orders*. Number 5 in London Mathematical Society Monographs,. Academic Press, 1975.

[34] L.H. Rowen. *Polynomial Identities in Ring Theory*. Number 84 in Pure and Applied Mathematics. Academic Press, 1980.

[35] D. Rumynin. Hopf-Galois extensions with central invariants and their geometric properties. *Algebra and Rep. Theory*, 1:353–381, 1998.

[36] J. R. Schue. Symmetry for the enveloping algebra of a restricted Lie algebra. *Proc. Amer. Math. Soc.*, 16:1123–1124, 1965.

[37] B. Ju. Veisfeiler and V.G. Kac. The irreducible representations of Lie p-algebras. *Funkcional. Anal. i Priložen.*, 5(2):28–36, 1971.

THE YANG-BAXTER EQUATION FOR OPERATORS ON FUNCTION FIELDS

JINTAI DING AND TIMOTHY J. HODGES

INTRODUCTION

In this note we show how infinite dimensional versions of the Cremmer-Gervais and Jordanian R-matrices occur as operators on rational function fields. Moreover we classify certain types of solutions of the Yang-Baxter equation on rational function fields and show that the above operators are essentially the only solutions.

Recall that if A is an integral domain and σ is an automorphism of A, then σ extends naturally to the field of rational functions $A(x)$ by acting on the coefficients. Denote by $\mathbb{F}(z_1, z_2)$ the field of rational functions in the variables z_1 and z_2. Then for any $\sigma \in \operatorname{Aut} \mathbb{F}(z_1, z_2)$, and any $i, j \in \{1, 2, 3\}$, we may define $\sigma_{ij} \in \operatorname{Aut} \mathbb{F}(z_1, z_2, z_3)$ by realizing $\mathbb{F}(z_1, z_2, z_3)$ as $\mathbb{F}(z_i, z_j)(z_k)$. Set $\Gamma = \operatorname{Aut} \mathbb{F}(z_1, z_2)$. Elements $R = \sum \alpha_i(z_1, z_2)\sigma_i$ of the group algebra $\mathbb{F}(z_1, z_2)[\Gamma]$ act as linear operators on $\mathbb{F}(z_1, z_2)$ and we may define in this way R_{ij} as linear operators on $\mathbb{F}(z_1, z_2, z_3)$. Thus we may look for solutions of the Yang-Baxter equation $R_{12}R_{13}R_{23} = R_{23}R_{13}R_{12}$ amongst such operators. Denote by P the operator $P \cdot f(z_1, z_2) = f(z_2, z_1)$.

First we look for operators of the form

$$R \cdot f(z_1, z_2) = \alpha(z_1/z_2)f(z_1, z_2) + \beta(z_1/z_2)f(z_2, z_1)$$

where $\alpha(x), \beta(x) \in \mathbb{F}(x)$. One solution of this is an infinite dimensional version of one member of the two parameter family of Cremmer-Gervais R-matrices [2, 12]. Moreover up to a certain natural equivalence, this is essentially the only such solution. This approach explains and clarifies the results on generating functions for the Cremmer-Gervais R-matrices in [11].

Next we consider the additive version; that is, operators of the form

$$R \cdot f(z_1, z_2) = \alpha(z_1 - z_2)f(z_1, z_2) + \beta(z_1 - z_2)f(z_2, z_1)$$

where $\alpha(x), \beta(x) \in \mathbb{F}(x)$. In this case the solution is again essentially unique. It is again a kind of direct limit of finite dimensional olutions of the Yang-Baxter equation. However this time the solutions are less familiar. They turn out (when twisted appropriately) to be generalizations of the Jordanian R-matrix [4] which may be viewed as quantizations of certain skew-symmetric solutions of the classical Yang-Baxter equation [5, 9, 10].

This work is an analog for the constant Yang-Baxter equation of work of Shibukawa and Ueno. In [14], Shibukawa and Ueno construct solutions of the

The second author was supported in part by NSA grant MDA904-99-1-0026 and by the Charles P. Taft Foundation.

Yang-Baxter equation with spectral parameter on meromorphic functions of two variables. They show that operators of the form

$$R(\lambda) = G(z_1 - z_2, \lambda)P - G(z_1 - z_2, \kappa)I$$

satisfy the equation

$$R_{12}(\lambda_1)R_{13}(\lambda_1 + \lambda_2)R_{23}(\lambda_2) = R_{23}(\lambda_2)R_{13}(\lambda_1 + \lambda_2)R_{12}(\lambda_1)$$

for any $\kappa \in \mathbb{F}$ if G is of the form

$$G(z, \lambda) = \frac{\theta'(0)\theta(\lambda + z)}{\theta(\lambda)\theta(z)}$$

and θ satisfies the equation

$$\theta(x + y)\theta(x - y)\theta(z + w)\theta(z - w) + \theta(x + z)\theta(x - z)\theta(y + w)\theta(w - y)$$
$$+ \theta(x + w)\theta(x - w)\theta(y + z)\theta(y - z) = 0$$

The solutions of this equation are well known (see for instance, [15]). They are of three types: elliptic, trigonometric and rational. These lead to operators of the respective types which we shall denote $R_e(\lambda)$, $R_t(\lambda)$, and $R_r(\lambda)$. Felder and Pasquier showed that the elliptic solutions when twisted and restricted to certain finite dimensional spaces yield the Belavin R-matrices. A similar procedure when applied to $R_t(\lambda)$, and $R_r(\lambda)$ yields trigonometric and rational degenerations of the Belavin R-matrices. In these two degenerate cases it is possible to take the limit as $\lambda \to \infty$ to obtain solutions of the constant Yang-Baxter equation For $R_t(\lambda)$ this yields (after the usual twisting and restriction) the Cremmer-Gervais R-matrices. For $R_r(\lambda)$ it yields the generalized Jordanian R-matrices referred to above. This situation can be summed up in the following way.

$$
\begin{array}{ccc}
R_e(\lambda) & & R_B(\lambda) \\
\downarrow & & \downarrow \\
R_t(\lambda) \longrightarrow R_t^\infty & \qquad & \hat{R}_{CG}(\lambda) \longrightarrow R_{CG} \\
\downarrow \qquad\quad \downarrow & & \downarrow \qquad\quad \downarrow \\
R_r(\lambda) \longrightarrow R_r^\infty & & R_{B,r}(\lambda) \longrightarrow R_p
\end{array}
$$

The matrices on the left are the operators on meromorphic functions. The vertical arrows denote the degeneration of θ from a true theta function to a trigonometric function and then a rational function. The horizontal arrows denote passing to the limit as the spectral parameter tends to infinity. The right hand diagram gives the corresponding finite dimensional R-matrices.

1. THE CONSTANT YANG-BAXTER EQUATION ON RATIONAL FUNCTION FIELDS

We begin by looking at operators of the form

$$R \cdot f(z_1, z_2) = \alpha(z_1/z_2)f(z_2, z_1) + \beta(z_1/z_2)f(z_1, z_2)$$

where $\alpha(x), \beta(x) \in \mathbb{F}(x)$. We wish to determine for which functions α and β the operator R satisfies the Yang-Baxter equation. For a rational function $\alpha(x) \in \mathbb{F}(x)$, we shall denote by $\tilde{\alpha}$ the function $\alpha(z_1/z_2) \in \mathbb{F}(z_1, z_2)$. With this notation we may write R as $\tilde{\alpha} P + \tilde{\beta} I$.

Lemma 1.1. *The operator $R = \tilde{\alpha} P + \tilde{\beta} I$ satisfies the Yang-Baxter equation if and only if the following two equations are satisfied (in the rational function field $\mathbb{F}(x, y)$)*

$$(1.1) \qquad \alpha(x)\alpha(y) = \alpha(xy^{-1})\alpha(y) + \alpha(x)\alpha(yx^{-1})$$

$$(1.2) \qquad \alpha(x)\alpha(y)^2 + \beta(y)\beta(y^{-1})\alpha(xy) = \alpha(x)^2\alpha(y) + \beta(x)\beta(x^{-1})\alpha(xy)$$

Proof. Applying both sides of the Yang-Baxter equation to an arbitrary function $f(z_1, z_2, z_3)$ and comparing coefficients of $f(z_{\sigma(1)}, z_{\sigma(2)}, z_{\sigma(3)})$ for $\sigma \in \Sigma_3$ shows that the Yang-Baxter equation holds if and only if the following equations are satisfied.

$$\alpha(z_1/z_2)\alpha(z_2/z_3)^2 + \beta(z_2/z_3)\alpha(z_1/z_3)\beta(z_3/z_2) =$$
$$\alpha(z_1/z_2)^2\alpha(z_2/z_3) + \beta(z_1/z_2)\alpha(z_1/z_3)\beta(z_2/z_1)$$

$$\alpha(z_2/z_3)\beta(z_1/z_2)\alpha(z_1/z_3) = \beta(z_1/z_2)\alpha(z_1/z_3)\alpha(z_2/z_1)$$
$$+ \alpha(z_1/z_2)\alpha(z_2/z_3)\beta(z_1/z_2)$$

$$\beta(z_2/z_3)\alpha(z_1/z_3)\alpha(z_3/z_2) + \beta(z_2/z_3)\alpha(z_2/z_3)\alpha(z_1/z_2)$$
$$= \beta(z_2/z_3)\alpha(z_1/z_2)\alpha(z_1/z_3)$$

The last two equations are both equivalent to 1.1 and the first one to 1.2. $\square$

Notice that if $\alpha(x)$ and $\beta(x)$ are solutions of the above equations then so are $\lambda\alpha(x^n)$ and $\lambda\beta(x^n)$ for any integer n and any $\lambda \in \mathbb{F}$. Moreover we may clearly multiply $\beta(x)$ by any member of $\Gamma = \{g(x) \in \mathbb{F}(x) \mid g(x)g(x^{-1}) = 1\}$ and $\beta(x)$ will still satisfy 1.2. Surprisingly, the solutions to 1.1 and 1.2 turn out to be almost unique modulo these adjustments.

Theorem 1.2. *If $\alpha \neq 0$ the operator $R = \tilde{\alpha} P + \tilde{\beta} I$ satisfies the Yang-Baxter equation if and only if, there exist $\lambda \in \mathbb{F}\backslash 0$, $\mu \in \mathbb{F}$, $k \in \mathbb{Z}$ and $g(x) \in \Gamma$ such that*

$$\alpha(x) = \frac{\lambda}{1 - x^k} \quad and \quad \beta(x) = \left(\frac{\lambda}{1 - x^k} + \mu\right) g(x).$$

Proof. The verification that α and β satisfy the hypotheses of the lemma is elementary. Conversely, suppose that α satisfies 1.1. If a is a zero of α then equation 1.1 implies that $\alpha(ay^{-1})\alpha(y) = 0$ which is impossible unless $a = 0$. Thus we may write α in the form

$$\alpha(x) = \frac{x^m}{p(x)}$$

where $p(x)$ is a polynomial with $p(0) \neq 0$. If b is a root of p then equation 1.1 implies that

$$x^m p(by^{-1})[b^m p(yb^{-1}) - p(y)] = 0$$

whence $p(y) = b^m p(yb^{-1})$. Thus if a and b are roots of p so is ab^{-1} and the roots form a group H of order, say, k. Since $p(0) \neq 0$, we must have $b^m = 1$ for all $b \in H$. Hence $k|m$ and $p(x) = p(xb)$ for all $b \in H$. This implies that $p(x) = \lambda(1 - x^k)^l$ for some integer l. Setting $y = x^{-1}$ in equation 1.1 and substituting yields $x^{m-lk} + x^m = (1 + x^k)^l$ so $l = 1$ and $m = 0$ or k. This implies that α has the required form.

If α and β satisfy equations 1.1 and 1.2 then it is easily seen that

$$\beta(x)\beta(x^{-1}) = \alpha(x)\alpha(x^{-1}) + \gamma$$

for some scalar γ. Moreover for any fixed γ, any two solutions for β differ by a factor $g(x) \in \Gamma$. If α and β are as in the statement of the Theorem, then

$$\beta(x)\beta(x^{-1}) = \alpha(x)\alpha(x^{-1}) + \mu(\alpha(x) + \alpha(x^{-1})) + \mu^2$$
$$= \alpha(x)\alpha(x^{-1}) + \mu\lambda + \mu^2.$$

Thus for any α the β satisfying 1.2 are precisely the functions $\beta(x) = (\alpha(x) + \mu)g(x)$. $\qquad\square$

Remark 1. This result remains true in the case of meromorphic functions.

The most interesting example is the simplest one; when $k = 1$, $g(x) = 1$ and the constants are chosen so that RP satisfies the Hecke condition $(RP - q)(RP + q^{-1}) = 0$. That is,

$$\alpha(x) = \frac{\hat{q}}{1 - x} \quad \text{and} \quad \beta(x) = \frac{q^{-1} - qx}{1 - x}.$$

In this case R restricts to one of the Cremmer-Gervais R-matrices on certain finite dimensional subspaces.

Recall that the Cremmer-Gervais R-matrices are a two-parameter family of solutions to the constant Yang-Baxter equation. If V is an n-dimensional vector space with basis $\{e_1, e_2, \ldots, e_n\}$, the Cremmer-Gervais operators are defined by [2, 12]

$$\rho_p(e_j \otimes e_i) = qp^{i-j}e_j \otimes e_i + \sum_k \hat{q}p^{i-k}\eta(i, j, k)e_k \otimes e_{i+j-k}$$

where q and p are non-zero elements of the base field $\mathbb{F}$, $\hat{q} = q - q^{-1}$ and

$$\eta(i, j, k) = \begin{cases} 1 & \text{if } i \leq k < j \\ -1 & \text{if } j \leq k < i \\ 0 & \text{otherwise} \end{cases}$$

Proposition 1.3. *Let R be the operator $\tilde{\alpha}P + \tilde{\beta}I$ where $\alpha(x) = \hat{q}/(1 - x)$ and $\beta(x) = (q^{-1} - qx)/(1 - x)$. Let V be an n-dimensional vector space and identify $V \otimes V$ with the subspace of $\mathbb{F}(z_1, z_2)$ of polynomials of degree less than or equal to n in both variables. Then R leaves $V \otimes V$ invariant and the matrix representing R with respect to the basis $z_1^j z_2^i$ is ρ_1.*

Proof. Notice that for any i and j,

$$\sum_k \eta(i,j,k)x^k = \frac{x^i - x^j}{1-x}$$

From this it follows easily that

$$R \cdot z_1^j z_2^i = q z_1^j z_2^i + \sum_k \eta(i,j,k) z_1^k z_2^{i+j-k}$$

as required.   □

Just as the more general Cremmer-Gervais operators ρ_p may be obtained from ρ_1 by a simple twisting method, so we may twist this operator R to obtain more general matrices R_p which restrict to ρ_p in the same way as R restricts to ρ_1. Fix $p \in \mathbb{F}\backslash 0$, and define an invertible operator F on $\mathbb{F}(z_1, z_2)$ by

$$F \cdot f(z_1, z_2) = f(p^{-1}z_1, pz_2)$$

Theorem 1.4. *Let $R = \tilde{\alpha}P + \tilde{\beta}I$ be a solution of the Yang-Baxter equation on $\mathbb{F}(z_1, z_2)$. Then $R_F = F_{21}^{-1}RF_{12}$ is also a solution of the Yang-Baxter equation. Moreover, when $\alpha(x) = \hat{q}/(1-x)$ and $\beta(x) = (q^{-1} - qx)/(1-x)$, then R_F again leaves $V \otimes V$ invariant and the matrix representing R_F with respect to the basis $z_1^j z_2^i$ is ρ_p.*

Proof. For the first assertion it suffices to check the relations

(1) $F_{12}F_{13}F_{23} = F_{23}F_{13}F_{12}$
(2) $R_{12}F_{23}F_{13} = F_{13}F_{23}R_{12}$
(3) $R_{23}F_{12}F_{13} = F_{13}F_{12}R_{23}$

The second assertion is verified exactly as in the previous proposition.   □

We now turn beiefly to consider the "additive" version of this construction. Namely we can look for operators of the form

$$R \cdot f(z_1, z_2) = \alpha(z_1 - z_2)f(z_2, z_1) + \beta(z_1 - z_2)f(z_1, z_2)$$

satisfying the Yang-Baxter equation. One sees analogously that R satisfies the Yang-Baxter equation if and only if

$$\alpha(x)\alpha(y) = \alpha(x-y)\alpha(y) + \alpha(x)\alpha(y-x)$$

$$\alpha(x)\alpha(y)^2 + \beta(y)\beta(-y)\alpha(x+y) = \alpha(x)^2\alpha(y) + \beta(x)\beta(-x)\alpha(x+y)$$

Theorem 1.5. *The operator $R = \check{\alpha}P + \check{\beta}I$ satisfies the Yang-Baxter equation if and only if there exist λ, $\mu \in \mathbb{F}$, and $g(x) \in \Gamma'$ such that*

$$\alpha(x) = -\frac{\lambda}{x} \quad and \quad \beta(x) = \left(\mu + \frac{\lambda}{x}\right)g(x).$$

For this operator,

$$R \cdot z_1^j z_2^i = \mu z_1^j z_2^i + \sum_k \eta(i,j,k) z_1^k z_2^{i+j-k-1}$$

Hence R again leaves the subspaces $V \otimes V$ invariant and on these spaces becomes the operator

$$R(e_j \otimes e_i) = \mu e_j \otimes e_i + \sum_k \lambda \eta(i, j, k) e_k \otimes e_{i+j-k-1}.$$

One can again twist using an additive version of the twist described in 1.4. Restricting these matrices to finite dimensional subspaces yields quantizations of the boundary classical r-matrices discovered by Gerstenhaber and Giaquinto [9, 10] (see [5] for details).

If we replace $\mathbb{F}(z_1, z_2)$ by meromorphic functions in two variables, then replacing x by $e^{2\pi i z}$ in the funcions of Theorem 1.2 clearly leads to solutions of the additive form. In particular the simplest such solution yields

$$\alpha(z) = \frac{\sin \pi h}{\sin \pi z}, \quad \beta(z) = \frac{\sin \pi(h + z)}{\sin \pi z}.$$

Because R is Hecke, The family of operators

$$R(\lambda) = e^{-\pi i \lambda} R - e^{\pi i \lambda} R^{-1}$$

will satisfy the general Yang-Baxter equation. After multiplying by a suitable scalar, R will be of the particularly symmetric form $R(\lambda) = \check{\alpha} I + \check{\beta} P$ where

$$\alpha(z) = \frac{\sin \pi(z + \lambda)}{\sin \pi z} \sin \pi \lambda, \quad \beta(z) = \frac{\sin \pi(z + h)}{\sin \pi z} \sin \pi h.$$

This is the trigonometric version of the R operator of Shibukawa and Ueno discussed in the introduction.

Finaly let us note an interesting connection between the symmetric algebra associated to infinite dimensional Cremmer-Gervais operator and the affine quantum universal enveloping algebra $U_q(\hat{\mathfrak{sl}}(2))$. Let $V = \mathbb{F}[z]$ and let R be the operator on $V \otimes V$ described in 1.3. As usual we can form the associated q-symmetric algebra as $S_{R,q}(V) = T(V)/((RP - q)(V \otimes V))$. Now for any $f \in V \otimes V = \mathbb{F}[z_1, z_2]$, we have $(R - q)f = \beta(P - I)f$. From this it follows that the elements of $(R - q)(V \otimes V)$ are of the form $(qz_2 - q^{-1}z_1)g$ where $g(z_1, z_2)$ is symmetric (in the usual sense). Now,

$$(qz_2 - q^{-1}z_1)(z_1^k z_2^l + z_1^l z_2^k) = qz_1^k z_2^{l+1} - q^{-1}z_1^{k+1} z_2^l + qz_1^l z_2^{k+1} - q^{-1}z_1^{l+1} z_2^k.$$

Hence if X_k denotes the image of z^k in $S_{R,q}(\mathbb{F}[z])$, then the relations defining the symmetric algebra are

$$X_{k+1}X_l - q^2 X_k X_{l+1} = q^2 X_l X_{k+1} - X_{l+1}X_k$$

These are precisely the relations defining the subalgebra U^+ of $U_q(\hat{\mathfrak{sl}}(2))$ [3]. Hence we have $S_{R,q}(\mathbb{F}[z]) \cong U^+$. It would be interesting to have a more abstract explanation for this isomorphism.

REFERENCES

[1] A. Bilal and J.-L. Gervais, Systematic constructions of conformal theories with higher spin Virasoro symmetries, *Nucl. Phys. B.*, 318 (1989).

[2] E. Cremmer and J.-L. Gervais, The quantum group structure associated with non-linearly extended Virasoro algebras, *Comm. Math. Phys.*, 134 (1990), 619-632.

[3] V. G. Drinfeld, New realization of Yangians and affine algebras, Soviet Math. Dokl., 36 (1988), 212-216.

[4] E. Demidov, Y. I. Manin, E. E. Mukhin and D. V. Zhdanovich, Non-standard quantum deformations of $GL(n)$ and constant solutions of the Yang-Baxter equation, *Progr. Theor. Phys. Suppl.* 102 (1990), 203-218.

[5] R. Endelman and T. J. Hodges, Generalizations of the Jordanian R-matrix of Cremmer-Gervais type, preprint.

[6] P. Etingof and A. Varchenko, Solutions of the quantum dynamical Yang-Baxter equation and dynamical quantum groups, *Comm. Math. Phys.* 196 (1998), 591–640.

[7] P. Etingof and A. Varchenko, Exchange dynamical quantum groups, q-alg/9801135.

[8] G. Felder and V. Pasquier, A simple construction of elliptic R-matrices, *Lett. Math. Phys.*, 32 (1994), 167-171.

[9] M. Gerstenhaber and A. Giaquinto, Boundary solutions of the classical Yang-Baxter equation.

[10] M. Gerstenhaber and A. Giaquinto, Boundary solutions of the quantum Yang-Baxter equation and solutions in three dimensions.

[11] T. J. Hodges, The Cremmer-Gervais solution of the Yang Baxter equation, *Proc. Amer. Math. Soc.*, 127 (1999), 1819-1826.

[12] T. J. Hodges, On the Cremmer Gervais quantizations of $SL(n)$, *Int. Math. Res. Notices*, 10 (1995), 465-481.

[13] M. Jimbo, H. Konno, S. Odake and J. Shiraishi, Quasi-Hopf twistors for elliptic quantum groups, q-alg/9712029.

[14] Y. Shibukawa and K. Ueno, Completely symmetric R matrix, *Lett. Math. Phys.*, 25 (1992), 239-248.

[15] E. T. Whittaker and G. N. Watson, *A Course of Modern Analysis*, Cambridge, 1999.

UNIVERSITY OF CINCINNATI, CINCINNATI, OH 45221-0025, U.S.A.
E-mail address: jintai.ding@uc.edu

UNIVERSITY OF CINCINNATI, CINCINNATI, OH 45221-0025, U.S.A.
E-mail address: timothy.hodges@uc.edu

Noncommutative Differential Geometry and Twisting of Quantum Groups

Shahn Majid

School of Mathematical Sciences, Queen Mary and Westfield College
University of London, Mile End Rd, London E1 4NS[1]

Abstract We outline the recent classification of differential structures for all main classes of quantum groups. We also outline the algebraic notion of 'quantum manifold' and 'quantum Riemannian manifold' based on quantum group principal bundles, a formulation that works over general unital algebras.

1 Introduction

There have been many attempts in the last decades to arrive at a theory of noncommutative geometry applicable to 'coordinate' algebras that are not necessarily commutative, notably that of A. Connes coming out of abstract C^*-algebra theory in the light of the Gelfand-Naimark and Serre-Swan theorems. One has tools such as cyclic cohomology and examples such as the noncommutative torus and other foliation C^*-algebras. Another 'bottom up' approach, which we outline, is based on the idea that the theory should be guided by the inclusion of the large vein of 'naturally occuring' examples, the coordinate algebras of the quantum groups $U_q(g)$ in particular, and Hopf algebras in general, whose validity for several branches of mathematics has already been established. This is similar to the key role that Lie groups played in the development of modern differential geometry. Much progress has been made in recent years and there is by now (at least at the algebraic level) a more or less clear formulation of 'quantum manifold' suggested by this approach. After being validated on the q-deformation examples such as quantum groups, quantum homogeneous spaces etc, one can eventually apply the theory quite broadly to a wide range of unital algebras. The approach will be algebraic, although not incompatible with C^* completions at a later stage. In particular, as a bonus, one can apply the theory to finite-dimensional algebra, i.e. to discrete classical and quantum systems.

An outline of the paper is the following. We start with the lowest level structure which (in our approach) is the choice of differential structure. This

---

[1]Reader and Royal Society University Research Fellow

is the topic of Section 2 where we outline the recently achieved more or less complete classification results. In Section 3 we develop the notion of 'quantum manifold' [1] based on noncommutative frame bundles with quantum group fibre. Usual ideas of 'sheaf theory' and 'local trivialisations' do not work in this setting, but from [2] one has global algebraic replacements. There is also an appropriate notion of automorphism or diffeomorphism quantum groups [3].

2   Quantum differential forms

Let M be a unital algebra, which we consider as playing the role of 'co-ordinates' in algebraic geometry, except that we do not require the algebra to be commutative. The appropriate notion of cotangent space or differential 1-forms in this case is

1. Ω^1 an M-bimodule
2. $\mathrm{d} : M \to \Omega^1$ a linear map obeying the Leibniz rule $\mathrm{d}(ab) = a\mathrm{d}b + (\mathrm{d}a)b$ for all $a, b \in M$.
3. The map $M \otimes M \to \Omega^1$, $a \otimes b \mapsto a\mathrm{d}b$ is surjective.

Differential structures are not unique even classically, and even more non-unique in the quantum case. There is, however, one universal example of which others are quotients. This is

$$\Omega^1_{\mathrm{univ}} = \ker \cdot \subset M \otimes M, \quad \mathrm{d}a = a \otimes 1 - 1 \otimes a. \tag{1}$$

This is common to more or less all approaches to noncommutative geometry.

The main feature here is that, in usual algebraic geometry, the multiplication of forms Ω^1 by 'functions' M is the same from the left or from the right. However, if $a\mathrm{d}b = (\mathrm{d}b)a$ then by axiom 2. we have $\mathrm{d}(ab - ba) = 0$, i.e. we cannot naturally suppose this when M is noncommutative. We say that a differential calculus is noncommutative or 'quantum' if the left and right multiplication of forms by functions do not coincide.

When M has a Hopf algebra structure with coproduct $\Delta : M \to M \otimes M$ and counit $\epsilon : M \to k$ (k the ground field), we say that Ω^1 is *bicovariant* if [4]

4. Ω^1 is a bicomodule with coactions $\Delta_L : \Omega^1 \to M \otimes \Omega^1, \Delta_R : \Omega^1 \to \Omega^1 \otimes M$ bimodule maps (with the tensor product bimodule structure on the target spaces, where M is a bimodule by left and right multiplication).
5. d is a bicomodule map with the left and right regular coactions on M provided by Δ.

A morphism of calculi means a bimodule and bicomodule map forming a commuting triangle with the respective d maps. One says [5] that a calculus is *coirreducible* if it has no proper quotients. Whereas the translation-invariant calculus is unique classically, in the quantum group case we have at least complete classification results in terms of representation theory [5]. The dimension of a calculus is that of its space of (say) left-invariant 1-forms, which can be viewed as generating the rest of the calculus as a right M-module. Similarly with left and right interchanged.

We note also that in the bicovariant case there is a natural extension [4] from Ω^1 to Ω^n with $d^2 = 0$. This is defined as the tensor algebra over M generated by Ω^1 modulo relations defined by a braiding which acts by a simple transposition on left and right invariant differential 1-forms. Other extensions are also possible and in general the differential structure can be specified order by order. Given the extension, one has a quantum DeRahm cohomology defined in the usual way as closed forms modulo exact ones. Apart from cohomology one can also start to do '$U(1)$' gauge theory with trivial bundles, where a gauge field is just a differential form $\alpha \in \Omega^1$ and its curvature is $F = d\alpha + \alpha \wedge \alpha$, etc. A gauge transform is

$$\alpha^\gamma = \gamma^{-1}\alpha\gamma + \gamma^{-1}d\gamma, \quad F^\gamma = \gamma^{-1}F\gamma \tag{2}$$

for any invertible 'function' $\gamma \in M$, and so on. One can define then the space of flat connections as those with $F = 0$ modulo gauge transformation. This gives two examples of 'geometric' invariants which work therefore for general algebras equipped with differential structure.

2.1 $M = k[x]$

For polynomials in one variable the coirreducible calculi have the form [6]

$$\Omega^1 = k_\lambda[x], \quad df(x) = \frac{f(x + \lambda) - f(x)}{\lambda},$$

$$f(x) \cdot g(\lambda, x) = f(x + \lambda)g(\lambda, x), \quad g(\lambda, x) \cdot f(x) = g(\lambda, x)f(x)$$

for functions f and one-forms g. Here k_λ is a field extension of the form $k[\lambda]$ modulo $m(\lambda) = 0$ and m is an irreducible monic polynomial. The dimension of the calculus is the order of the field extension or the degree of m.

For example, the calculi on $\mathbb{C}[x]$ are classified by $\lambda_0 \in \mathbb{C}$ (here $m(\lambda) = \lambda - \lambda_0$) and one has

$$\Omega^1 = dx\mathbb{C}[x], \quad df = dx\frac{f(x + \lambda_0) - f(x)}{\lambda_0}, \quad xdx = (dx)x + \lambda_0.$$

We see that the Newtonian case $\lambda_0 = 0$ is only one special point in the moduli space of quantum differential calculi. But if Newton had not supposed that differentials and forms commute he would have had no need to take this limit. What one finds with noncommutative geometry is that there is no need to take this limit at all. It is also interesting that the most important field extension in physics, $\mathbb{R} \subset \mathbb{C}$, can be viewed noncommutative-geometrically with complex functions $\mathbb{C}[x]$ the quantum 1-forms on the algebra of real functions $\mathbb{R}[x]$. There is nontrivial quantum DeRahm cohomology in this case.

2.2 $M = \mathbb{C}[G]$

For the coordinate algebra of a finite group G (for convenience we work over $\mathbb{C}$) the coirreducible calculi correspond to nontrivial conjugacy classes $\mathcal{C} \subset G$ and have the form

$$\Omega^1 = \mathcal{C} \cdot \mathbb{C}[G], \quad \mathrm{d}f = \sum_{g \in \mathcal{C}} g \cdot (L_g(f) - f), \quad f \cdot g = g \cdot L_g(f)$$

where $L_g(f) = f(g \cdot)$ is the translate of $f \in \mathbb{C}[G]$. The dimension of the calculus is the order of the conjugacy class.

For the coordinate algebra $\mathbb{C}[G]$ of a Lie group with Lie algebra $\mathfrak{g}$ the coirreducible calculi correspond to maximal ideals in $\ker \epsilon$ stable under the adjoint coaction. Or in a natural reformulation [5] in terms of quantum tangent spaces the correspondence is with irreducible Ad-invariant subspaces of the enveloping algebra $\ker \epsilon \subset U(\mathfrak{g})$ which are stable under the coaction $\Delta_L = \Delta - \mathrm{id} \otimes 1$ of $U(\mathfrak{g})$. For example $\mathfrak{g}$ itself defines the standard translation-invariant calculus and this is coirreducible when $\mathfrak{g}$ is semisimple.

2.3 $M = \mathbb{C}G$

For the group algebra of a nonAbelian finite group G, we definitely need the machinery of noncommutative geometry since M itself is noncommutative. We regard these group algebras 'up side down' as if coordinates, i.e. we describe the geometry of the noncommutative space $\hat{G}$ in some sense. The above definitions make sense and differential structures abound. The coirreducible calculi correspond to pairs (V, ρ, λ) where (V, ρ) is a nontrivial irreducible representation and $\lambda \in V/\mathbb{C}$ [5]. They have the form

$$\Omega^1 = V \cdot \mathbb{C}G, \quad \mathrm{d}g = ((\rho(g) - 1)\lambda) \cdot g, \quad g \cdot v = (\rho(g)v) \cdot g$$

where $g \in G$ is regarded as a 'function'. The dimension of the calculus is that of V. The minimum assumption for merely a differential calculus is that λ should be cyclic.

For $M = U(\mathfrak{g})$ (the Kirillov-Kostant quantisation of $\mathfrak{g}^*$) one has a similar construction for any irreducible representation V of the Lie algebra $\mathfrak{g}$ and choice of ray λ in it. Then

$$\Omega^1 = V \cdot U(\mathfrak{g}), \quad \mathrm{d}\xi = \rho(\xi)\lambda, \quad \xi \cdot v = \rho(\xi)v + v \cdot \xi$$

where $\xi \in \mathfrak{g}$ is regarded as a 'function'. The dimension is again that of V.

For example, let $\mathfrak{g} = b_+$ be the 2-dimensional Lie algebra with $[x, t] = x$. Let V be the 2-dimensional representation with matrix and ray vector

$$\rho(t) = \begin{pmatrix} 0 & 0 \\ 0 & 1 \end{pmatrix}, \quad \rho(x) = \begin{pmatrix} 0 & 1 \\ 0 & 0 \end{pmatrix}, \quad \lambda = \begin{pmatrix} 0 \\ 1 \end{pmatrix}.$$

Then $dt = \lambda$ and dx are the usual basis of V and obey

$$[t, dx] = [x, dx] = 0, \quad [t, dt] = dt, \quad [x, dt] = dx.$$

Replacing x trivially by a vector x_i, $i = 1, 2, 3$ one obtains similarly a natural candidate for noncommutative Minkowski space along with its differential structure. It has measurable astronomical predictions [7].

This covers the classical objects or their duals viewed 'up side down' as noncommutative spaces. For a finite-group bicrossproduct $\mathbb{C}[M] \blacktriangleright\!\!\triangleleft \mathbb{C}G$ the classification is a mixture of the two cases above and is given in [8]. The Lie version remains to be worked out in detail. The important example of the Planck scale quantum group $\mathbb{C}[x] \blacktriangleright\!\!\triangleleft \mathbb{C}[p]$, however, is a twisting by a cocycle of its classical limit $\mathbb{C}[\mathbb{R} \bowtie \mathbb{R}]$ and is therefore covered by a later subsection.

2.4 Proofs

The above cases are all sufficiently elementary that they can be easily worked out using the following simple observations known essentially (in some form or other) since [4]. We suppose for convenience that H has invertible antipode.

1. $\ker \epsilon \subset M$ is an object in the braided category of left crossed M-modules (i.e. modules over the quantum double $D(M)$ in the finite-dimensional case) by multiplication and the left adjoint coaction.

2. The isomorphism $M \otimes M \cong M \otimes M$ given by $a \otimes b \mapsto (\Delta a)b$ restricts to an isomorphism $\Omega^1_{\text{Univ}} \cong \ker \epsilon \otimes M$ of bimodules and of bicomodules, where the right hand side is a right (co)module by the (co)product of M and a left (co)module by the tensor product of the (co)action on $\ker \epsilon$ and the (co)product of M.

This implies that every other bicovariant Ω^1 is of the form $\Omega^1 \cong \Omega_0 \otimes M$ where Ω_0 is a quotient object of $\ker \epsilon$ in the category of crossed M-modules. I.e. the calculi correspond to ideals in $\ker \epsilon$ stable under the adjoint coaction. Given Ω^1 the space Ω_0 is given by the right-invariant differentials. In categorical terms the braided category of bicovariant M-bimodules as featuring above (i.e. bimodules which are also bicomodules with structure maps being bimodule maps) can be identified with that of crossed M-modules, under which the Hopf module for the universal calculus corresponds to $\ker \epsilon$.

When this is combined with the notion of coirreducibility and with the Peter-Weyl decomposition of an appropriate type for $\ker \epsilon$, one obtains the classification results above. These latter steps have been introduced by the author [5] (before that one found only sporadic examples of calculi on particular quantum groups, usually close to the unique classical calculus.)

We also note that for any finite-dimensional bicovariant calculus the map $d : M \to \Omega_0 \otimes M$ can be viewed as a 'partial derivative' $\partial_x : M \to M$ for each $x \in \Omega_0^*$. The space Ω_0^* is called the invariant 'quantum tangent space' and is often more important than the 1-forms in applications. These ∂_x are not derivations but together form a braided derivation in the braided category of M-crossed modules (there is a braiding as $x \in \Omega_0^*$ passes $a \in M$) [5].

2.5 Cotriangular quantum groups and twisting of calculi

We recall [9] that if M is a quantum group and $\chi : M \otimes M \to k$ a cocycle in the sense

$$\chi(b_{(1)} \otimes c_{(1)})\chi(a \otimes b_{(2)}c_{(2)}) = \chi(a_{(1)} \otimes b_{(1)})\chi(a_{(2)}b_{(2)} \otimes c),$$

$$\chi(1 \otimes a) = \epsilon(a), \quad \forall a, b, c \in M$$

then there is a 'twisted' quantum group M^χ with product

$$a \cdot_\chi b = \chi(a_{(1)} \otimes b_{(1)})a_{(2)}b_{(2)}\chi^{-1}(a_{(3)} \otimes b_{(3)})$$

and unchanged unit, counit and coproduct. Here χ^{-1} is the inverse in $(M \otimes M)^*$, which we assume, and $\Delta a = a_{(1)} \otimes a_{(2)}$, etc., is a notation.

Theorem 1 *[10] The bicovariant differentials $\Omega^1(M^\chi)$ are in 1-1 correspondence with those of M.*

In fact the entire exterior algbera in the bicovariant case is known to be a super-Hopf algebra (Brzezinski's theorem) and that of M^χ is the twist of that of M when χ is trivially extended to a cocycle on the latter. The more direct proof involves the following:

Theorem 2 *[10] There is an equivalence $\mathcal{F}$ of braided monoidal categories from left M-crossed modules to left M^χ-crossed modules given by the functor*

$$\mathcal{F}(V, \triangleright, \Delta_L) = (V, \triangleright^\chi, \Delta_L),$$

$$a \triangleright^\chi v = \chi(a_{(1)} \otimes v^{(\bar{1})})(a_{(2)} \triangleright v^{(\bar{2})})^{(\bar{2})}\chi^{-1}((a_{(2)} \triangleright v^{(\bar{2})})^{(\bar{1})} \otimes a_{(3)}), \quad \forall a \in M, \ v \in V,$$

where $\triangleright$ denotes the action and $\Delta_L v = v^{(\bar{1})} \otimes v^{(\bar{2})}$ is a notation. There is an associated natural transformation

$$c_{V,W} : \mathcal{F}(V) \otimes \mathcal{F}(W) \cong \mathcal{F}(V \otimes W), \quad c_{V,W}(v \otimes w) = \chi(v^{(\bar{1})} \otimes w^{(\bar{1})})v^{(\bar{2})} \otimes w^{(\bar{2})}.$$

As a corollary we deduce by Tannaka-Krein reconstruction arguments:

Corollary 3 *[10] When M is finite-dimensional the dual of the Drinfeld double $D(M^\times)$ is isomorphic to a twist of the dual of the Drinfeld double $D(M)$ by χ^{-1} viewed on $D(M) \otimes D(M)$.*

Here we use the theorem [11] that an equivalence of comodule categories respecting the forgetful functor corresponds to a twist of the underlying quantum groups. The corollary itself can then be verified directly at an algebraic level once the required (nontrivial) isomorphism has been found in this way. Of course one can state it also in terms of Drinfeld's coproduct twists.

Starting with a classical (commutative) Hopf algebra such a twist yields a cotriangular one and (from recent work of Etingof and Gelaki [12]) every finite-dimensional cotriangular Hopf algebra in the (co)semisimple case over k algebraically closed is of this form. Hence the differential calculus in this case reduces by the above theorem to the classification in the classical cases considered in previous sections. There are many other instances where an important quantum group is a twisting of another – the theorem provides its differential calculus from that of the other.

2.6 Factorisable quantum groups $M = \mathbb{C}_q[G]$

Finally, we come to the standard quantum groups $\mathbb{C}_q[G]$ dual to the Drinfeld-Jimbo $U_q(\mathfrak{g})$. Here [5] the coirreducible calculi are essentially provided by nontrivial finite-dimensional irreducible right comodules V of the quantum group (i.e. essentially by the irreducible representations of the Lie algebra) and have the form

$$\Omega^1 = \mathrm{End}(V) \cdot \mathbb{C}_q[G], \qquad \mathrm{d}a = \rho_+(a_{(1)}) \circ \rho_-(Sa_{(2)}) \cdot a_{(3)} - \mathrm{id} \cdot a,$$
$$a \cdot \phi = \rho_+(a_{(1)}) \circ \phi \circ \rho_-(Sa_{(2)}) \cdot a_{(3)} \tag{3}$$

for all $\phi \in \mathrm{End}(V)$, where

$$\rho_+(a)v = v^{(1)} \mathcal{R}(a \otimes v^{(2)}), \quad \rho_-(a)v = v^{(1)} \mathcal{R}^{-1}(v^{(2)} \otimes a)$$

and $\mathcal{R} : \mathbb{C}_q[G] \otimes \mathbb{C}_q[G] \to \mathbb{C}$ is the dual-quasitriangular structure [9]. S denotes the antipode. The construction works for any dual-quasitriangular Hopf algebra with factorisable $\mathcal{R}$ (the minimum one needs for a differential calculus is that $\mathcal{Q}_\rho(a) = \rho_+(a_{(1)})\rho_-(Sa_{(2)})$ is surjective) and gives a classification of calculi if M has in addition the Peter-Weyl property that $M = \oplus_V V \otimes V^*$ as left and right M-comodules. This is the result in [5, Thm. 4.3] cast in a purely comodule form. Or in the original formulation in terms of quantum tangent spaces the correspondence is given in more familiar terms with representations ρ of the quantum enveloping algebra and

$$\rho_+(a) = (a \otimes \rho)(\mathcal{R}), \quad \rho_-(a) = (\rho \otimes a)(\mathcal{R}^{-1}),$$

where $\mathcal{R}$ is the quasitriangular structure or universal R-matrix of $U_q(\mathfrak{g})$. For finite-dimensional representations only a finite number of terms in its powerseries contribute here, i.e. there are no convergence problems.

The factorisability etc. hold formally for $\mathbb{C}_q[G]$ so that although one has one natural calculus for each irreducible representation there are also some 'shadows' or technical variants allowed according to the precise formulation of the relevant quantum groups and their duality (this is more a deficit in the technical definitions than anything else). The latter aspect has been subsequently clarified in [13] [14] following our basic result in [5].

For the sake of a sketch of the proof of the basic result assume that M is strictly factorisable dual-quasitriangular and has Peter-Weyl decomposition in terms of irreducible representations (V, ρ) of a suitable dual Hopf algebra U. Classifying the quotient M-crossed modules of $\ker \epsilon$ is equivalent essentially to classifying the subobjects of $\ker \epsilon \subset U$ as U-crossed modules (the quantum tangent spaces). When U is strictly factorisable its quantum double $D(U)$ is isomorphic to $U \bowtie U$ which, as an algebra, is a tensor product (the coproduct is twisted). Hence U-crossed modules are equivalent to two U-modules. Next, under the isomorphism of linear spaces $U \cong M$ assumed under strict factorisability, this is the same as classifying subobjects of $\ker \epsilon \subset M$. The U-crossed module structure on this final $\ker \epsilon \subset M$ under this chain of reasoning is simply evaluation against M coacting independently from the left and the right (viewed from the left via the antipode). This is just the action with respect to which the assumed Peter-Weyl decomposition $M = \oplus_V \mathrm{End}(V)$ is a decomposition into irreducibles as V runs over the irreducible representations of U. One may make a similar proof working only with M-crossed modules throughout and the corresponding comodule Peter-Weyl decomposition.

Proposition 4 *[5] The quantum tangent spaces $\Omega_0^* = V^* \otimes V$ for the above differential calculi on $\mathbb{C}_q[G]$ are braided-Lie algebras in the sense of [15]. The action of basis element $f^i \otimes e_i$ is*

$$\partial_{x^i{}_j}(a) = \langle x^i{}_j, a_{(1)} \rangle a_{(2)}, \quad x^i{}_j = \mathcal{Q}(\rho^i{}_j) - \delta^i{}_j$$

where $\mathcal{Q} : \mathbb{C}_q[G] \to U_q(\mathfrak{g})$ is defined by $\mathcal{R}_{21}\mathcal{R}$ and $\rho^i{}_j \in \mathbb{C}_q[G]$ are the matrix elements in the representation V with basis $\{e_i\}$ and dual basis $\{f^i\}$.

Recall that the dual of any quantum group acts on the quantum group by the 'coregular representation' in the manner shown, in our case by the $x^i{}_j$. These and 1 together form a braided-Lie algebra. This is described by a system of axioms in any braided category including a pentagonal 'braided-Jacobi' identity. Moreover, such objects have braided enveloping algebras which, for usual Lie algebras $\mathfrak{g}$, returns a homogenized version of $U(\mathfrak{g})$. In our case it returns a quadratic and braided version of $U_q(\mathfrak{g})$, i.e. this solved (some years ago [15]) the Lie problem for such quantum groups. The above gives its geometric interpretation.

For the example of $\mathbb{C}_q[SU_2]$ there is basically one bicovariant calculus for each spin j with dimension $(2j+1)^2$. The lowest corresponds to the 4-dimensional braided-Lie algebra $gl_{q,2}$ spanned by

$$x = \begin{pmatrix} q^H - 1 & q^{-\frac{1}{2}}(q - q^{-1})q^{\frac{H}{2}}X_- \\ q^{-\frac{1}{2}}(q - q^{-1})X_+q^{\frac{H}{2}} & q^{-H} - 1 + q^{-1}(q - q^{-1})^2 X_+ X_- \end{pmatrix}$$

in the usual notations for $U_q(su_2)$. This braided-Lie algebra is irreducible for generic q but as $q \to 1$ it degenerates into $su_2 \oplus u(1)$. The partial derivatives degenerate into the usual invariant vector fields on SU_2 and an additional 2nd order operator which turns out to be the Casimir or Laplacian.

2.7 Discrete manifolds

To close with one non-quantum group example, consider any actual manifold with a finite good cover $\{U_i\}_{i\in I}$. Instead of building geometric invariants on a manifold and studying them modulo diffeomorphisms we can use the methods above to first pass to the skeleton of the manifold defined by its open set structure and do differential geometry directly on this indexing set I. Thus we take $M = \mathbb{C}[I]$ which just means collections $\{f_i \in \mathbb{C}\}$. The universal Ω^1 is just matrices $\{f_{ij}\}$ vanishing on the diagonal. We use the intersection data for the open sets to set some of these to zero. Similarly for higher forms. Thus [16]

$$\Omega^1 = \{f_{ij}|\ U_i \cap U_j \neq \emptyset\}, \quad \Omega^2 = \{f_{ijk}|\ U_i \cap U_j \cap U_k \neq \emptyset\}$$

$$(df)_{ij} = f_i - f_j, \quad (df)_{ijk} = f_{ij} - f_{ik} + f_{jk}$$

and so on. Then one has that the quantum cohomology is just the additive Cech cohomology of the original manifold. Similarly, one has that the zero curvature gauge fields modulo gauge transformations recovers again the first Cech cohomology, but now in a multiplicative form.

3 Bundles and connections

The next layer of differential geometry is bundles, connections, etc. Usually in physics one needs only the local picture with trivial bundles in each open set – but for a general noncommutative algebra M there may be no reasonable 'open sets' and one has therefore to develop the global picture from the start. This is needed for example to describe the frame bundle of a topologically nontrivial 'manifold'. It also turns out to be rather easier to do the gauge theory beyond the '$U(1)$' case (i.e. with a nontrivial quantum structure group and nonuniversal calculus on it) if one takes the global point of view, even if the bundle itself is trivial. We take a Hopf algebra H in the role of 'functions' on the structure group of the bundle. To keep things simple we concentrate on the universal differential calculus but it is important that the general case is also covered by making suitable quotients. Recall that a classical bundle has a free action of a group on the total space P and a local triviality property. In our algebraic terms we need [2]:

1. An algebra P and a coaction $\Delta_R : P \to P \otimes H$ of the quantum group H such that the fixed subalgebra is M,

$$M = P^H = \{p \in P| \; \Delta_R p = p \otimes 1\}. \tag{4}$$

We assume that P is flat as an M-module.

2. The sequence

$$0 \to P(\Omega^1 M)P \to \Omega^1 P \xrightarrow{\text{ver}} P \otimes \ker \epsilon \to 0 \tag{5}$$

is exact, where $\text{ver} = (\cdot \otimes \text{id})\Delta_R$.

The map ver plays the role of generator of the vertical vector fields corresponding classically to the action of the group (for each element of H^* it maps $\Omega^1 P \to P$ like a vector field). Exactness on the left says that the one-forms $P(\Omega^1 M)P$ lifted from the base are exactly the ones annihilated by the vertical vector fields. In the universal calculus case this can be formulated as a Hopf-Galois extension, a condition arising in other contexts in Hopf algebra theory also. The differential geometric picture is more powerful and includes general calculi when we use the right-handed version of Ω_0 in place of $\ker \epsilon$.

One can then define a connection as an equivariant splitting

$$\Omega^1 P = P(\Omega^1 M)P \oplus \text{complement} \tag{6}$$

i.e. an equivariant projection Π on $\Omega^1 P$. One can show [2] the required analogue of the usual theory, i.e. that such a projection corresponds to a connection form such that

$$\omega : \ker \epsilon \to \Omega^1 P, \quad \text{ver} \circ \omega = 1 \otimes \text{id} \tag{7}$$

where ω intertwines with the adjoint coaction of H on itself. Finally, if V is a vector space on which H coacts then we define the associated 'bundles' $E^* = (P \otimes V)^H$ and $E = \text{hom}^H(V, P)$, the space of intertwiners. The two bundles should be viewed geometrically as 'sections' in classical geometry of bundles associated to V and V^*. Given a suitable (so-called strong) connection one has a covariant derivative [2]

$$D_\omega : E \to \Omega^1 M \underset{M}{\otimes} E, \quad D_\omega = (\text{id} - \Pi) \circ \text{d}. \tag{8}$$

All of this can be checked out for the example of the q-monopole bundle over the q-sphere [2]. Recall that classically the inclusion $U(1) \subset SU_2$ in the diagonal has coset space S^2 and defines the $U(1)$ bundle over the sphere on which the monopole lives. In our case the coordinate algebra of $U(1)$ is the polynomials $\mathbb{C}[g, g^{-1}]$ and the classical inclusion becomes the projection

$$\pi : \mathbb{C}_q[SU_2] \to \mathbb{C}[g, g^{-1}], \quad \pi \begin{pmatrix} a & b \\ c & d \end{pmatrix} = \begin{pmatrix} g & 0 \\ 0 & g^{-1} \end{pmatrix}$$

Its induced coaction $\Delta_R = (\text{id} \otimes \pi)\Delta$ is by the degree defined as the number of a, c minus the number of b, d in an expression. The quantum sphere $\mathbb{C}_q[S^2]$ is the fixed subalgebra i.e. the degree zero part. Explicitly, it is generated by $b_3 = ad$, $b_+ = cd$, $b_- = ab$ with q-commutativity relations

$$b_\pm b_3 = q^{\pm 2} b_3 b_\pm + (1 - q^{\pm 2}) b_\pm, \quad q^2 b_- b_+ = q^{-2} b_+ b_- + (q - q^{-1})(b_3 - 1)$$

and the sphere equation $b_3^2 = b_3 + q b_- b_+$. When $q \to 1$ we can write $b_\pm = \pm(x \pm iy)$, $b_3 = z + \frac{1}{2}$ and the sphere equation becomes $x^2 + y^2 + z^2 = \frac{1}{4}$ while the others become that x, y, z commute. One may verify that we have a quantum bundle in the sense above and that there is a connection $\omega(g-1) = dda - q b dc$ which, as $q \to 1$, becomes the usual Dirac monopole constructed algebraically. If we take $V = k$ with coaction $1 \mapsto 1 \otimes g^n$, the sections of the associated vector bundles E_n for each charge n are just the degree n parts of $\mathbb{C}_q[SU_2]$. The associated covariant derivative acts on these.

This example demonstrates compatibility with the more traditional C^*-algebra approach of A. Connes [17] and others. Traditionally a vector bundle over any algebra is defined as a finitely generated projective module. However, there was no notion of quantum principal bundle before quantum groups.

Proposition 5 *[18] The associated bundles E_n for the q-monopole bundle are finitely generated projective modules, i.e. there exist*

$$e_n \in M_{|n|+1}(\mathbb{C}_q[S^2]), \quad e_n^2 = e_n, \quad E_n = e_n \mathbb{C}_q[S^2]^{|n|+1}.$$

The covariant derivative for the monopole has the form $e_n d e_n$. The classes $[e_n]$ are elements of the noncommutative K-theory $K_0(\mathbb{C}_q[S^2])$ and have nontrivial duality pairing with cyclic cohomology, hence the q-monopole bundle is nontrivial.

The potential applications of quantum group gauge theory hardly need to be elaborated. For example, for a classical manifold

$$\left\{ \begin{array}{c} \text{Flat connections on } G - \text{bundle} \\ \text{modulo gauge} \end{array} \right\} \cong \hom(\pi_1, G)/G \qquad (9).$$

using the holonomy. One can view this as a functor from groups to sets and the homotopy group π_1 as more or less the representing object in the category of groups. The same idea with quantum group gauge theory essentially defines a homotopy quantum group $\pi_1(M)$ for any algebra M as more or less the representing object of the functor that assigns to a quantum group H the set of zero-curvature gauge fields with this quantum structure group. This goes somewhat beyond vector bundles and K-theory alone. Although in principle defined, this idea has yet to be developed in a computable form.

Finally we mention that one needs to make a slight generalisation of the above to include other noncommutative examples of interest. In fact (and a little unexpectedly) the general theory above can be developed with only a coalgebra rather than a Hopf algebra H. Or dually it means only an algebra A in place of the enveloping algebra of a Lie algebra. This was achieved more recently, in [19] [20], and allows us to include the full 2-parameter quantum spheres as well as (in principle) all known q-deformed symmetric spaces. This setting of gauge theory based on inclusions of algebras could perhaps be viewed as an algebraic analogue of the notion of 'paragroup' in the theory of operator algebras. Also, in a different direction, one may do the quantum group gauge theory in any braided monoidal category at the level of braids and tangles [21] so that one has braided group gauge theory and in principle gauge theory for quasiassociative algebras such as [22] the octonions.

4   Quantum soldering forms and metrics

We are finally ready to take the plunge and offer at least a first definition of a 'quantum manifold'. The approach we take is basically that of the existence of a bein or, in global terms, a soldering form. The first step is to define a generalised frame bundle or *frame resolution* of our algebra M as [1]

1. A quantum principal bundle (P, H, Δ_R) over M.
2. A comodule V and an equivariant 'soldering form' $\theta : V \to P\Omega^1 M \subset \Omega^1 P$ such that the induced map

$$E^* \to \Omega^1 M, \quad p \otimes v \mapsto p\theta(v) \tag{10}$$

is an isomorphism.

What this does is to express the cotangent bundle as associated to a principal one. Other tensors are then similarly associated, for example vector fields are $E \cong \Omega^{-1} M$. Of course, all of this has to be done with suitable choices of differential calculi on M, P, H whereas we have been focusing for simplicity on the universal calculi. There are some technicalities here but more or less the same definitions work in general. The working definition [1] of a *quantum manifold* is simply this data $(M, \Omega^1, P, H, \Delta_R, V, \theta)$. The definition works in that one has analogues of many usual results. For example, a connection

ω on the frame bundle induces a covariant derivative D_ω on the associated bundle E^* which maps over under the soldering isomorphism to a covariant derivative

$$\nabla : \Omega^1 M \to \Omega^1 M \underset{M}{\otimes} \Omega^1 M. \qquad (11)$$

Its torsion is defined as corresponding similarly to $\bar{D}_\omega \theta$, where we use a suitable (right-handed) version of the covariant derivative.

Defining a Riemannian structure can be done in a 'self-dual' manner as follows. Given a framing, a 'generalised metric' isomorphism $\Omega^{-1} M \cong \Omega^1 M$ between vector fields is equivalent to [1]

3. Another framing $\theta^* : V^* \to (\Omega^1 M)P$, which we call the *coframing*, this time with V^*.

The associated quantum metric is

$$g = \theta^*(f^a)\theta(e_a) \in \Omega^1 M \underset{M}{\otimes} \Omega^1 M \qquad (12)$$

where $\{e_a\}$ is a basis of V and $\{f^a\}$ is a dual basis.

Now, this self-dual formulation of 'metric' as framing and coframing is symmetric between the two. One could regard the coframing as the framing and vice versa. From our original point of view its torsion tensor corresponding to $D_\omega \theta^*$ is some other tensor, which we call the *cotorsion tensor*. A natural proposal for a generalised Levi-Civita connection on a quantum Riemannian manifold is therefore [1]

4. A connection ω such that the torsion and cotorsion tensors both vanish.

There is a corresponding covariant derivative ∇. The Riemannian curvature of course corresponds to the curvature of ω, which is $d\omega + \omega \wedge \omega$, via the soldering form. I would not say that the Ricci tensor and Einstein tensor are understood abstractly enough in this formalism but of course one can just write down the relevant contractions and proceed blindly.

Theorem 6 *[1] Every quantum group M has a framing by $H = M$, $P = M \otimes M$, $V = \ker \epsilon$ and θ induced from the quantum group Maurer-Cartan form $e(v) = Sv_{(1)} \otimes v_{(2)}$. Likewise for all M equipped with a bicovariant differential calculus, with $V = \Omega_0$.*

In this construction one builds the framing from a V-bein e inducing the isomorphism $\Omega^1(M) = M \otimes \Omega_0$ as in Section 2 (in a right-handed setting). Moreover, for quantum groups such as $\mathbb{C}_q[SU_2]$ there is an Ad-invariant nondegenerate braided Killing form [15] on the underlying braided-Lie algebra, which provides a coframing from a framing – so that quantum groups such as $\mathbb{C}_q[SU_2]$ with the corresponding bicovariant differential calculi are quantum Riemannian manifolds in the required sense. The existence of a generalised Levi-Civita connection in such cases remains open and may require one to go beyond strong connections.

At least with the universal calculus every quantum homogeneous space is a quantum manifold too. That includes quantum spheres, quantum planes etc. In fact, there is a notion of comeasuring or quantum automorphism bialgebra [3] for practically any algebra M and when this has an antipode (which typically requires some form of completion) one can write M as a quantum homogeneous space. So almost any algebra M is more or less a quantum manifold for some principal bundle (at least rather formally). This is analogous to the idea that any classical manifold is, rather formally, a homogeneous space of diffeomorphisms modulo diffeomorphisms fixing a base point.

Finally, to get the physical meaning of the cotorsion tensor and other novel ideas coming out of this noncommutative Riemannian geometry, let us consider the semiclassical limit. What we find is that noncommutative geometry forces us to slightly generalise conventional Riemannian geometry itself [1]:

1. We should allow any group G in the 'frame bundle', hence the more general concept of a 'frame resolution' (P, G, V, θ^a_μ) or generalised manifold.

2. The generalised metric $g_{\mu\nu} = \theta^{*a}_\mu \theta_{\nu a}$ corresponding to a coframing θ^{*a}_μ is nondegenerate but need not be symmetric.

3. The generalised Levi-Civita connection defined as having vanishing torsion and vanishing cotorsion respects the metric only in a skew sense

$$\nabla_\mu g_{\nu\rho} - \nabla_\nu g_{\mu\rho} = 0 \tag{13}$$

and need not be uniquely determined.

This generalisation of Riemannian geometry includes special cases of symplectic geometry, where the generalised metric is totally antisymmetric. It is also remarkable that metrics with antisymmetric part are exactly what are needed in string theory to establish T-duality. In summary, one has on the table a general noncommutative Riemannian geometry to play with. It can be applied to a variety of algebras far removed from conventional geometry. Some finite dimensional examples will be presented elsewhere.

Acknowledgements

It is a pleasure to thank the organisers of the LMS symposium in Durham for a thoroughly enjoyable conference.

References

[1] S. Majid. Quantum and braided group Riemannian geometry. *J. Geom. Phys.*, 30:113–146, 1999.

[2] T. Brzeziński and S. Majid. Quantum group gauge theory on quantum spaces. *Commun. Math. Phys.*, 157:591–638, 1993. Erratum 167:235, 1995.

[3] S. Majid. Quantum and braided diffeomorphism groups. *J. Geom. Phys.*, 28:94–128, 1998.

[4] S.L. Woronowicz. Differential calculus on compact matrix pseudogroups (quantum groups). *Commun. Math. Phys.*, 122:125–170, 1989.

[5] S. Majid. Classification of bicovariant differential calculi. *J. Geom. Phys.*, 25:119–140, 1998.

[6] S. Majid. Quantum geometry of field extensions. *J. Math. Phys.*, 40:2311–2323, 1999.

[7] G. Amelino-Camelia and S. Majid. Waves on noncommutative spacetime and gamma-ray bursts. *Int. J. Mod. Phys. A* (In press.)

[8] E. Beggs and S. Majid. Quasitriangular and differential structures on bicrossproduct hopf algebras. *J. Algebra*, 219:682–727, 1999.

[9] S. Majid. *Foundations of Quantum Group Theory*. Cambridge Univeristy Press, 1995.

[10] S. Majid and R. Oeckl. Twisting of quantum differentials and the Planck scale Hopf algebra. *Commun. Math. Phys.*, 205:617–655, 1999.

[11] S. Majid. Tannaka-Krein theorem for quasiHopf algebras and other results. *Contemp. Maths*, 134:219–232, 1992.

[12] P. Etingof and S. Gelaki. The classification of triangular semisimple and cosemisimple Hopf algebras over an algebraically closed field. *Preprint*, 1999.

[13] P. Baumann and F. Schmitt. Classification of bicovariant differential calculi on quantum groups (a representation-theoretic approach). *Commun. Math. Phys.*, 194:71–86, 1998.

[14] I. Heckenberger and K. Schmudgen. Classification of bicovariant differential calculi on the quantum groups SLq(n+1) and Sp(q)(2n). J. Reine. Angewandte Mat., 502:141-162, 1998.

[15] S. Majid. Quantum and braided Lie algebras. *J. Geom. Phys.*, 13:307–356, 1994.

[16] T. Brzeziński and S. Majid. Quantum differentials and the q-monopole revisited. *Acta Appl. Math.*, 54:185–232, 1998.

[17] A. Connes. *Noncommutative Geometry*. Academic Press, 1994.

[18] P. Hajac and S. Majid. Projective module description of the q-monopole. *Commun. Math. Phys.*, 206:246–464, 1999.

[19] T. Brzezinski and S. Majid. Coalgebra bundles. *Commun. Math. Phys.*, 191, 1998. 467-492.

[20] T. Brzeziński and S. Majid. Quantum geometry of algebra factorisations and coalgebra bundles. *Commun. Math. Phys.* (In press.)

[21] S. Majid. Diagrammatics of braided group gauge theory. *J. Knot Theor. Ramif.*, 8:731–771, 1999.

[22] H. Albuquerque and S. Majid. Quasialgebra structure of the octonions. *J. Algebra*, 220:188–224, 1999.

Finite Quantum Groups and Pointed Hopf Algebras

Ian M. Musson[1]
Department of Mathematical Sciences
University of Wisconsin-Milwaukee

Abstract

We show that under certain conditions a finite dimensional graded
pointed Hopf algebra is an image of an algebra twist of a quantized
enveloping algebra $U_q(\mathfrak{b})$ when q is a root of unity. In addition we
obtain a classification of Hopf algebras H such that $G(H)$ has odd
prime order $p > 7$ and grH is of Cartan type.

Throughout this paper K will denote an algebraically closed base field of characteristic zero. Recently there has been considerable interest in the structure
of finite dimensional pointed Hopf algebras over K. For example if p is prime
all Hopf algebras of dimension p are group algebras. Also any pointed Hopf
algebra of dimension p^2 is either a group algebra or a Taft algebra, while those
of dimension p^3 have been classified [AS1], [CD], [SvO], [Z]. In addition there
are infinitely many isomorphism classes of pointed Hopf algebras of dimension
p^4, [AS1], [BDG], see also [G].

If H is pointed the coradical filtration $\{H_n\}$ on H is a Hopf algebra filtration and the associated graded algebra $grH = \oplus_{n \geq 0} H_n/H_{n-1}$ is a Hopf
algebra, see [M1], also [M2,Lemma 5.5.1]. In [AS2] pointed Hopf algebras H
such that $H \cong grH$ are studied using methods from Lie theory and quantum groups. These Hopf algebras are crossed products $H = R * G$ where
$G = G(H)$ is the group of grouplike elements in H, and R is an analog of the
infinitesimal part of H in the classical theory of cocommutative Hopf algebras. Another key ingredient in understanding the structure of H is a matrix
b known as the braiding matrix which is determined by the action of G on
R.

Our first task is to understand finite dimensional graded pointed Hopf
algebras in terms of something more familiar. In [AS2] this is done under
certain conditions by twisting the coalgebra structure of a Frobenius-Lusztig
kernel. We show that under similar conditions a finite dimensional graded

---

[1]Research partially supported by NSF grant DMS-9801151.

pointed Hopf algebra is an image of an algebra twist of a quantized enveloping algebra $U_q(\mathfrak{b})$ when q is a root of unity. This approach seems to have certain technical advantages, for example we don't use quantum antisymmetrizers.

We also consider the problem of determining the finite dimensional pointed Hopf algebras H for which grH is known. This is known as the lifting problem. In particular we show that if $G(H)$ has prime order $p > 7$ and grH is of Cartan type, then $H \cong grH$. We also obtain some results for low primes; in particular we construct some apparently new Hopf algebras H with $G(H)$ of order 3 and grH of type $A_2 \times A_2$. Since the first version of this paper was written, Andruskiewitsch and Schneider have obtained further results on the structure of pointed Hopf algebras H such that $G(H)$ is an elementary abelian p−group with $p > 17$, [AS4].

1. POINTED HOPF ALGEBRAS AS IMAGES OF QUANTIZED ENVELOPING ALGEBRAS.

1.1 Let $H = \oplus_{n\geq o}H(n)$ be a graded pointed Hopf algebra with coradical KG. We say that H is *coradically graded* if the coradical filtration of H is given by $H_n = \oplus_{m\leq n}H(m)$, [CM2].

If H is coradically graded, the projection $\pi : H \longrightarrow H(0)$ is a Hopf algebra map. If Δ is the coproduct on H then $\rho = (1 \otimes \pi)\Delta$ makes H into a left $H(0)$ - Hopf module. Since ρ is an algebra maps it follows that the coinvariants $R = \{h \in H|\rho(h) = h \otimes 1\}$ form a subalgebra of H which is invariant under conjugation by elements of G. It follows from [M2, Theorem 1.9.4] that $H = R * G$ is a crossed product of R by G.

If $R(n) = H(n)\cap R$ then $R = \oplus_{n\geq o}R(n)$ is a graded algebra with $R(0) = K$ by [AS1, Lemma 2.1].

If $x \in H$ and $\Delta x = g \otimes x + x \otimes h$ for grouplike elements $g, h \in G$ we say x is (h, g)-primitive. Since G is abelian, G acts on the space $P_g(H)$ of $(1, g)$-primitives by conjugation. For $\chi \in G^* = \mathrm{Hom}(G, K^*)$ we set

$$P_g^\chi = P_g^\chi(H) = \{x \in P_g(H)|hxh^{-1} = \chi(h)x \text{ for all } h \in G\}.$$

Note that if G is finite then $\chi(h)$ is a root of unity for all $\chi \in G^*$ and $h \in G$. It is frequently useful to make the following assumptions.

(1) $R(1)$ has a basis $x_1, \ldots, x_n$ such that $x_i \in P_{g_i}^{\chi_i}$ for some $g_i \in G$, $\chi_i \in G^*$.

(2) H is generated as an algebra by G and $R(1)$.

(3) $G =< g_1, \ldots, g_n >$ is abelian.

Note that if $R(1)$ is finite dimensional and (3) holds then (1) also holds.

The algebra R is an Hopf algebra in the category ${}_{KG}^{KG}\mathcal{YD}$ of left Yetter-Drinfeld modules (also known as a braided bialgebra) over KG and H can be reconstructed by bosonization as a biproduct $H = R \# KG$, see [Mj],[R]. In [AS2] the braided bialgebra R is christened a Nichols algebra since algebras of this form were first studied by Nichols, [N]. By [AG, Prop. 3.2.12] R is determined by the subspace V regarded as an object in $^{KG}_{KG}\mathcal{YD}$, and as in [AS2] we write $R = \mathcal{B}(V)$ in this situation. The structure of V as a Yetter-Drinfeld module is given as follows: V is a left KG-comodule via the map $x_i \longrightarrow g_i \otimes x_i$, and a left KG-module via $g.x_j = \chi_j(g)x_j$. It is easily seen that the Yetter-Drinfeld compatability condition ([M2, 10.6.11]) holds. We set $b_{ij} = \chi_j(g_i)$ and call the matrix $\mathbf{b} = (b_{ij})$ the *braiding* of H and dim R(1) the *rank* of H.

As we shall see the relations in R depend largely on the matrix $\mathbf{b}$ and are independent of the elements χ_j and g_i. Thus we also call an $n \times n$ matrix $\mathbf{b} = (b_{ij})$ a *braiding of rank n* if all the entries b_{ij} are nonzero.

Lemma. If $\mathbf{b}$ is the braiding matrix of a finite dimensional Hopf algebra, then $b_{ii} \neq 1$ for all i.

Proof. See [AS1, Lemma 3.1] or [N, page 1538].

1.2 Following [AS2] we say that a braiding matrix $\mathbf{b} = (b_{ij})$ is of *Cartan type* if $b_{ii} \neq 1$ and there exists $a_{ij} \in \mathbb{Z}$ such that

$$b_{ij}b_{ji} = b_{ii}^{a_{ij}} \tag{1}$$

where $a_{ii} = 2$ for all i. If b_{ii} is a root of unity we assume that a_{ij} is the unique integer such that $-ord(b_{ii}) < a_{ij} \leq 0$. With this choice of the a_{ij} we say that $\mathbf{b}$ is of Cartan type (a_{ij}).

If $\mathbf{b}$ is a braiding of Cartan type (a_{ij}), we say that $\mathbf{b}$ is *admissible* if b_{ij} is a root of unity of odd order and a_{ij} is either zero or relatively prime to the order of b_{ij} for all i, j. Observe that if (a_{ij}) is the Cartan matrix of a simple Lie algebra $\mathfrak{g}$ then this condition simply means that the order of each b_{ij} is odd and, if $\mathfrak{g}$ is of type G_2 not divisible by 3.

Also we say that $\mathbf{b}$ is a braiding of *FL-type* (a_{ij}) if there exist positive integers $d_1, \ldots, d_n$ such that for all i, j

$$d_i a_{ij} = d_j a_{ji} \tag{2}$$

and

$$\text{there exists } q \in K^* \text{ such that } b_{ij} = q^{d_i a_{ij}} \tag{3}$$

Finally we say that a braiding matrix $\mathbf{b} = (b_{ij})$ has *exponent e* if e is the least positive integer such that $b_{ij}^e = 1$ for all i, j.

1.3 Let L be a free abelian group with basis $e_1, \ldots, e_n$. If H is an L-graded Hopf algebra and $p : L \times L \to K^*$ an antisymmetric bicharacter, we obtain a new Hopf algebra by "twisting" H by p as follows. Let H' be an isomorphic copy of H as a coalgebra with canonical isomorphism $h \to h'$. The new algebra structure on H' is defined by

$$a'.b' = p(\alpha, \beta)(ab)'$$

for $a \in H_\alpha, b \in H_\beta$. By [HLT, Theorem 2.1] H' is a Hopf algebra.

Suppose in addition that $H = \oplus H(n)$ is a coradically graded Hopf algebra such that assumptions $(1)-(3)$ of 1.1 hold, and let $\mathbf{b} = (b_{ij})$ be the braiding of H. Assume also that H is an L-graded Hopf algebra with $\deg x_i = \deg g_i = e_i$. Suppose that $p : L \times L \to K^*$ is an antisymmetric bicharacter with $p(e_i, e_j) = p_{ij}$ for $i < j$. A short calculation shows that in the twisted algebra H' we have

$$g_i x_j g_i^{-1} = b_{ij} p_{ij}^2 x_j.$$

Thus the braiding $\mathbf{b}'$ in H' satisfies

$$b'_{ij} = b_{ij} p_{ij}^2.$$

Henceforth we denote the twisted Hopf algebra H' by $H^{(p)}$.

Motivated by the above calculation we say that two braidings $\mathbf{b}, \mathbf{b}'$ of rank n are *twist equivalent* if there exists $p_{ij} \in K^*$ for $1 \leq i, j \leq n$ such that

$$p_{ij} p_{ji} = 1 \text{ and } b'_{ij} = b_{ij} p_{ij}^2$$

for all i, j.

Obviously any braiding of FL-type is a braiding of Cartan type. Conversely we have the following result.

Lemma. If (a_{ij}) be a symmetrizable Cartan matrix and $\mathbf{b}$ an admissible braiding of Cartan type (a_{ij}), then $\mathbf{b}$ is twist equivalent to a braiding of FL-type.

Proof. There is a root of unity q of odd order and integers e_{ij} such that $b_{ij} = q^{e_{ij}}$ for all i, j. Thus q has a unique fourth root and we set $p_{ij} = q^{(e_{ji} - e_{ij})/4}$. Then $b'_{ij} = b_{ij} p_{ij}^2 = q^{(e_{ij} + e_{ji})/2} = b'_{ji}$ for all i, j. Therefore the result follows from [AS2, Lemma 4.3].

1.4. Let (a_{ij}) be a generalized Cartan matrix and suppose that there exist relatively prime positive integers d_i such that $d_i a_{ij} = d_j a_{ji}$ for all i, j.

Define $[a] = (v^a - v^{-a})/(v - v^{-1})$, $[a]! = \Pi_{i=1}^a [i]$ and $\begin{bmatrix} a \\ i \end{bmatrix} = [a]!/[i]![a-i]!$.

The result of substituting q for v in $\begin{bmatrix} a \\ i \end{bmatrix}$ is denoted $\begin{bmatrix} a \\ i \end{bmatrix}_q$. Let $v_i = v^{d_i}$.

We define the quantized enveloping algebra $\mathbf{U} = U_v(\mathfrak{b})$ to be the $\mathbb{Q}(v)$-algebra with generators $K_1^{\pm 1}, \ldots, K_n^{\pm 1}$ and $E_1, \ldots, E_n$, subject to the relations

$$K_i K_i^{-1} = K_i^{-1} K_i = 1 \tag{1}$$

$$K_j E_i K_j^{-1} = v^{d_i a_{ij}} E_i \tag{2}$$

$$\sum_{r=0}^{1-a_{ij}} (-1)^r \begin{bmatrix} 1 - a_{ij} \\ r \end{bmatrix}_{v_i} E_i^{1-a_{ij}-r} F_j E_i^r = 0. \tag{3}$$

Then $\mathbf{U}$ is a Hopf algebra whose coproduct satisfies

$$\Delta K_i^{\pm 1} = K_i^{\pm 1} \otimes K_i^{\pm 1}$$

$$\Delta E_i = K_i \otimes E_i + E_i \otimes 1.$$

Note that $\mathbf{U}$ is a graded Hopf algebra if we set $\deg K_i = 0$ and $\deg E_i = 1$ for all i. It follows from [CM2, Theorem B] that $\mathbf{U}$ is coradically graded. Let $\Gamma = <K_1, \ldots, K_n>$ and define $\psi_j \in \Gamma^*$ by $\psi_j(K_i) = v^{d_i a_{ij}}$. If R is the subalgebra of $\mathbf{U}$ generated by $E_1, \ldots, E_n$ then $\mathbf{U} = R * \Gamma$ as in 1.1 and $E_i \in P_{K_i}^{\psi_i}$ for $1 \leq i \leq n$.

As in [J, 4.3] it is useful to consider also the Hopf algebra $\widetilde{\mathbf{U}}$ over $\mathbb{Q}(v)$ with generators $K_1^{\pm 1}, \ldots, K_n^{\pm 1}$, $E_1, \ldots, E_n$ satisfying only relations (1) and (2) above.

Let $\mathcal{A} = \mathbb{Q}[v, v^{-1}]$. We define a "nonrestricted $\mathcal{A}$-form" of $\mathbf{U}$ (c.f. [CP, 9.2]). This is the $\mathcal{A}$-subalgebra U of $\mathbf{U}$ generated by the elements $E_i, K_i^{\pm 1}$, and

$$[K_i : 0] = (K_i - K_i^{-1})/(v_i - v_i^{-1}).$$

The $\mathcal{A}$-subalgebra $\widetilde{U}$ of $\widetilde{\mathbf{U}}$ is defined similarly. The defining relations for U are the same as those for $\mathbf{U}$ except that we have the additional relations

$$K_i - K_i^{-1} = (v_i - v_i^{-1})[K_i : 0]$$

in U. Note that we can regard U and $\widetilde{U}$ as L-graded Hopf algebras ($L = \mathbb{Z}^n$) by setting $\deg K_i = \deg E_i = e_i$. If p is an antisymmetric bicharacter on L, then $U^{(p)}$ denotes the twisted algebra as in 1.3. Finally if $q \in K$, we set $U_q^{(p)} = U^{(p)} \otimes_{\mathbb{Q}[v^{\pm 1}]} K$ where K is a $\mathbb{Q}[v^{\pm 1}]$-algebra with v acting as q. The Hopf algebras $\widetilde{U}^{(p)}$ and $\widetilde{U}_q^{(p)}$ are defined analogously.

1.5 Recall that if H is a Hopf algebra with antipode S, the adjoint action is defined by

$$(ada)(b) = \sum a_1 b S(a_2).$$

The next result is a key ingredient in our work. In the case of quantized enveloping algebras it says that the rather complex Serre relations of 1.4 (3) follow from the relatively simple relations in 1.4 (2) together with a mild assumption on the skew primitives. This might seem rather remarkable, although a similar situation obtains for Kac-Moody algebras, see [Kac,3.3].

Lemma. If $x_i \in P_{g_i}^{\chi_i}(H)(i = 1,2)$ and r is a positive integer such that

$$\chi_2(g_1)\chi_1(g_2)\chi_1(g_1)^{r-1} = 1$$

then $(adx_1)^r(x_2)$ is $(1, g_1^r g_2)$-primitive.

Proof. A result similar to this is proved in [AS2, Lemma A.1.] for a braided adjoint action. The result in [AS2] translates into the Lemma after bosonization. Set $b_{ij} = \chi_j(g_i)$. Because of the importance of the result we indicate another proof in the case where **b** is an admissible rank 2 braiding of Cartan type (a_{ij}) (this is the only case that we shall need). By specialization it suffices to prove the result in the case where $H = \widetilde{U}$ as in 1.4 and $x_i = E_i, g_i = K_i$ and $\chi_i = \psi_i$ for i = 1,2. If **b** is of FL-type this follows from [J, Lemma 4.10]. (Note that it is assumed in [J, Chapter 4] that (a_{ij}) has finite type, but this is not necessary for the proof of [J, Lemma 4.10]). In general since **b** is twist equivalent to a braiding of FL-type, the result follows from [CM1, Lemma 3.2].

1.6 Theorem. Let H be a finite dimensional coradically graded pointed Hopf algebra with braiding **b**. Assume that **b** is admissible of Cartan type (a_{ij}) and that H satisfies (1)-(3) of Section 1.1. Let U be the $\mathcal{A}$-form of $U_v(\mathbf{b})$ described in 1.4. Then there is a surjective map of Hopf algebras $U_q^{(p)} \longrightarrow H$ for some twist $U_q^{(p)}$ of U_q, and some $q \in K$.

Proof. By Lemma 1.3 and its proof there exist roots of unity p_{ij} such that the braiding $\mathbf{b}'$ given by $b'_{ij} = b_{ij}p_{ij}^{-2}$ is of FL-type. Hence there exists a root of unity q such that $p_{ij} \in (q)$ and $b'_{ij} = q^{d_i a_{ij}}$ for all i, j. Let L be a free abelian group with basis $e_1, \ldots, e_n$ and define $p : L \times L \to \mathbb{Q}[v^{\pm 1}]$ by $p(e_i, e_j) = v^{e_{ij}}$, where $p_{ij} = q^{e_{ij}}$ and $0 \le e_{ij} < ord(q)$.

By our assumptions, H is generated by $G = < g_1, \ldots, g_n >$ and $x_i \in P_{g_i}^{\chi_i}(H), 1 \le i \le n$. We claim there is a surjective algebra map $\theta : \widetilde{U}^{(p)} \longrightarrow H$ sending K_i to g_i and E_i to x_i. Clearly θ preserves relation (1) in 1.4. Since the braiding in $\widetilde{U}_q^{(p)}$ satisfies

$$\psi_j(K_i) = q^{d_i a_{ij} + 2e_{ij}}$$

and the braiding in H satisfies $b_{ij} = b'_{ij}p_{ij}^2 = q^{d_i a_{ij} + 2e_{ij}}$, it follows that the twisted version of relation (2) in 1.4 is preserved by θ and the claim follows. The relation (3) in 1.4 can be written in the form $(adE_i)^{1-a_{ij}}(E_j) = 0$ and by [CM1, Lemma 3.2] the same relation holds in $U_q^{(p)}$. Since $\chi_j(g_i)\chi_i(g_j)\chi_i(g_i)^{-a_{ij}} = b_{ij}b_{ji}b_{ii}^{-a_{ij}} = 1$ we have $(adx_i)^{1-a_{ij}}(x_j) = 0$ in H since H is graded so θ descends to an algebra map $U_q^{(p)} \longrightarrow H$.

To see that θ is a map of bialgebras note that the set

$$\{a \in U_q^{(p)} | \Delta\theta(a) = (\theta \otimes \theta)\Delta(a)\}$$

is a subalgebra of $U_p^{(p)}$ which contains the generators $K_i^{\pm 1}$ and E_i. Similarly θ is a map of Hopf algebras.

1.7. We describe the kernel of the Hopf algebra map in the previous theorem. Let Φ^+ be the set of positive roots for the semisimple Lie algebra associated to a Cartan matrix (a_{ij}) of finite type. Using a reduced expression for the longest element in the Weyl group and Lusztig's braid group action we can construct root vectors $E_\alpha \in U_q$ for $\alpha \in \Phi^+$. Suitably ordered monomials in the E_α form a PBW basis for the subalgebra U_q^+ of U_q generated by $E_1, \ldots, E_n$, see [L] or [J, Theorem 8.24]. The same monomials form a basis for the twisted subalgebra $(U_q^+)^{(p)}$ of $U_q^{(p)}$.

Proposition. Suppose that H is a finite dimensional coradically graded pointed Hopf algebra of finite and indecomposable type $A = (a_{ij})$ such that $G(H) \cong (\mathbb{Z}/N)^s$ for some odd N with $N \neq 3$ if A has type G_2. Let $\theta : U_q^{(p)} \longrightarrow H$ be the surjective map of Hopf algebras described in Theorem 1.6 and $L = Ker\theta|_\Gamma$. Then $Ker\theta$ is the ideal generated by the elements $g - 1, g \in L$ and $E_\alpha^N, \alpha \in \Phi^+$.
Proof. This follows from [AS4, Theorem 4.2].

1.8. We illustrate Theorem 1.6 in case where $H = R * G$ is a graded Hopf algebra with $G = (g)$ a group of prime order p, $\dim R(1) = 2$ and $\mathbf{b}$ a braiding of finite Cartan type (a_{ij}). These possibilities are described in [AS 2, Section 5]. Let $d \in \{1, 2, 3\}$ and let p be an odd prime ($p > 3$ if $d = 3$). Suppose that q is a primitive p^{th} root of unity in K. Then q has a unique $2d^{th}$ root in K. Define

$$A = \begin{pmatrix} 2 & -1 \\ -d & 2 \end{pmatrix}, \quad D = \begin{pmatrix} d & 0 \\ 0 & 1 \end{pmatrix}$$

Then DA is symmetric and any twist of the braiding $\mathbf{b}'$ for $U_v(\mathfrak{b})$ has the form

$$\begin{pmatrix} v^{2d} & v^{a-d} \\ v^{-a-d} & v^2 \end{pmatrix}.$$

Now specialize v to $q^{1/2d}$ and set $b = a/2d - 1/2$, $c = -a/2d - 1/2$. This means that $b + c + 1 \equiv 0 \pmod{p}$. Also the braiding matrix of the twisted specialization $U_q^{(p)}$ has the form

$$\begin{pmatrix} q & q^b \\ q^c & q^{1/d} \end{pmatrix}$$

Now consider a homomorphism from $U_q^{(p)}$ to H. We can assume that K_1 maps to g and K_2 to g^c. Thus such a map exists if and only if $dbc \equiv 1 \bmod p$ or equivalently $db(b + 1) \equiv -1 \bmod p$. The last congruence imposes some conditions on p which can be found using quadratic reciprocity, see [AS2, Section 5].

By [AS2, Theorem 1.3] the number of isomorphism types of Nichols algebras with coradical of prime dimension p is equal to $(p-1)$ for type A_2 and $2(p-1)$ for type B_2 and G_2. The factor $p - 1$ comes from the choice of a pth root of unity q. With q fixed the two roots of the congruence $db(b+1) + 1 \equiv 0 \bmod p$ give rise to 2 nonisomorphic Hopf algebras of types B_2 or G_2. There is a unique Hopf algebra of type A_2 because of the diagram automorphism in this case.

2. THE LIFTING PROBLEM.

2.1. Let H be a pointed Hopf algebra with coradical filtration $\{H_n\}$, and $H_0 = KG$. Then the graded Hopf algebra $grH = \oplus_{n \geq 0} H_n / H_{n-1}$ is coradically graded. Assuming the structure of grH is known we investigate the possibilities for H.

We assume that G is abelian and that $grH = R * G$ as before. By [M2, 5.4.1], we can find skew primitive elements $y_1, \ldots, y_n$ in H_1 such that the images x_i of these elements in grH form a basis for $R(1)$. By considering the action of G by conjugation we can assume further that $y_i \in P_{g_i}^{\chi_i}(H)$ for suitable g_i, χ_i. As before we set $b_{ij} = \chi_j(g_i)$.

Lemma. Suppose that $y_i \in P_{g_i}^{\chi_i}(H)$ for $i = 1, 2$, and that r is a positive integer such that
$$\chi_2(g_1)\chi_1(g_2)\chi_1(g_1)^{r-1} = 1. \tag{1}$$
If $(ady_1)^r(y_2) = a(g_1^r g_2 - 1)$ with $a \neq 0$ then $\chi_2 = \chi_1^{-r}$ and $\chi_1(g_1) = \chi_1(g_2)$

Proof. If $k \in G$ then k commutes with $(ady_1)^r(y_2)$ since G is abelian and $a \neq 0$. This forces $(\chi_1^r \chi_2)(k) = 1$. In particular $(\chi_1^{r-1}\chi_2)(g_1) = \chi_1^{-1}(g_1)$ and then (1) implies that $\chi_1(g_1) = \chi_1(g_2)$.

Corollary. With the hypothesis of the Lemma suppose that the braiding **b** in grH has rank 2 and finite Cartan type $A = (a_{ij})$ as in 2.2. Assume that H is finite dimensional and that the exponent of G is an odd prime p which is different from 3 if $d = 3$. Then either

　　1) **b** has exponent dividing $2d + 1$

or

　　2) $(ad\, y_1)^2(y_2) = (ad\, y_2)^{d+1}(y_1) = 0$.

Proof. By Lemma 1.5 $z_1 = (ady_1)^2(y_2)$ is $(1, g_1^2 g_2)$-primitive and $z_2 =$

$(ad\, y_2)^{d+1}(y_1)$ is $(1, g_1 g_2^{d+1})$-primitive. We show first that for $g = g_1 g_2^{d+1}$ all $(1, g)$-primitives are trivial. There are two cases to consider as follows. Note that $g_1 \neq 1 \neq g_2$.

1) If $g_1 g_2^{d+1} = g_1$ then $p | d + 1$. The only possibility is $d + 1 = p = 3$, but then the congruence $2b^2 + 2b + 1 \equiv 0 \mod 3$ from section 2.2 has no solution.

2) If $g_1 g_2^{d+1} = g_2$ then for $j = 1, 2$

$$b_{1j} = \chi_j(g_1) = \chi_j(g_2)^{-d} = b_{2j}^{-d}.$$

Thus

$$b_{22}^{-d} b_{21} = b_{12} b_{21} = b_{22}^{a_{21}} = b_{22}^{-d}.$$

Hence $b_{21} = 1$ and $b_{11} = b_{21}^{-d} = 1$. This is impossible by Lemma 1.1. Similarly all $(1, g_1^2 g_2)$-primitives are trivial.

Now suppose $z_1 = a(g_1^2 g_2 - 1)$ with $a \in K$, $a \neq 0$. By Lemma 2.1 $\chi_2 = \chi_1^{-2}$ and $\chi_1(g_1) = \chi_1(g_2)$. Set $q = b_{11} = b_{21}$. Then for $j = 1, 2$.

$$b_{j2} = \chi_2(g_j) = q^{-2},$$

and $b_{12} b_{21} = b_{22}^{a_{21}}$ gives $q^{-1} = q^{2d}$.

Similarly if $z_2 = b(g_1 g_2^{d+1} - 1)$ with $b \neq 0$ we have $\chi_1 = \chi_2^{-(d+1)}$ and $\chi_2(g_1) = \chi_2(g_2)$. Then if $q = b_{12} = b_{22}$ we have

$$b_{j1} = \chi_1(g_j) = b_{j2}^{-(d+1)} = q^{-(d+1)}$$

and $b_{12} b_{21} = b_{11}^{a_{12}}$ gives $q^{-d} = q^{d+1}$ and hence the result.

2.2. We apply our results to pointed Hopf algebras H such that $G = G(H)$ has odd prime order p and $gr\, H$ is a Nichols algebra of finite Cartan type. We first discuss the indecomposable case.

Theorem. Let H be a finite dimensional pointed Hopf algebra such that $G(H) = (g)$ has odd prime order p and $gr\, H$ is of finite indecomposable Cartan type. Assume that
1) If $p = 3$ or 7 then $gr\, H$ is not of type G_2
2) If $p = 5$ then $gr\, H$ is not of type B_2.
Then $H \cong gr\, H$.

Proof. By [AS2,Section 5], $gr\, H$ has type A_1, A_2, B_2 or G_2. For type A_1 the only possibility is the Taft algebra which has no nongraded analog. We assume $gr\, H$ has rank 2 and that the Cartan matrix A is as described in Section 1.8. In particular, $gr\, H$ has generators g, x_1, x_2 satisfying

$$(adx_1)^2(x_2) = (adx_2)^{d+1}(x_1) = 0.$$

By [Mo, Theorem 5.4.1] we can choose $y_i \in P_{g_i}^{\chi_i}(i = 1, 2)$ such that the image of y_i in grH is x_i. If $d \neq 1$ or $p \neq 3$, Corollary 2.1 implies

$$(ady_1)^2(y_2) = (ady_2)^{d+1}(y_1) = 0. \tag{1}$$

If $d = 1$ and $p = 3$ we can assume

$$g_1 = g, \quad g_2 = g^b, \quad \chi_1 = \chi, \quad \chi_2 = \chi^c$$

where $\chi(g) = q$ is a primitive cube root of unity. From section 1.8 we have

$$b + c + 1 \equiv b(b + 1) \equiv 0 \bmod 3.$$

The only solution is $b = c = 1$. Then Lemma 1.5 implies that (1) holds in this case also.

To see this we modify the proof of [AS4, Lemma 6.9]. Let $U_q^{(p)}$ be the twisted quantized enveloping algebra used in the proof of Theorem 1.6. By the choice of the twist p and (1) there is a surjective Hopf algebra map $\phi : U_q^{(p)} \longrightarrow H$ such that $\phi(K_i) = g_i$ and $\phi(E_i) = y_i$. Set $y_\alpha = \phi(E_\alpha)$. We claim that $y_\alpha^p = 0$. To see this we modify the proof of [AS4, Lemma 6.9]. By [DCP, Section 19] the elements $E_\alpha^p, K_i^{\pm p}$ generate a Hopf subalgebra L of $U_q^{(p)}$. By [Mo, Cor. 5.3.5], $\phi(L)$ is a finite dimensional pointed Hopf algebra with trivial coradical. Thus $\phi(L) = 0$ and $y_\alpha^p = 0$.

2.3. We next extend Theorem 2.2 to the case where the Cartan matrix (a_{ij}) is decomposable. By [AS2, Section 5] the only new root systems that arise are subsystems of $A_2 \times A_2$.

We construct some examples of pointed Hopf algebras H such that grH has Cartan type $A_2 \times A_2$. Let q be a primitive cube root of unity and let $K < x_1, x_2 >$ be the free algebra on x_1, x_2. Consider the crossed product $\tilde{B} = K < x_1, x_2 > *(g)$ where g has order 3 and $gx_ig^{-1} = qx_i$ for $i = 1, 2$.

Now let I be the ideal of $\tilde{B}$ generated by the elements

$$x_i^3 \quad i = 1, 2$$

$$(x_1x_2 - qx_2x_1)^3$$

$$x_i^2x_j + x_ix_jx_i + x_jx_i^2 \quad i \neq j.$$

Similarly let $\tilde{C} = K < y_1, y_2 > *(\chi)$ where χ has order 3 and $\chi y_i \chi^{-1} = q^{-1}y_i$ for $i = 1, 2$. Let J be the ideal of $\tilde{C}$ generated by the elements

$$y_i^3 \quad i = 1, 2$$

$$(y_2 y_1 - q y_1 y_2)^3$$

$$y_i^2 y_j + y_i y_j y_i + y_j y_i^2 \quad i \neq j.$$

Set $B = \tilde{B}/I$ and $C = \tilde{C}/J$. We denote the images of elements of $\tilde{B}, \tilde{C}$ in the factor algebras by the same symbol. We make $B, C, \tilde{B}, \tilde{C}$ into Hopf algebras via the coproducts

$$\Delta g = g \otimes g, \quad \Delta \chi = \chi \otimes \chi$$

$$\Delta x_i = g \otimes x_i + x_i \otimes 1$$

$$\Delta y_i = 1 \otimes y_i + y_i \otimes \chi^{-1}.$$

Thus B, C are coradically graded pointed Hopf algebras of type A_2 and $dim B = dim C = 3^4$. It is easy to check that there is a Hopf algebra isomorphism $\psi : \tilde{B} \longrightarrow \tilde{C}^{opp}$ defined by $\psi(g) = \chi$ and $\psi(x_i) = y_i \chi$. Note that $\psi(I) = J$.

Lemma. Given a 2×2 matrix $\Lambda = (\lambda_{ij})$ there are unique linear maps $\delta_i \in \tilde{C}^*$ such that

(1) $\delta_i(ab) = \epsilon(a)\delta_i(b) + \delta_i(a)\gamma(b)$ for all $a, b \in \tilde{C}$

(2) $\delta_i(y_j) = \lambda_{ij}$.

Furthermore $\delta_i(J) = 0$.

Proof. This is similar to [AS4, Lemma 5.19 (b)]. It suffices to show that there are algebra maps $T_i : C \longrightarrow M_2(K)$ satisfying

$$T_i(g) = \begin{pmatrix} 1 & 0 \\ 0 & q \end{pmatrix}, \quad T_i(x_j) = \begin{pmatrix} 0 & \lambda_{ij} \\ 0 & 0 \end{pmatrix}.$$

Then T_i will have the form

$$T_i(c) = \begin{pmatrix} \epsilon(c) & \delta_i(c) \\ 0 & \gamma(c) \end{pmatrix}.$$

We leave the details to the reader.

Now it is easy to see that there is an algebra map $\phi_\Lambda : \tilde{B} \longrightarrow \tilde{C}^*$ defined by

$$\phi_\Lambda(g) = \gamma, \quad \phi_\Lambda(x_i) = \delta_i, \quad i = 1, 2$$

It follows that there is a pairing of Hopf algebras $(\, , \,)_\Lambda : \tilde{C}^{opp} \times \tilde{B} \to K$ defined by $(c, b)_\Lambda = \phi_\Lambda(b)(c)$ for all $b \in \tilde{B}, c \in \tilde{C}$. The pairing is determined by the rules

$$(y_j, x_i)_\Lambda = \lambda_{ij} \quad i, j = 1, 2$$

$$(y_i, g)_\Lambda = (\chi, x_i)_\Lambda = 0$$

$$(\chi, g)_\Lambda = q.$$

Let Λ^{tr} be the transpose of Λ. Then

$$(c, b)_\Lambda = (\psi(b), \psi^{-1}(c))_{\Lambda^{tr}}$$

for all $b \in \tilde{B}, c \in \tilde{C}$. Since $\delta_i(J) = 0$ it follows that

$$(J, \tilde{B})_\Lambda = (\tilde{C}, I)_\Lambda = 0.$$

Thus $(\ ,\ )_\Lambda$ induces a pairing of Hopf algebras $(\ ,\ ) : C^{opp} \times B \to K$.

As in [HLT, Section 2] we can form the Drinfeld double $D(B, C)$ of the pair (B, C). As a coalgebra $D(B, C) = C \otimes B$. The algebra structure is determined by the requirements that $1 \otimes B$ and $C \otimes 1$ are subalgebras of $D(B, C)$ and that

$$b \otimes c = (c_1, Sb_1)(c_3, b_3)c_2 \otimes b_2.$$

Here we use the abbreviated summation notation $\Delta b = b_1 \otimes b_2$. This easily gives

$$\chi x_i = q x_i \chi, \quad y_i g = q g y_i$$

$$x_i y_j - y_j x_i = \lambda_{ij}(g - \chi^{-1})$$

for $i, j = 1, 2$. It follows that $g\chi^{-1}$ is a central grouplike in $D(B, C)$. We denote the Hopf algebra obtained from $D(B, C)$ by factoring the ideal generated by $g\chi^{-1} - 1$ by $H(q, \Lambda)$. Finally let Λ_1 be the identity matrix and Λ_0 the matrix $\Lambda_0 = \begin{pmatrix} 1 & 0 \\ 0 & 0 \end{pmatrix}$ and set $H(q, \epsilon) = H(q, \Lambda_\epsilon)$.

Theorem. Let H be a finite dimensional pointed Hopf algebra such that $G(H)$ is cyclic of prime order p and grH has Cartan type $A_2 \times A_2$. Then either $H \cong grH$ or $H \cong H(q, \epsilon)$ for some primitive cube root q and $\epsilon = 0, 1$. Also

$$H(q, \epsilon) \cong H(q', \epsilon')$$

if and only if $q = q'$ and $\epsilon = \epsilon'$.

Proof. By [AS2, Theorem 1.3] graded pointed Hopf algebras of type $A_2 \times A_2$ and coradical of prime dimension p exist only if $p = 3$. Furthermore when $p = 3$ there are 4 isomorphism classes of such algebras. They are denoted $R(q, e) \# KG$ for $e = 1, 2$ and q a primitive cube root of 1 in [AS2, Section 6]. Suppose now that H is a pointed Hopf algebra such that

$$grH \cong R(q, e) \# KG.$$

By [AS2, (5.7)] we can assume there exist $x_i \in P_{g_i}^{\chi_i}(H)$, $(1 \le i \le 4)$ such that the images of $x_1, \ldots, x_4$ in grH from a basis for $R(1)$ and

$$g_1 = g_2 = g \quad , \quad g_3 = g_4 = g^e,$$

$$\chi_1 = \chi_2 = \chi \quad , \quad \chi_3 = \chi_4 = \chi^{-e}$$

where $\chi(g) = q$.

Set $y_{i-2} = x_i g_i^{-1}$ for $i = 3, 4$. The subalgebra B generated by g, x_1, x_2 is a Hopf algebra such that $G(B)$ has order 3 and grH is of type A_2. It follows from Theorem 2.2 that $H \cong grH$ and hence

$$(adx_1)^2(x_2) = (adx_2^2)(x_1) = x_1^3 = x_2^3 = (x_1x_2 - qx_2x_1)^3 = 0.$$

Similarly

$$(ady_1)^2(y_2) = (ady_2)^2(y_1) = y_1^3 = y_2^3 = (y_2y_1 - qy_1y_2)^3 = 0.$$

Also since $\chi_1(g_3)\chi_3(g_1) = 1$ it follows from Lemma 1.5 with $r = 1$ or by a direct calculation that $[x_j, y_k]$ is (g, g^{-1})-primitive for all j, k. If $[x_j, y_k] = 0$ for all j, k then $H \cong grH$. Otherwise $e = 1$ by Lemma 2.1 and we have

$$[x_j, y_k] = \lambda_{jk}(g - g^{-1})$$

for some 2×2 matrix Λ. Now suppose $P, Q \in GL_2(K)$ and define $x_i' = \sum_j p_{ij}x_j, y_\ell' = \sum_k q_{k\ell}y_k$. Then

$$[x_i', y_\ell'] = \lambda_{i\ell}'(g - g^{-1})$$

where $\Lambda' = P\Lambda Q$. Since $\Lambda \ne 0$ we can choose P, Q such that $\Lambda' = \Lambda_\epsilon$ for $\epsilon = 0$ or 1. Then replacing x_i' by x_i and y_i' by y_i we have the first claim in the Theorem.

If $\phi : H = H(q, \epsilon) \to H' = H(q', \epsilon')$ is an isomorphism then $q = q'$ by passing to the graded algebras and using [AS2, Lemma 6.5]. Denote the generators of $H(q', \epsilon')$ by $g, x_1', x_2', y_1', y_2'$ Since $\phi(x_i)$ is a nontrivial $(1, \phi(g))$-primitive in H', we have $\phi(g) = g$ and $\phi(x_i) \in span\{g - 1, x_1', x_2', y_1', y_2'\}$. Applying ϕ to the equation $gx_ig^{-1} = qx_i$ we see that ϕ maps $span\{x_1, x_2\}$ onto $span\{x_1, x_2\}$. Similarly ϕ maps $span\{y_1, y_2\}$ onto $span\{y_1', y_2'\}$. The result follows easily.

2.4 Now we discuss decomposable case.

Theorem. Let H be a finite dimensional pointed Hopf algebra such that $G(H) = (g)$ has odd prime order p and grH is of finite decomposable Cartan type. Then either $H \cong grH$ or one of the following holds

(1) H is the Frobenius-Lusztig kernels $u_q(sl(2))$ where q is a pth root of unity.

(2) $p = 3$ and H is one of the Hopf algebras described $H(q, \epsilon)$ in Theorem 2.2.

(3) $p = 3$ and H is the subalgebra of $H(q, 1)$ generated by x_1, x_2, y_1 and g.

Proof. By [AS2, Proposition 5,1] $gr(H)$ has type Cartan type $A_1 \times A_1, A_2 \times A_1$ or $A_2 \times A_2$ and in the last two cases $p = 3$. For type $A_1 \times A_1$ the only non-graded examples are the algebras $u_q(sl(2))$ by for example [AS1, Section 1]. For type $A_2 \times A_2$ the result follows from Theorem 2.2. The result for type $A_2 \times A_1$ is easily deduced from the proof of Theorem 2.2 and we leave the details to the reader.

References

[AG] N. Andruskiewitsch and M. Graña, Braided Hopf algebras over non-abelian groups, Bol. Acad. Nacional de Ciencias, (Córdoba) 63, (1999), 45-78.

[AS1] N. Andruskiewitsch and H.-J. Schneider, Lifting of quantum linear spaces and pointed Hopf algebras of order p^3, J. Alg. 209 (1998), 658-691.

[AS2] _____, Finite quantum groups and Cartan matrices, Adv. in Math., to appear.

[AS3] _____, Lifting of Nichols algebras of type A_2 and pointed Hopf algebras of order p^4, in "Hopf algebras and quantum groups", Proceedings of the colloquium in Brussels, 1998, ed. S. Caenepeel, 1-18.

[AS4] _____, Finite quantum groups over abelian groups of prime exponent , preprint, 1999.

[BDG] M. Beattie, S. Dăscălescu and L. Grunenfelder, On the number of types of finite dimensional Hopf algebras, Invent. Math. 136 (1999), 1-7.

[CD] S. Caenepeel and S. Dăscălescu, Pointed Hopf algebras of dimension p^3, J. Algebra 209 (1998) 622-634.

[CM1] W. Chin and I.M. Musson, Multiparameter quantum enveloping algebras, J. Pure and Applied Algebra, 107 (1996), 171-191.

[CM2] _____, The coradical filtration for quantized enveloping algebras, J. London Math. Soc 53 (1996), 50-62, Corrig. to appear.

[CP] V. Chari and A. Pressley, A Guide to Quantum Groups, Cambridge University Press, 1994, Cambridge.

[DCP] C. De Concini and C. Procesi, Quantum Groups, in "D-modules, Representation Theory and Quantum Groups", pages 31-140, Lecture Notes in Math. 1565, Springer-Verlag, 1993, Berlin.

[G] S. Gelaki, Pointed Hopf algebras and Kaplansky's 10th conjecture, J. Algebra 209 (1998), 635-657.

[HLT] T. J. Hodges, T. Levasseur and M. Toro, Algebraic structure of multi-parameter quantum groups, Adv. in Math. 126 (1997), 52-92.

[J] J. C. Jantzen, Lectures on Quantum Groups, Graduate Studies in Mathematics, vol. 6, Amer. Math. Soc. Providence, 1996.

[K] V. G. Kac, Infinite Dimensional Lie Algebras, Cambridge University Press, Cambridge, 1990.

[L] G. Lusztig, Quantum groups at roots of 1, Geom. Dedicata 35 (1990), 89-114.

[M1] S. Montgomery, Some remarks on filtrations of Hopf algebras, Comm. Alg. 21 (1993), 999-1007.

[M2] S. Montgomery, Hopf Algebras and Their Actions on Rings, CBMS Regional Conference Series No. 82, Amer. Math. Soc. Providence, second printing 1995.

[Mj] S. Majid, Crossed products by braided groups and bosonization, J. Algebra 163 (1994), 165-190.

[N] W. D. Nichols, Bialgebras of type one, Comm. Alg 6 (1978), 1521-1552.

[R] D. Radford, Hopf algebras with projection, J.Algebra 92 (1985), 322-347.

[SvO] D. Stefan and F. van Oystaeyen, Hochschild cohomology and coradical filtration of pointed Hopf algebras, J. Alg 210 (1998), 535-556.

[Z] Y.Zhu, Hopf algebras of prime dimension, Int. Math. Res. Notices 1 (1994), 53-59.

ON SOME TWO PARAMETER QUANTUM AND JORDANIAN DEFORMATIONS, AND THEIR COLOURED EXTENSIONS*

Deepak Parashar[†] and Roger J. McDermott[‡]

School of Computer and Mathematical Sciences
The Robert Gordon University, St. Andrew Street
Aberdeen AB25 1HG, United Kingdom

Abstract

This paper suveys some recent algebraic developments in two parameter Quantum deformations and their Nonstandard (or Jordanian) counterparts. In particular, we discuss the contraction procedure and the quantum group homomorphisms associated to these deformations. The scheme is then set in the wider context of the coloured extensions of these deformations, namely, the so-called Coloured Quantum Groups.

I Introduction

Recent years have witnessed considerable development in the study of multi-parameter quantum deformations from both, the algebraic as well the differential geometric point of view. These have also found profound applications in many diverse areas of Mathematical Physics. Despite of the intensive and successful development of the mathematical theory of multiparameter quantum deformations or quantum groups, various important aspects still need thorough investigation. Besides, all quantum groups seem to have a natural coloured extension thereby defining corresponding coloured quantum groups. It is the aim of this paper to address some of the key issues involved.

Two parameter deformations provide an obvious step in constructing generalisations of single parameter deformations. Besides being mathematically interesting in their own right, two parameter quantum groups serve as very good examples in generalising physical theories based on the quantum group symmetry. $GL_{p,q}(2)$ and $GL_{h,h'}(2)$ are well known examples of two parameter Quantum and Jordanian deformations of the space of 2×2 matrices. Just as both these quantum groups are of great significance in building up various

---

*Presented by D.P. at the LMS - Durham Symposium on Quantum Groups, Durham, 19-29 July 1999, to appear in the Proceedings.

[†]email: deeps@scms.rgu.ac.uk
[‡]email: rm@scms.rgu.ac.uk

mathematical and physical theories, it is worthwhile to look for other possible examples, including the 'coloured' ones, which might play a fundamental role in future researches. We wish to focus our attention on a new two parameter quantum group [1], $G_{r,s}$ which sheds light on some of the above mentioned issues. $G_{r,s}$ is a quasitriangular Hopf algebra generated by five elements, four of which form a Hopf subalgebra ismomorphic to $GL_q(2)$, while the fifth generator relates $G_{r,s}$ to $GL_{p,q}(2)$.

The $G_{r,s}$ quantum group, which is the basis of our investigation, is defined in Section II. In Section III, we give a new Jordanian analogue of $G_{r,s}$, denoted $G_{m,k}$ and establish a homomorphism with $GL_{h,h'}(2)$. Both $G_{r,s}$ and $G_{m,k}$ admit a natural coloured extension and this is given in Section IV. Section V generalises the contraction procedure to the case of coloured quantum groups and discusses various homomorphisms. In section VI, we make concluding remarks and give possible physical significance of our results. Throughout this paper, we shall endeavour to refrain from too much of technical details, which can be found in the appropriate references.

II   Two parameter q- deformations

The quantum group $G_{r,s}$ was defined in [1] as a quasitriangular Hopf algebra with two deformation parameters r and s, and generated by five elements a,b,c,d and f. The generators a,b,c,d of this Hopf algebra form a subalgebra, infact a Hopf subalgebra, which coincides exactly with the single parameter dependent $GL_q(2)$ quantum group when $q = r^{-1}$. Moreover, the two parameter dependent $GL_{p,q}(2)$ can also be realised through the generators of this $G_{r,s}$ Hopf algebra, provided the sets of deformation parameters (p,q) and (r,s) are related to each other in a particular fashion. This new algebra can, therefore, be used to realise both $GL_q(2)$ and $GL_{p,q}(2)$ quantum groups. Alternatively, this $G_{r,s}$ structure can be considered as a two parameter quantisation of the classical $GL(2) \otimes GL(1)$ group. The first four generators of $G_{r,s}$, i.e. a, b, c, d correspond to $GL(2)$ group at the classical level and the remaining generator f is related to $GL(1)$. In fact, $G_{r,s}$ can also be interpreted as a quotient of multiparameter q- deformation of $GL(3)$.

The elements of $G_{r,s}$ can be conveniently arranged in the matrix $T = \left(\begin{smallmatrix} a & b & 0 \\ c & d & 0 \\ 0 & 0 & f \end{smallmatrix} \right)$ and the coalgebra and counit are $\Delta(T) = T \dot{\otimes} T$, $\varepsilon(T) = \mathbf{1}$. It should be mentioned that the quantum determinant $\delta = Df$ (where $D = ad - r^{-1}bc$) is group-like but not central. The above block diagonal form of the T- matrix is particularly convenient to understand the related schematics. The $G_{r,s}$ R- matrix is given in [1,2], and the most general Hopf algebra generated by this R- matrix is multiparameter $GL(3)$ with the T-matrix of the form $\left(\begin{smallmatrix} a & b & x_1 \\ c & d & x_2 \\ v_1 & v_2 & f \end{smallmatrix} \right)$. It can be shown that the two-sided Hopf ideal generated by x_1, x_2 when factored out yields the Inhomogenous multiparameter $IGL(2)$. Furthermore, if one factors out yet another two-sided Hopf ideal generated by elements v_1, v_2, what one obtains is precisely the $G_{r,s}$ Hopf algebra. The relation of $G_{r,s}$ with various known q-deformed groups can be exhibited as

$$GL_Q(3)$$
$$\downarrow Q$$
$$IGL_Q(2)$$
$$\downarrow Q$$
$$GL(2) \otimes GL(1) \xleftarrow{\mathcal{L}} G_{r,s} \xrightarrow{\mathcal{F}} GL_{p,q}(2)$$
$$\downarrow \mathcal{S}$$
$$GL_q(2)$$

where $\mathcal{Q}$, $\mathcal{F}$, $\mathcal{S}$ and $\mathcal{L}$ denote the Quotient, Hopf algebra homomorphism, (Hopf)Subalgebra and (classical) Limit respectively. $GL_Q(3)$ denotes the multiparameter q- deformed $GL(3)$ and $IGL_Q(2)$ is the inhomogenous multiparameter q- deformation of $GL(2)$. Motivated by the rich structure of $G_{r,s}$, this quantum group has recently been studied by the authors in detail [2]. As an intial step in the further understanding of $G_{r,s}$, the authors have derived explicitly the dual algebra and showed that it is isomorphic to the single parameter deformation of $gl(2) \oplus gl(1)$, with the second parameter appearing in the costructure. In [2], the authors have also constructed a differential calculus on $G_{r,s}$, which inturn provides a realisation of the calculus on $GL_{p,q}(2)$.

III   Two parameter h- deformations

Jordanian deformations (also known as h-deformations) of Lie groups and Lie algebras have attracted a lot of attention in recent years. A peculiar feature of this deformation is that the corresponding R-matrix is triangular i.e. $R_{12}R_{21} = 1$. These deformations are called 'Jordanian' due to the Jordan normal form of the R-matrix. It was shown in [3] that upto isomorphism, $GL_q(2)$ and $GL_h(2)$ are the only possible distinct deformations (with central determinant) of the group $GL(2)$. In [4], an interesting observation was made that the h- deformations could be obtained by a singular limit of a similarity transformation from the q- deformations, and this was generalised to multiparameter deformations as well as to higher dimensions i.e space of $n \times n$ quantum matrices [5]. For the purpose of current investigation, the authors have applied the contraction procedure to $G_{r,s}$ to obtain a new Jordanian quantum group $G_{m,k}$ [6]. It turns out that this new structure is also related to other known Jordanian quantum groups.

The $G_{m,k}$ quantum group can be defined as a triangular Hopf algebra generated by the T- matrix $\left(\begin{smallmatrix} a & b & 0 \\ c & d & 0 \\ 0 & 0 & f \end{smallmatrix} \right)$. The set of commutation relations consisting of elements a,b,c,d form a subalgebra that coincides exactly with the single parameter Jordanian $GL_h(2)$ for $m = h$. This is exactly analogous to the q-deformed case where the first four elements of $G_{r,s}$ form the $GL_q(2)$ Hopf subalgebra. Again, the remaining fifth element f generates the $GL(1)$ group, as it did in the q-deformed case, and the second parameter appears only through the cross commutation relations between $GL_m(2)$ and $GL(1)$ elements. Therefore, $G_{m,k}$ can also be considered as a two parameter Jordanian deformation of classical $GL(2) \otimes GL(1)$ group. Furthermore, $G_{m,k}$ also provides a realisation of the two parameter Jordanian $GL_{h,h'}(2)$. Besides, it may be interpreted as a quotient of the multiparameter Jordanian deformation of $GL(3)$, denoted $GL_J(3)$ as well as that of inhomogenous $IGL(2)$, denoted $IGL_J(2)$ quantum groups. This can be represented as follows

$$GL_J(3)$$
$$\downarrow \mathcal{Q}$$
$$IGL_J(2)$$
$$\downarrow \mathcal{Q}$$
$$GL(2) \otimes GL(1) \xleftarrow{\mathcal{L}} G_{m,k} \xrightarrow{\mathcal{F}} GL_{h,h'}(2)$$
$$\downarrow \mathcal{S}$$
$$GL_h(2)$$

where the maps $\mathcal{Q}$, $\mathcal{F}$, $\mathcal{S}$ and $\mathcal{L}$ are as before.

IV   Coloured Extensions

The standard quantum group relations can be extended by parametrising the corresponding generators using some continuous 'colour' variables and redefining the associated algebra and coalgebra in a way that all Hopf algebraic properties remain preserved [1,7,8]. For the case of a single parameter quantum deformation of $GL(2)$ (with deformation parameter r), its 'coloured' version [1] is given by the R-matrix, denoted $R_r^{\lambda,\mu}$ which satisfies

$$R_{12}^{\lambda,\mu} R_{13}^{\lambda,\nu} R_{23}^{\mu,\nu} = R_{23}^{\mu,\nu} R_{13}^{\lambda,\nu} R_{12}^{\lambda,\mu}$$

the so-called 'Coloured' Quantum Yang Baxter Equation (CQYBE). It should be stressed at this point that the coloured R- matrix provides a nonadditive-type solution $R^{\lambda,\mu} \neq R(\lambda - \mu)$ of the Yang-Baxter equation, which is in general multicomponent and the parameters λ, μ, ν are considered as 'colour' parameters. Such solutions were first discovered in the study of integrable models [9]. This gives rise to the coloured RTT relations

$$R_r^{\lambda,\mu} T_{1\lambda} T_{2\mu} = T_{2\mu} T_{1\lambda} R_r^{\lambda,\mu}$$

(where $T_{1\lambda} = T_\lambda \dot{\otimes} \mathbf{1}$ and $T_{2\mu} = \mathbf{1} \dot{\otimes} T_\mu$) in which the entries of the T matrices carry colour dependence. The coproduct and counit for the coalgebra structure are given by $\Delta(T_\lambda) = T_\lambda \dot{\otimes} T_\lambda$, $\varepsilon(T_\lambda) = \mathbf{1}$ and depend only on one colour parameter. By contrast, the algebra structure is more complicated with generators of two different colours appearing simultaneously in the algebraic relations. The full Hopf algebraic structure can be constructed and results in a coloured extension of the quantum group. Since λ and μ are continuous variables, this implies the coloured quantum group has an infinite number of generators.

The above coloured generalisation of the FRT formalism was given by Kundu and Basu-Mallick [1,10] and that of the Drinfeld-Jimbo formulation of quantised universal enveloping algebras has been given by Bonatos, Quesne *et al* [11]. In the context of knot theory, Ohtsuki [12] introduced some coloured quasitriangular Hopf algebras, which are characterised by the existence of a coloured universal R- matrix, and he applied his theory to $U_q sl(2)$. Coloured generalisations of quantum groups can also be understood as an application of the twisting procedure, in a manner similar to the multiparameter generalisation of quantum groups. Jordanian deformations also admit coloured extensions [7]. The associated R-matrix satisfies the CQYBE and is 'colour' triangular i.e. $R_{12}^{\lambda,\mu} = (R_{21}^{\mu,\lambda})^{-1}$, a coloured extension of the notion of triangularity.

Coloured Extension of $G_{r,s}$: $G_r^{s,s'}$

The coloured extension of $G_{r,s}$ proposed in [1] has only one deformation parameter r and two colour paramters s and s'. The second deformation parameter of the uncoloured case now plays the role of a colour parameter. In such a coloured extension, the first four generators a,b,c,d are kept independent of the colour parameters while the fifth generator f is now paramterised by s and s'. The matrices of generators are

$$T_s = \begin{pmatrix} a & b & 0 \\ c & d & 0 \\ 0 & 0 & f_s \end{pmatrix} \quad , \quad T_{s'} = \begin{pmatrix} a & b & 0 \\ c & d & 0 \\ 0 & 0 & f_{s'} \end{pmatrix}$$

From the RTT relations, one observes that the commutation relations between a,b,c,d are as before but f_s and $f_{s'}$ now satisfy two colour copies of the relations satisfied by f of the uncloured $G_{r,s}$. In addition, the relation $[f_s, f_{s'}] = 0$ holds. The associated coloured R-matrix, denoted $R_r^{s,s'}$ satisfies the CQYBE

$$R_{12}(r; s, s')R_{13}(r; s, s'')R_{23}(r; s', s'') = R_{23}(r; s', s'')R_{13}(r; s, s'')R_{12}(r; s, s')$$

and the corresponding coloured quantum group is denoted $G_r^{s,s'}$.

Coloured Extension of $G_{m,k}$: $G_m^{k,k'}$

Similar to the case of $G_{r,s}$, we have proposed [13] a coloured extension of the Jordanian quantum group $G_{m,k}$. The first four generators remain independent of the colour parameters k and k' whereas the generator f is parameterised by k and k'. Again, the second deformation parameter k of the uncoloured case now plays the role of a colour parameter and the T-matrices are

$$T_k = \begin{pmatrix} a & b & 0 \\ c & d & 0 \\ 0 & 0 & f_k \end{pmatrix} \quad , \quad T_{k'} = \begin{pmatrix} a & b & 0 \\ c & d & 0 \\ 0 & 0 & f_{k'} \end{pmatrix}$$

The commutation relations between a,b,c,d remain unchanged whereas f_k and $f_{k'}$ satisfy two colour copies of the relations satisfied by f of the uncloured $G_{m,k}$. In addition, the relation $[f_k, f_{k'}] = 0$ holds. The associated coloured R-matrix, denoted $R_m^{k,k'}$, is a solution of the CQYBE

$$R_{12}(m; k, k')R_{13}(m; k, k'')R_{23}(m; k', k'')$$

$$= R_{23}(m; k', k'')R_{13}(m; k, k'')R_{12}(m; k, k')$$

and is colour triangular. The corresponding coloured Jordanian quantum group is denoted $G_m^{k,k'}$.

V   Contractions and Homomorphisms

The R-matrix of the Jordanian (or h-deformation) can be viewed as a singular limit of a similarity transformation on the q-deformed R-matrix [4]. Let $g(\eta)$ be a matrix dependent on a contraction parameter η which is itself a function of one of the deformation parameters of the q-deformed algebra. This can be used to define a transformed q-deformed R-matrix

$$R_h = (g^{-1} \otimes g^{-1})R_q(g \otimes g)$$

The R-matrix of the Jordanian deformation is then obtained by taking a limiting value of the parameter η. Even though the contraction parameter η is undefined in this limit, the new R-matrix is finite and gives rise to a new quantum group structure through the RTT-relations. For example, in the contraction process which takes $GL_q(2)$ to $GL_h(2)$, the contraction matrix is

$$g(\eta) = \begin{pmatrix} 1 & \eta \\ 0 & 1 \end{pmatrix}$$

where $\eta = \frac{h}{1-q}$ with h a new free parameter. Such transformations have proved to be powerful tools in establishing various connections between the q- and the h- deformed quantum groups, which were previously obscure. In the context of the quantum groups under consideration in the present paper, the contraction procedure was successfully applied [6] to the $G_{r,s}$ quantum group of Section II to obtain the Jordanian $G_{m,k}$ given in Section III. Furthermore, the multiparameter Jordanian $GL_J(3)$ and hence the multiparamter

Inhomogeneous $IGL_J(2)$ were also obtained by contracting their respective q- deformed counterparts [14].

The Hopf algebra homomorphism $\mathcal{F}$ from $G_{r,s}$ to $GL_{p,q}(2)$, which provides a realisation of the latter, is given by

$$\mathcal{F} : G_{r,s} \longmapsto GL_{p,q}(2)$$

$$\mathcal{F}\begin{pmatrix} a & b \\ c & d \end{pmatrix} \longmapsto \begin{pmatrix} a' & b' \\ c' & d' \end{pmatrix} = f^N \begin{pmatrix} a & b \\ c & d \end{pmatrix}$$

The elements a', b', c' and d' are the generators of $GL_{p,q}(2)$ and N is a fixed non-zero integer. The relation between the deformation parameters (p, q) and (r, s) is given by

$$p = r^{-1}s^N \quad , \qquad q = r^{-1}s^{-N}$$

A Hopf algebra homomorphism

$$\mathcal{F} : G_{m,k} \longmapsto GL_{h,h'}(2)$$

of exactly the same form as in the q-deformed case, exists between the generators of $G_{m,k}$ and $GL_{h,h'}(2)$ provided that the two sets of deformation parameters (h, h') and (m, k) are related via the equation

$$h = -m + Nk \quad , \qquad h' = -m - Nk$$

Note that for vanishing k, one gets the one parameter case. In addition, using the above realisation together with the coproduct, counit and antipode axioms for the $G_{m,k}$ algebra and the respective homeomorphism properties, one can easily recover the standard coproduct, counit and antipode for $GL_{h,h'}(2)$. Thus, the Jordanian $GL_{h,h'}(2)$ group can in fact be reproduced from the newly defined Jordanian $G_{m,k}$. It is curious to note that if we write $p = e^h$, $q = e^{h'}$, $r = e^m$ and $s = e^k$, then the relations between the parameters in the q-deformed case and the h-deformed case are identical. The systematics of the uncloured quantum groups discussed here can be summarised in the following commutative diagram

$$
\begin{array}{ccccccc}
GL_Q(3) & \xrightarrow{\ \mathcal{Q}\ } & IGL_Q(2) & \xrightarrow{\ \mathcal{Q}\ } & G_{r,s} & \xrightarrow{\ \mathcal{F}\ } & GL_{p,q}(2) \\
\Big\downarrow{\scriptstyle c} & & \Big\downarrow{\scriptstyle c} & & \Big\downarrow{\scriptstyle c} & & \Big\downarrow{\scriptstyle c} \\
GL_J(3) & \xrightarrow[\ \mathcal{Q}\ ]{} & IGL_J(2) & \xrightarrow[\ \mathcal{Q}\ ]{} & G_{m,k} & \xrightarrow[\ \mathcal{F}\ ]{} & GL_{h,h'}(2)
\end{array}
$$

where $\mathcal{Q}$, $\mathcal{C}$ and $\mathcal{F}$ denote the quotient, contraction and the Hopf algebra homomorphism. The contraction procedure discussed above has been successfully applied [13] to the case of coloured quantum groups yielding new coloured Jordanian deformations. We apply to $R_r^{\lambda,\mu}$, the coloured R-matrix for q-deformed $GL(2)$, the transformation

$$(g \otimes g)^{-1} R_r^{\lambda,\mu} (g \otimes g)$$

where g is the two dimensional transformation matrix $\begin{pmatrix} 1 & \eta \\ 0 & 1 \end{pmatrix}$ and η is chosen to be $\eta = \frac{m}{1-r}$. In the limit $r \to 1$, we obtain a new R-matrix, $R_m^{\lambda,\mu}$ which is a coloured R-matrix for a Jordanian deformation of $GL(2)$. The contraction is then also used to obtain the coloured extension $G_m^{k,k'}$ of $G_{m,k}$, from the coloured extension $G_r^{s,s'}$ of $G_{r,s}$. The R-matrix $R_m^{k,k'}$ is obtained as the contraction limit of the R-matrix for the coloured extension of $G_{r,s}$ via the transformation

$$R_m^{k,k'} = \lim_{r \to 1}(G \otimes G)^{-1} R_r^{s,s'}(G \otimes G)$$

where

$$G = \begin{pmatrix} g & 0 \\ 0 & 1 \end{pmatrix}; \quad g = \begin{pmatrix} 1 & \eta \\ 0 & 1 \end{pmatrix}, \quad \eta = \frac{m}{r-1}$$

The Hopf algebra homomorphism from $G_r^{s,s'}$ to $GL_r^{\lambda,\mu}(2)$

$$\mathcal{F}_N : G_r^{s,s'} \longmapsto GL_r^{\lambda,\mu}(2)$$

is given by

$$\mathcal{F}_N : \begin{pmatrix} a & b \\ c & d \end{pmatrix} \longmapsto \begin{pmatrix} a'_\lambda & b'_\lambda \\ c'_\lambda & d'_\lambda \end{pmatrix} = f_s^N \begin{pmatrix} a & b \\ c & d \end{pmatrix}$$

$$\mathcal{F}_N : \begin{pmatrix} a & b \\ c & d \end{pmatrix} \longmapsto \begin{pmatrix} a'_\mu & b'_\mu \\ c'_\mu & d'_\mu \end{pmatrix} = f_{s'}^N \begin{pmatrix} a & b \\ c & d \end{pmatrix}$$

where N is a fixed non-zero integer and the sets of colour parameters (s, s') and (λ, μ) are related through quantum deformation parameter r by

$$s = r^{2N\lambda} \quad , \quad s' = r^{2N\mu}$$

The primed generators $a'_\lambda, b'_\lambda, c'_\lambda, d'_\lambda$ and $a'_\mu, b'_\mu, c'_\mu, d'_\mu$ belong to $GL_r^{\lambda,\mu}(2)$ whereas the unprimed ones a, b, c, d, f_s and $f_{s'}$ are generators of $G_r^{s,s'}$. If we now denote the generators of $GL_m^{\lambda,\mu}(2)$ by $a'_\lambda, b'_\lambda, c'_\lambda, d'_\lambda$ and $a'_\mu, b'_\mu, c'_\mu, d'_\mu$ and the generators of $G_m^{k,k'}$ by a, b, c, d, f_k and $f_{k'}$ then a Hopf algebra homomorphism from $G_m^{k,k'}$ to $GL_m^{\lambda,\mu}(2)$

$$\mathcal{F}_N : G_m^{k,k'} \longmapsto GL_m^{\lambda,\mu}(2)$$

is of exactly the same form

$$\mathcal{F}_N : \begin{pmatrix} a & b \\ c & d \end{pmatrix} \longmapsto \begin{pmatrix} a'_\lambda & b'_\lambda \\ c'_\lambda & d'_\lambda \end{pmatrix} = f_k^N \begin{pmatrix} a & b \\ c & d \end{pmatrix}$$

$$\mathcal{F}_N : \begin{pmatrix} a & b \\ c & d \end{pmatrix} \longmapsto \begin{pmatrix} a'_\mu & b'_\mu \\ c'_\mu & d'_\mu \end{pmatrix} = f_{k'}^N \begin{pmatrix} a & b \\ c & d \end{pmatrix}$$

The sets of colour parameters (k, k') and (λ, μ) are related to the Jordanian deformation parameter m by

$$Nk = -2m\lambda \quad , \quad Nk' = -2m\mu$$

and N, again, is a fixed non-zero integer. The schematics of our analysis for the coloured quantum groups is represented in the diagram

$$
\begin{array}{ccc}
G_{r,s} & \xrightarrow{\;\mathcal{E}\;} & G_r^{s,s'} & \xrightarrow{\;\mathcal{F}\;} & GL_r^{\lambda,\mu}(2) \\
\mathcal{C}\downarrow & & \downarrow\mathcal{C} & & \downarrow\mathcal{C} \\
G_{m,k} & \xrightarrow[\;\mathcal{E}\;]{} & G_m^{k,k'} & \xrightarrow[\;\mathcal{F}\;]{} & GL_m^{\lambda,\mu}(2)
\end{array}
$$

where $\mathcal{C}$, $\mathcal{F}$ and $\mathcal{E}$ denote the contraction, Hopf algebra homomorphism and coloured extension respectively. In both of the commutative diagrams above, the objects at the top level are the q deformed ones and the corresponding Jordanian counterparts are shown at the bottom level.

VI   Conclusions

In the present work, we have obtained a new Jordanian quantum group $G_{m,k}$ by contraction of the q- deformed quantum group $G_{r,s}$. We then used this new structure to establish quantum group homomorphisms with other known two parameter quantum groups at the Jordanian level. At the same time we also showed that such homomorphisms commute with the contraction procedure. Our analysis is then set in the wider context of coloured quantum groups. We give a coloured generalisation of the contraction procedure and obtain new coloured Jordanian quantum groups. A careful study of the properties of both $G_{r,s}$ and $G_{m,k}$ lead to their respective coloured extensions. Furthermore, we show that the homomorphisms of the uncoloured case naturally extend to the coloured case.

The physical interest in studying $G_{r,s}$ lies in the observation that when endowed with a $*$- structure, this quantum group specialises to a two parameter quantum deformation of $SU(2) \otimes U(1)$ which is precisely the gauge group for the theory of electroweak interactions. Since gauge theories have an obvious differential geometric description, the study of differential calculus [2] provides insights in constructing a q-gauge theory based on $G_{r,s}$. It would also be of significance to generalise the formalism of differential calculus to the case of coloured quantum groups and explore possible physical applications.

Acknowledgments

D.P. is grateful to the organisers of the Symposium and would like to thank Prof. David Radford and Dr. Gustav Delius for useful comments. The authors have also benefited by discussions with Prof. Vlado Dobrev and Dr. Preeti Parashar.

References

[1] B. Basu-Mallick, hep-th/9402142; *Intl. J. Mod. Phys.* **A10**, 2851 (1995).

[2] D. Parashar and R. J. McDermott, Kyoto University preprint RIMS - 1260 (1999), math.QA/9901132.

[3] B. A. Kupershmidt, *J. Phys.* **A25**, L1239 (1992).

[4] A. Aghamohammadi, M. Khorrami and A. Shariati, *J. Phys.* **A28**, L225 (1995).

[5] M. Alishahiha, *J. Phys.* **A28**, 6187 (1995).

[6] D. Parashar and R. J. McDermott, math.QA/9909001, *Czech. J. Phys.*, in press.

[7] P. Parashar, *Lett. Math. Phys.* **45**, 105 (1998).

[8] C. Quesne, *J. Math. Phys.* **38**, 6018 (1997); *ibid* **39**, 1199 (1998).

[9] V. V. Bazhanov and Yu. G. Stroganov, *Theor. Math. Phys.* **62**, 253 (1985).

[10] A. Kundu and B. Basu-Mallick, *J. Phys.* **A27**, 3091 (1994); B. Basu-Mallick, *Mod. Phys. Lett.* **A9**, 2733 (1994).

[11] D. Bonatos *et. al.*, *J. Math. Phys.* **38**, 369 (1997); C. Quesne, q-alg/9705022.

[12] T. Ohtsuki, *J. Knot Theor. Its Rami.* **2**, 211 (1993).

[13] D. Parashar and R. J. McDermott, math.QA/9911194, *J. Math. Phys.*, in press.

[14] R. J. McDermott and D. Parashar, math.QA/9909045, *Czech. J. Phys.*, in press.

TENSOR CATEGORIES AND BRAID REPRESENTATIONS

HANS WENZL

ABSTRACT. Representations of braid groups of types A and B are discussed which are useful for the purpose of extending an analog of Schur duality to quantum groups also of Lie type other than A. In particular, this contains some new results in connection with spinor representations and and with exceptional Lie types. As a possible application, a reconstruction technique is described which so far has been successfully applied to classify tensor categories whose Grothendieck semiring is equal to $Rep(SU(N))$.

Classically, the symmetric groups were used to describe representations of the general linear group $Gl(N)$ by decomposing tensor products of its vector representations. For other Lie types this approach becomes much more complicated or is not available. For quantum groups, the symmetric groups are replaced by braid groups which have a far richer representation theory. The purpose of this article is to describe how braid groups can be used to decompose tensor products of representations of quantum groups and the corresponding objects in fusion categories.

In Section 1 we review Schur duality, and its analog for quantum groups of type A discovered by Jimbo. Using representations of Hecke algebras of type A, one can completely classify any rigid semisimple monoidal tensor category whose Grothendieck semiring coincides with the one of $Rep(SU(N))$. This result also extends to the associated fusion categories.

In Section 2, braid representations are considered which can be regarded as a q-analog of Brauer's centralizer algebra for orthogonal and symplectic groups. Moreover, in a similar approach, using representations of braid groups of type B, one can describe the decomposition of tensor products of the form $V_\varepsilon \otimes V^{\otimes n}$, with V_ε a spinor representation and V the vector representation of $U_q so_N$.

In the 3rd section, we describe braid representations in connection with representations of quantum groups of exceptional Lie type. In particular, we obtain a nontrivial generalization of the braid representations considered in Section 2. This can be regarded as a version of an exceptional series in the sense of Vogel and Deligne. However we consider the representations of smallest dimensions instead of the adjoint representations. The uniform behaviour is only observed in the new part of tensor powers of these representations, i.e. that part of the tensor powers whose simple components have not already appeared in smaller tensor powers (see Remark 3.11). At least for E_6 and E_7

Supported in part by NSF grant #DMS 9706839.

it is not difficult to get from this the decomposition of the complete tensor product.

As mentioned before, the focus of this review is rather limited, namely the use of braid representations for the decomposition of tensor products of representations of quantum groups; work of this nature has also appeared in the work of a number of authors e.g. [12] [26] [19]. Many important and closely related results have been omitted due to lack of time and knowledge. So in particular, very little is said about general reconstruction results for categories and the connection of this work with low dimensional topology (see e.g. [14] [15] [30] [37]), operator algebras (see [13] [33]) and mathematical physics (see e.g. [23] [37]).

1. LIE TYPE A

1.1. Schur duality.

To motivate the constructions used later, let us review Schur duality. Let V be the N-dimensional vector representation of $SU(N)$. Then we have an action of $SU(N)$ on $V^{\otimes n}$ defined in the usual way, as well as an action of the symmetric group S_n, defined via permuting the tensor factors in $V^{\otimes n}$. Schur duality then says

Theorem 1.1. *The algebras generated by the actions of $SU(N)$ and by S_n are commutants of each other.*

Schur duality allows us to describe the tensor category $Rep(SU(N))$ of representations of $SU(N)$ only in terms of the representation theory of symmetric groups. We shall sketch this construction here in some detail as it serves as a model for the constructions done later in this section. In the following, we shall denote the space of linear maps between 2 $SU(N)$-modules V and W which intertwine the group action by $\mathrm{Hom}(V, W)$, with a similar notation for endomorphisms.

Observe that any irreducible representation of $SU(N)$ appears in some tensor power of the vector representation V. It therefore suffices to consider the subcategory whose objects are $\{V^{\otimes n}, n = 0, 1, 2, \ldots\}$, where $V^{\otimes 0}$ is defined to be the trivial representation $\mathbf{1}$. Schur duality tells us that $\mathrm{End}(V^{\otimes n})$ is isomorphic to a quotient of $\mathbb{C}S_n$ which we denote by $\overline{\mathbb{C}S_n} = \overline{\mathbb{C}S_n}^{(N)}$; it is isomorphic to the direct sum of the irreducible representations labeled by Young diagrams with N rows at the most. Observe that the standard embedding of $S_n \times S_m$ into S_{n+m} leads to an embedding of $\mathbb{C}S_n \otimes \mathbb{C}S_m$ into $\mathbb{C}S_{n+m}$, which also factors over the quotients just defined. Hence we get from this the definition of the tensor product $f \otimes g$ for $f \in \mathrm{End}(V^{\otimes n}) \cong \overline{\mathbb{C}S_n}$ and $g \in \mathrm{End}(V^{\otimes m}) \cong \overline{\mathbb{C}S_m}$.

This still leaves us with the description of the tensor product for $f \in \mathrm{Hom}(V^{\otimes n_1}, V^{\otimes m_1})$ and $g \in \mathrm{Hom}(V^{\otimes n_2}, V^{\otimes m_2})$ in general. It is well-known that $\mathrm{Hom}(V^{\otimes n}, V^{\otimes m}) \neq 0$ only if $m - n$ is divisible by N. Recall that the trivial representation of $SU(N)$ appears with multiplicity 1 in $V^{\otimes N}$. Let $i : \mathbf{1} \to V^{\otimes N}$ and $c : V^{\otimes N} \to \mathbf{1}$ be morphisms such that $c \circ i = 1$. Assume that $m < n$ and $n - m = kN$. Then the map

$$\Phi : h \in \mathrm{Hom}(V^{\otimes n}, V^{\otimes m}) \mapsto h \otimes i^{\otimes k} \in \mathrm{End}(V^{\otimes n}) \cong \overline{\mathbb{C}S_n}$$

defines a linear embedding of $\mathrm{Hom}(V^{\otimes n}, V^{\otimes m})$ into $\overline{\mathbb{C}S_n}$. A similar map can be defined for $\mathrm{Hom}(V^{\otimes m}, V^{\otimes n})$, where one tensors with $c^{\otimes k}$. These embeddings are not functorial for $n \neq m$. However, there exist permutations w_1 and w_2 such that

(1.1) $$\Phi(f \otimes g) = w_1 \circ (\Phi(f) \otimes \Phi(g)) \circ w_2;$$

e.g. if $m_1 < n_1$ and $n_1 - m_1 = kN$, w_1 permutes the tensor factors of the target space of $i^{\otimes k}$ to the right of the factors belonging to the image of g, and w_2 is the identity.

1.2. Hecke algebras.

Let $H_n(q)$ denote the Hecke algebra of type A_{n-1}. It can be described as the quotient of the group algebra $\mathbb{C}\mathcal{B}_n$ of Artin's braid group $\mathcal{B}_n$ via the additional relation $(\sigma_i - q)(\sigma_i + 1) = 0$; here σ_i, $i = 1, 2, \ldots n-1$ are the standard generators of $\mathcal{B}_n$ satisfying $\sigma_i \sigma_{i+1} \sigma_i = \sigma_{i+1} \sigma_i \sigma_{i+1}$ and $\sigma_i \sigma_j = \sigma_j \sigma_i$ if $|i - j| \geq 2$. Let us denote the image of σ_i in $H_n(q)$ by g_i. For any $w \in S_n$ we define the element $g_w \in H_n$ by $g_w = g_{i_1} \cdots g_{i_r}$ if $s_{i_1} \cdots s_{i_r}$ is a reduced expression of w in terms of simple reflections s_j corresponding to the permutations $(j, j+1)$, $1 \leq j \leq n-1$. It is well-known that for q not a root of unity $H_n(q) \cong \mathbb{C}S_n$ as algebras. Now it is easy to mimick the construction of the previous subsection, replacing $\mathbb{C}S_n$ by $H_n(q)$ at each step; one only needs to replace w_1 by g_{w_1} and w_2 by $g_{w_2^{-1}}^{-1}$ in equation 1.1. In particular, we can obtain a category whose Grothendieck semiring is isomorphic to the one of $Rep(SU(N))$. Using Jimbo's observation of the analogy of Schur-Weyl duality between the quantum group $U_q sl_N$ and Hecke algebras, it is easy to see that we obtain the representation category of $U_q sl_N$ this way; here $q = v^2$ in the notation of [21].

Let us examine under which conditions we can get a functor between these 2 categories mapping the vector representation of $SU(N)$ to the vector representation of $U_q sl_N$: If such a functor, say F, exists, it would map the projection e corresponding to the antisymmetrization of $V^{\otimes 2}$ (or, in terms of the symmetric group, the eigenprojection of the permutation $(1,2)$ with eigenvalue -1) to a projection $f \in H_2$, i.e. to an eigenprojection of the image of g_1. We shall more generally denote the corresponding eigenprojections of s_i and g_i by e_i and f_i respectively. Functoriality of $\otimes$ then implies that

(1.2)
$$F((e\otimes 1)(1\otimes e)(e\otimes 1)) = (F(e)\otimes 1)(1\otimes F(e))(F(e)\otimes 1) = (f\otimes 1)(1\otimes f)(1\otimes f);$$

here 1 stands as the identity of V. We will use the identifications $e_1 = e \otimes 1$ and $e_2 = 1 \otimes e$, and similar identifications for f_1 and f_2. Observe that f_i is an eigenprojection of g_i, and e_i is an eigenprojection of s_i. It is now an easy exercise to check that $e_1 e_2 e_1$ has nonzero eigenvalues 1 and 1/4, while $f_1 f_2 f_1$ has nonzero eigenvalues 1 and $q/(1 + q)^2$. One concludes that a functor F

can only exist if $1/4 = q/(1+q)^2$, or $q = 1$. Doing this construction for 2 different Hecke algebras with parameters q_1 and q_2, one concludes in exactly the same way that the corresponding categories can only be equivalent if $q_1/(1+q_1)^2 = q_2/(1+q_2)^2$, or equivalently if $q_1 = q_2^{\pm 1}$.

1.3. Reconstruction. Let us now assume that we are given a rigid monoidal semisimple tensor category $\mathcal{C}$ whose Grothendieck semiring is isomorphic to the one of $Rep(SU(N))$. Observe that this isomorphism statement implies a 1-1 correspondence between isomorphism classes of simple objects of $\mathcal{C}$ and of $Rep(SU(N))$. Let X be an object in $\mathcal{C}$ which corresponds to the vector representation V under this correspondence. Let f be the projection onto the subobject of $X^{\otimes 2}$ corresponding to the antisymmetrization of $V^{\otimes 2}$. Using notation from the last subsection, one shows that the Grothendieck semiring structure implies that $f_1 f_2 f_1$ has eigenvalues 0, 1 and $\gamma \neq 0$; from these relations one also deduces that one obtains a homomorphism from the Hecke algebra H_n into $\mathrm{End}(X^{\otimes n})$. Using rigidity and a detailed analysis of certain traces, called Markov traces, this homomorphism can be shown to be surjective. This characterizes $\mathrm{End}(X^{\otimes n})$ for all $n \in \mathbb{N}$ in terms of quotients of group algebras of braid groups, which factor through Hecke algebras. Similarly, one reconstructs tensor products for morphisms in $\mathrm{Hom}(X^{\otimes n}, X^{\otimes m})$, using formula 1.1 and the constructions in subsection 1.2. However, the elements g_{w_1} and g_{w_2} can be varied, where the variation is induced by substituting each generator g_i by $\tilde{\theta} g_i$ with $\tilde{\theta}$ an N^2-th root of unity in the definition of g_{w_1} and g_{w_2}. It turns out that for choices $\tilde{\theta}_1$ and $\tilde{\theta}_2$ of N^2-th roots of unity, the resulting categories are equivalent if and only if $\theta_1 = \theta_2$, where $\theta_i = \tilde{\theta}_i^N$ for $i = 1, 2$. We denote by $Rep(U_q sl_N)^\theta$ the representation category of $U_q sl_N$ twisted by the N-th root of unity θ. We have given an outline of the proof of the following result, due to Kazhdan and the author, where the treatment of the twisting is based on the notes [6] by Bruguières.

Theorem 1.2. *([17]) Let $\mathcal{C}$ be a monoidal rigid semsimple tensor category whose Grothen- dieck semiring is isomorphic to the one of $Rep(SU(N))$. Then $\mathcal{C}$ is equivalent to $Rep(U_q sl_N)^\theta$ for q not a root of unity.*

1.4. Fusion Categories. There are important examples of tensor categories which are not the usual representation categories of groups or Hopf algebras. Due to their origin in the physics literature they are often referred to as fusion categories. They are braided tensor categories whose Grothendieck semiring is a quotient of the one of a compact Lie group, with only finitely many isomorphism classes of simple objects depending on an integer parameter k. E.g. for $G = SU(N)$ the objects are labeled by dominant weights λ such that $\lambda_1 - \lambda_N \leq k$, where we use the usual notation for the weight lattice of $SU(N)$. We shall denote the fusion category corresponding to $SU(N)$ at level k by $\mathcal{C}(SU(N), l)$, where $l = k + N$ (this notation is explained in (a) below). There are a number of ways of constructing fusion categories:

(a) The perhaps easiest way uses the concept of tilting modules, and is due to Andersen (and Paradowski) [1] [2]. It is well-known that a quantum group $U_q\mathfrak{g}$ is no longer semisimple if q is a root of unity. Andersen introduced the category $\mathcal{T}$ of tilting modules which in many ways behaves like a semisimple category. However, those tilting modules whose q-dimension is equal to 0 form a tensor ideal $\mathcal{I}$; this means if N is an object in $\mathcal{I}$ and T is an object in $\mathcal{T}$, any direct summand in $T \otimes N$ has q-dimension 0, i.e. it is an object in $\mathcal{I}$. The quotient category turns out to be a semisimple tensor category with only finitely many isomorphism classes of simple objects. The fusion category $\mathcal{C}(SU(N), l)$ is obtained from $U_q sl_N$ with q a primitive l-th root of unity (in our notation, q would correspond to v^2 in Lusztig's notation [21]).

(b) A highly nontrivial tensor product was defined by Kazhdan and Lusztig for representations of affine Kac-Moody algebras (see [16]). It was then shown by Finkelberg [10] that one can similarly define a fusion tensor category as a quotient for integer values of the level, and that this category is equivalent to the one constructed from quantum groups.

(c) On the level of loop groups, a fusion tensor product was defined by Wassermann by a completely different method for Lie type A see [32]; it is currently being worked out for the other cases. His method has the advantage that it works in the category of Hilbert spaces where various positivity properties required for physics and for applications in operator algebras appear naturally (see also Theorem 1.4).

(d) The last construction of fusion categories mentioned here is probably the least known but is very close to our context. Recall that we have described the representation categories of $SU(N)$ and $U_q sl_N$ via idempotents in certain quotients of $\mathbb{C}S_n$ and of the Hecke algebra H_n. If q is a root of unity, one gets smaller quotients of the Hecke algebras (see [33]), but the same constructions still work. Similarly, this can be done for certain other quotients of $\mathbb{C}B_n$ to be discussed in the next section, which are closely related to the representation theory of orthogonal and symplectic quantum groups. This construction has been carried out in [29] for the latter case (for the subcategory generated by the vector representation), based on the analysis of braid representations in [34] and for type A in [4]. Hence at least for classical Lie types, fusion categories can be constructed using braid representations.

Moreover, the methods used in [17] can also be employed in the context of fusion categories (for type A so far). More precisely we have

Theorem 1.3. *Let $\mathcal{C}$ be a monoidal rigid tensor category whose Grothendieck semiring is isomorphic to the one of a fusion category $\mathcal{C}(SU(N), l)$. Then it is equivalent to $\mathcal{C}(SU(N, l))^\theta$, i.e. a twist of $\mathcal{C}(SU(N), l)$ determined by an N-th root of unity θ.*

Finally, let us also mention another result in connection with fusion categories. In the context of physics, and for applications for Jones' theory of subfactors, we also require that the Hom-spaces are Hilbert spaces with the inner product having the usual functorial properties with respect to $\otimes$. While this is manifest in Wassermann's loop group construction (at the expense of the rather involved construction of the fusion tensor product) it is not so obvious in the context of quantum groups. A suitable Hermitian form was constructed by Kirillov jr [18] and shown to be positive definite for type A_1 for certain roots of unity; positive definiteness in general was proved in [35]. More precisely, we have

Theorem 1.4. *Let C be a fusion category constructed from a quantum group at a root of unity q, where in the non-simply laced case the degree l of the root of unity q^2 is divisible by the ratio d of the square lengths of a long with a short root. Then C admits functorial positive definite inner products on its Hom-spaces provided that $q = e^{\pm \pi i/l}$.*

2. CENTRALIZER ALGEBRAS FOR LIE TYPES B, C AND D

2.1. Brauer algebras and their q-deformations.
Let now V be the vector representation of G, where G is either an orthogonal group $O(N)$ or a symplectic group $Sp(2N)$. In this case, G leaves invariant a bilinear form, and hence acts trivially on the canonical vector $v_{(2)} \in V^{\otimes 2}$ corresponding to this form; if $\{v_i\}$ and $\{v^j\}$ are dual bases with respect to this form the element $v_{(2)}$ can be written as $v_{(2)} = \sum_i v_i \otimes v^i$. It has been shown by Brauer that $\mathrm{End}_G(V^{\otimes n})$ is generated by the action of S_n together with the orthogonal projection onto $\{v_{(2)}\} \otimes V^{\otimes(n-2)}$ (see [5]).

Brauer also showed that $\mathrm{End}_G(V^{\otimes n})$ is again the quotient of an abstract algebra, Brauer's centralizer algebra $D_n(x)$, which can be defined over any ring containing the polynomial ring $\mathbb{Z}[x]$. It has an integral basis given by all graphs with $2n$ vertices, arranged in 2 rows, and n edges connecting exactly 2 vertices, with each vertex belonging to exactly one edge. It is easy to check that the dimension of $D_n(x)$ is $1 \cdot 3 \cdot \ldots \cdot (2n-1)$. The multiplication is given by concatenation, as with braids, with a disconnected loop in a product removed at the expense of multiplying the remaining expression by x. This algebra could also be described via generators and relations, however this becomes somewhat messy.

There is a simpler description in terms of generators and relations if one considers the quantum case. Originally motivated by the definition of a link invariant, the Kauffman polynomial, algebras $C_n(r, q)$ were defined as follows (see [3] and [22]; the original definition was slightly different): Let K be a field of characteristic 0 with nonzero elements r and q such that the element $(r - r^{-1})/(q - q^{-1})$ is well-defined. Then $C_n(r, q)$ is a quotient algebra of $K\mathcal{B}_n$, where we denote the image of the braid generator σ_i in that quotient by g_i. The quotient is defined by requiring that the g_i satisfy a cubic polynomial and one additional relation. More precisely, the relations are

(R1) $(g_i - q)(g_i + q^{-1})(g_i - r^{-1}) = 0$, for $i = 1, 2, \ldots n-1$,

(R2) $e_i g_{i-1}^{\pm 1} e_i = r^{\pm 1} e_i$, for $i = 2, 3, \ldots n-1$, where $e_i = xp_i$ with p_i the eigenprojection of g_i corresponding to the eigenvalue r^{-1} of g_i, and where the parameter x is defined by

$$x = (r - r^{-1})/(q - q^{-1}) + 1.$$

The algebra $C_n(r, q)$ is not well-defined for $q = 1$. In order to get the classical limit, one needs to use the parametrization in terms of q, x and an additional variable z satisfying $z^2 = (q-q^{-1})^2(x-1)^2+4$. With these notations, one sets $r = ((q - q^{-1})(x - 1) + z)/2$. Also, in the limiting case, one has to introduce e_1 as an additional generator together with relations which can be deduced from $(R1)$ and $(R2)$ in the quantum case (see e.g. [34] p. 400).

Proposition 2.1. *Let V be the vector representation of $U_q\mathfrak{g}$ with $\mathfrak{g} = so_M$ or $\mathfrak{g} = sp_{2M}$, and where $q = v^2$ in types B and C, and $q = v$ for type D in the notation of [21].*

(a) (see also [26]) There exists an algebra homomorphism Φ from $C_n(q^{M-1}, q)$ (for $\mathfrak{g} = so_M$) resp. from $C_n(-q^{2M+1}, q)$ (for $\mathfrak{g} = sp_{2M}$) into $End_{U_q\mathfrak{g}}(V^{\otimes n})$ defined by

$$\Phi : g_i \mapsto \check{R}_{V,i} = 1 \otimes \ \dots \ \otimes \check{R}_V \otimes \ \dots \ \otimes 1,$$

where $\check{R}_V \in End_{U_q\mathfrak{g}}(V^{\otimes 2})$ acts on the i-th and $(i + 1)$-st copy of $(V^{\otimes n})$ via $P_V R$; here P_V interchanges the factors of $V^{\otimes 2}$ and R is Drinfeld's universal R-matrix.

(b) The homomorphism Φ in (a) is surjective for Lie types B_N and C_N for all $N \in \mathbb{N}$, and for D_N if $N > n$.

Remark 2.2. The fact that the homomorphism Φ is not surjective for type D can be explained by the fact that in the classical case the Brauer algebra describes the commutant of the full orthogonal group $O(N)$. So the natural choice of labeling the simple components of the image of Φ would be via the simple representations of $O(N)$; classically one chooses Young diagrams whose first 2 columns contain at most N boxes (see [36]). For N odd, any simple $O(N)$ representation remains simple if restricted to $SO(N)$, hence duality is preserved. For N even, simple $O(N)$-modules labeled by Young diagrams with exactly $N/2$ rows split into the direct sum of 2 simple $SO(N)$-modules. Here one obtains duality if one enlarges the Brauer algebra by an additional generator which acts in the $N/2$-th antisymmetrization of the vector representation.

2.2. Algebraic structure of $C_n(r, q)$.

Proposition 2.1 plays a crucial role in determining the structure of $C_n(r, q)$. These results can be summarized as follows

Theorem 2.3. *([3]) Assume the algebra C_n to be defined over the field $\mathbb{Q}(r, q)$ of rational functions in 2 variables over $\mathbb{Q}$.*

(a) The algebra C_n is a direct sum of full matrix algebras.

(b) Its simple components are labeled by the Young diagrams with n, $n - 2$, $n - 4$, ... boxes. A simple C_n-module $V_{n,\lambda}$, considered as a C_{n-1}-module is

isomorphic to $\oplus_\mu V_{n-1,\mu}$ where the summation goes over all Young diagrams μ which can be obtained from λ by adding or subtracting a box to/from λ.

(c) There exists a trace tr on the inductive limit of the algebras $C_n(r,q)$ which is defined inductively by $tr(1) = 1$ and by $tr(ag_n^{\pm1}b) = tr(g_n^{\pm1})tr(ab)$, for $a,b \in C_n$.

This theorem was proved in [3] with the proof relying on the existence of a certain link invariant, the Kauffman polynomial. It is possible to give a purely algebraic proof of this and the following theorem, using the representation theory of quantum groups, by rearranging the arguments and results in [34]. Details will appear in [25].

The statements of the last theorem do not hold for C_n in general, except for part (c). Hence we can define the annihilator ideal I_n of tr in general. Let $\bar{C}_n = C_n/I_n$. The techniques of the proof of Theorem 2.3 can now also be used to determine the structure of $\bar{C}_n$.

Theorem 2.4. *([34]) With the notations introduced above we have*

(a) The algebra C_n is a direct sum of full matrix algebras as in the generic case (Theorem 2.3) if $r \neq \pm q^N$ for $N \in \mathbb{Z}$, $|N| \leq n$ and if q is not a root of unity.

(b) The algebra $\bar{C}_n$ is semisimple for $r = \pm q^N$, $N \in \mathbb{N}$ even if q is a root of unity; if q is not a root of unity, it coincides with the image of the homomorphism defined in Proposition 2.1.

(c) The simple components of $\bar{C}_n$ are again labeled by Young diagrams with $n, n-2, \ldots$ boxes taken from a certain set $\Lambda^\pm(N,l)$ of Young diagrams which depends on the specialization of $r = \pm q^N$ and on the degree l of the root of unity q^2, with $l = \infty$ if $q = \pm 1$ or q not a root of unity (see [34], Table 1, p. 429).

(d) The dimension of a simple $\bar{C}_{n,\lambda}$ module can be computed inductively by the restriction rule as in Theorem 2.3, where now only diagrams in $\Lambda^\pm(N,l)$ occur.

Remark 2.5. 1. The quotients $\bar{C}_n$ can be used to construct subfactors (see [34]) and fusion tensor categories (see [29]). If q is not a root of unity and $r = \pm q^N$ for some $N \in \mathbb{Z}$, one obtains tensor categories whose Grothendieck semiring coincide with the one of a representation category of a full orthogonal or symplectic group (see also Remark 2.2).

2. An alternative definition of the algebras $C_n(r,q)$ can be given in terms of canonical bases which correspond to Brauer's basis of graphs in the classical case. As to my knowledge, this was done first in great detail in [24] which, however, has not been generally available. Similar work was also done by a number of authors independently. This alternative definition has the advantage that the dimension of the algebra remains stable also for the interesting values of the parameters when they are no longer semisimple. However, for our purposes, the quotients described in the last theorem are enough.

2.3. Littelmann paths. The dimensions of the simple braid modules we have studied so far coincide with multiplicities of irreducible representations of classical Lie groups in tensor powers of the vector representations. In order to deal with the remaining cases, we will need to understand the decomposition of tensor powers of minuscule and adjoint representations in general. This can be done easily and was already known to Brauer (see e.g. [27], Chapter 7).

A convenient way how to describe these rules is using Littelmann's path algorithm. Moreover, this approach is also useful for computing weight multiplicities and tensor product multiplicities for Kac-Moody algebras. Let $\mathfrak{g}$ be a symmetrizable Kac-Moody algebra with root lattice R, and let $F = R \otimes_{\mathbb{Z}} \mathbb{R}$. Moreover, let λ be a dominant integral weight for $\mathfrak{g}$ and let V_λ be the corresponding irreducible highest weight module. Then Littelmann showed that there exists a labeling set for a basis for V_λ consisting of paths which can be obtained by applying certain root operators to the straight line from 0 to λ. For our purposes, it will suffice to consider the following examples. We define for a weight ω of V_λ the path π_ω to be the straight line from 0 to ω.

(a) If V is a minuscule module, i.e. all the weights are conjugate to the highest weight, the basis is labeled by $\{\pi_\omega\}$ with ω running through the weights of V.

(b) If V is the adjoint representation of a simply laced Lie algebra, then the basis is labeled by $\{\pi_\alpha\} \cup \{\pi_i\}$; here α runs through the roots of $\mathfrak{g}$, and π_i is the piecewise linear path going from 0 to $-(1/2)\alpha_i$ and back to 0, where α_i is a simple root.

For 2 given piecewise linear paths π_1 and π_2 starting from 0 we define the concatenation $\pi_1 * \pi_2$ as usual by

$$(2.1) \qquad \pi_1 * \pi_2(t) = \begin{cases} \pi_1(2t) & \text{if } 0 \leq t \leq 1/2, \\ \pi_1(1) + \pi_2(2t-1) & \text{if } 1/2 \leq t \leq 1. \end{cases}$$

For a given module V, we define a path of length n to be any piecewise linear path which can be obtained by concatenating n basis paths. Then we have

Theorem 2.6. *([20]) The multiplicity of a highest weight module V_λ in $V^{\otimes n}$ is equal to the number of paths of length n which are entirely in the closure of the dominant Weyl chamber.*

Example Let $(\epsilon_i)_{i=1}^N$ be the standard basis for $\mathbb{R}^N$. The vector representation of D_N is minuscule, hence a labeling path just consists of a straight line from 0 to a weight of V, i.e. to $\pm\epsilon_i$ with $1 \leq i \leq N$. Then the only path of length 1 is the one leading into the dominant weight ϵ_1. Using the identification of dominant weights of so_{2N} with Young diagrams of $\leq N$ rows (where the last row is allowed to have a negative number of boxes), it is easy to check that extending a permissible path of length $(n-1)$ to one of length n corresponds to adding or removing a box from the Young diagram such that the resulting shape is still a Young diagram.

2.4. **Spinor representations.** The approach in the previous subsections does not cover the half integer spin representations in types B and D. One way to remedy this would be to consider the decomposition of tensor powers of spinor representations. We will not consider this here, as in this case the Grothendieck semirings do not show a similarly uniform behaviour as in the cases studied so far. Instead we shall consider the decomposition of tensor products of the form $V_\varepsilon \otimes V^{\otimes n}$, where V_ε is a spinor representation. This can be done by considering braid groups of type B.

The braid group BB_n corresponding to the Dynkin diagram B_n is given by generators $\tau, \sigma_1, ..., \sigma_{n-1}$; here the last $n-1$ generators generate a braid group of type A_{n-1} (the usual Artin braid group B_n of type A_{n-1}), and we have the additional relations $\tau\sigma_1\tau\sigma_1 = \sigma_1\tau\sigma_1\tau$ as well as $\tau\sigma_j = \sigma_j\tau$ for $j > 1$.

Häring-Oldenburg defined an algebra G_n in [11], which he called the $B-BMW$-algebra. It is a quotient of KBB_n, where K is a field of characteristic 0 containing invertible elements q, r, Q. Let t be the image of τ in this quotient, and let as before g_i be the image of σ_i in G_n. Then $g_1, g_2, ... g_{n-1}$ satisfy the relations of the q-Brauer algebra C_n, and we have the additional relations

(RB1) $t^2 = (Q-1)t + Q$,

(RB2) $tg_1te_1 = e_1$,

(RB3) $e_1te_1 = \frac{(Q-1)(r-q^{-1})}{q-q^{-1}}e_1$.

It is easy to show that this algebra has as a quotient the Hecke algebra of type B_n. It turns out that the structure analysis can be done quite analoguous to the one of the q-Brauer algebra. This was done in the generic case by Häring-Oldenburg; the connection with representation theory was observed in [25].

Theorem 2.7. *([11], except for part (a)) Let $G_n = G_n(q, r, Q)$ be the quotient algebra of KBB_n as defined above.*

(a) The algebra G_n is semisimple if q^2 is not a root of unity, and if r and Q are not $\pm$ powers of q.

(b) In the semisimple case, the simple components of G_n are labelled by ordered pairs of Young diagrams (α, β) such that $n - |\alpha| - |\beta| \in 2\mathbb{N}$.

(c) If G_n is generically semsimple, and $V_{(\alpha,\beta;n)}$ is a simple G_n-module, we have the following isomorphism of G_{n-1}-modules:

$$V_{(\alpha,\beta;n)} \cong \bigoplus V_{(\alpha',\beta';n-1)},$$

where the summation goes over all triples $(\alpha', \beta'; n-1)$ which can be obtained from $(\alpha, \beta; n)$ by removing or adding one box from/to α or β; adding a box is only possible if $|\alpha| + |\beta| < n$.

Theorem 2.8. *([25]) (a) There exists a homomorphism*

$$\Phi : G_n(q, q^{2N-1}, -q^{m(n-1)}) \rightarrow \mathrm{End}_{U_q so_{2N}}(V_{m\varepsilon} \otimes V^{\otimes n})$$

which maps t to $\check{R}^2_{V_\varepsilon, V} \otimes 1_{n-1}$ and which extends the homomorphism of Prop. 2.1.

(b) There exists a Markov trace tr defined inductively on G_{n+1} by $tr(a\chi b) = tr(\chi)tr(ab)$ for $\chi \in \{g_n^{\pm 1}, 1\}$ and $a, b \in G_n$. For the specializations of part (a), it can be derived from the q-dimensions of $U_q so_{2N}$.

(c) The image of the homomorphism in (a) coincides with $\mathrm{End}_{U_q so_{2N}}(V_{m\varepsilon} \otimes V^{\otimes n})$ if m is odd and q^2 is not a root of unity.

Remarks 1. A similar statement is true for the odd-dimensional orthogonal case for $m = 1$ which we do not list here; for $m > 1$ the situation is more complicated.

2. The last theorem again gives an intrinsic characterization of endomorphisms of certain modules of a quantum group in terms of quotients of braid groups which can be described explicitly. Moroever, this can also be done at roots of unity where one obtains endomorphisms of fusion categories.

3. The results of the last theorem also turn out to be useful in the description of the decomposition of tensor products of exceptional Lie groups, to be discussed in the next section.

3. Exceptional Groups

3.1. An E_N series. We consider Coxeter graphs of type E_N, with $N > 5$ and $N \neq 9$, and with the vertices labeled as below.

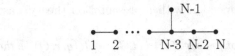

Figure 1

We can give an explicit construction of the root system for E_N as follows. Let (ϵ_i) be the standard basis for $\mathbb{R}^N$, which we supply with a symmetric bilinear form for which $\langle \epsilon_i, \epsilon_j \rangle = \delta_{ij}$ for $i \neq N$ and $\langle \epsilon_N, \epsilon_N \rangle = 1/(9 - N)$. Hence the form is positive definite for $N < 9$ and has signature $(N - 1, 1)$ for $N > 9$. The simple roots are given by

$$(3.1) \qquad \alpha_i = \begin{cases} \epsilon_i - \epsilon_{i+1} & \text{if } i \leq N - 2, \\ \epsilon_i + \epsilon_{i+1} & \text{if } i = N - 1, \\ (1/2)((9 - N)\epsilon_N + \epsilon_{N-1} - \sum_{i=1}^{N-2} \epsilon_i) & \text{if } i = N, \end{cases}$$

Observe that removing the vertex N results in the diagram D_{N-1}; hence $(\alpha_i)_{i=1}^{N-1}$ defines a set of simple roots for type D_{N-1} which lives in the subspace of $\mathbb{R}^N$ generated by the first $N - 1$ basis vectors. For a weight $\omega = \sum_i \omega_i \epsilon_i$ of our root system of type E_N we will use the notation

$$\omega = (\omega^{(D)}; \omega_N),$$

where $\omega^{(D)} = \sum_i^{N-1} w_i \epsilon_i$. We denote by Λ_i the fundamental weight corresponding to the vertex i for the graph E_N. It is easy to see that $\Lambda_i^{(D)}$ then is a fundamental weight of our sub root system of type D_{N-1} for $i < N$. With this notations, it is also easy to write down general formulas for the Casimir. Let $\rho = (N-2, N-3, \ldots 1, 0; 2 + (N-1)(N-2)/2)$, and let $\lambda \in \mathbb{R}^N$. Then we have $\langle \rho, \alpha_i \rangle = 1$ for $i = 1, 2, \ldots N$, and

$$(3.2) \qquad c_\lambda = \langle \lambda + 2\rho, \lambda \rangle = c_\lambda^{(D)} - 2(N+6)\lambda_N + 2\lambda_N \frac{\lambda_N + 60}{9 - N},$$

where

$$(3.3) \qquad c_\lambda^{(D)} = \langle \lambda^{(D)} + 2\rho^{(D)}, \lambda^{(D)} \rangle = 2\|\lambda^{(D)}\|_2^2 + 2(N-1)|\lambda^{(D)}| - 2\sum_{i=1}^{N-1} \lambda_i i.$$

Let $\mathfrak{g}(E_N)$ be the Kac-Moody algebra corresponding to the graph E_N. The role of the vector representation will be played by the module $V = V_{\Lambda_1}$, the irreducible module with highest weight Λ_1. Observe that we obtain for any highest weight module M of $\mathfrak{g}(E_N)$ a gradation in terms of $\mathfrak{g}(D_{N-1}) = so_{2N-2}$-modules by defining

$$M(g) = span\{F_{i_1} \ldots F_{i_r} m, \ |\{i_j, \ i_j = N\}| = g\},$$

where m is the highest weight vector of M and $|S|$ denotes the number of elements in the set S. The following lemma is proved using Littelmann's root operators.

Lemma 3.1. *The gradation of $V = V_{\Lambda_1}$ satisfies the following properties:*

(a) $V(0) \cong V_{\Lambda_1^{(D)}}$, the vector representation of so_{2N-2},

(b) $V(1) \cong V_{\Lambda_{N-1}^{(D)}}$, a spinor representation of so_{2N-2},

(c) $V(2) \cong V_{\Lambda_{N-6}^{(D)}} \oplus V_{\Lambda_{N-8}^{(D)}} \oplus \ldots$, a direct sum of antisymmetrizations of the vector representation of so_{2N-2}.

(d) $V \cong V(0) \oplus V(1) \oplus V(2)$ for $N = 6, 7$, and $V \cong \oplus_{i=0}^4 V(i)$ for $N = 8$.

(e) The Littelmann paths corresponding to the weights in $V(0)$, $V(1)$ and for the weights conjugate to the highest weight in $V_{\Lambda_{N-6}^{(D)}}$ (with respect to the Weyl group of so_{2N-2}) are all straight lines. We shall call the corresponding weights the straight weights.

Recall that the dimension of the vector representation of so_{2N-2} is equal to $2N-2$, and that the dimension of each of its 2 spinor representations is equal to 2^{N-2}. Using this, one easily checks that the dimensions of V are equal to 27 for E_6 and 56 for E_7. With a little bit more work one can also compute that the dimension of V for E_8 is equal to 248.

In the following we will give a description of the dominant weights which will be convenient in studying in which tensor power $V^{\otimes n}$ the corresponding highest weight module will appear for the first time. We will refer to a D^{N-1}-weight $\nu^{(D)}$ with integer coefficients (with respect to the basis (ϵ_i)) as an *integer weight.* We also define $|\nu^{(D)}| = \sum_{i=1}^{N-1} |\nu_i|$. The different cases below come from the fact that the highest weight vectors appear in different parts of the gradation of $V^{\otimes n}$.

Lemma 3.2. *Let λ be a dominant weight for E_N. Then it can be written in exactly one of the following ways (where $\mathbb{N}$ includes 0):*

(a) $\lambda = (\nu^{(D)}; n)$, where $\nu^{(D)}$ is a dominant integer weight of so_{2N-2} and $n - |\nu^{(D)}| \in 2\mathbb{N}$,

(b1) $\lambda = \Lambda_{N-2} + (\nu^{(D)}; n-4)$, where $\nu^{(D)}$ is a dominant integer weight of so_{2N-2} with $\nu_{N-1}^{(D)} \leq 0$, and $n - 4 - |\nu^{(D)}| \in 2\mathbb{N}$,

(b2) $\lambda = \Lambda_{N-1} + (\nu^{(D)}; n-3)$, where $\nu^{(D)}$ is a dominant integer weight of so_{2N-2} with $\nu_{N-1}^{(D)} \geq 0$, and $n - 3 - |\nu^{(D)}| \in 2\mathbb{N}$,

(c̃) $\lambda = (\lambda^{(D)}; \lambda_N)$ with $k(\lambda) = |\lambda^{(D)}| - \lambda_N > 1$, in which case λ can be written as $\lambda = k(\lambda)\Lambda_{N-1} + (\nu^{(D)}, \nu_N)$, where ν is an integer weight with $\nu_{N-1} \geq 0$ and $\nu_N = |\nu^{(D)}|$.

3.2. A new 2-parameter series of braid representations. In the following we define a generalization of the q-Brauer algebra. We first need a suitable labeling set for its simple components.

Definition 3.3. 1. Assume $N > n$, so that any Young diagram with $\leq n$ boxes can be identified with a vector in $\mathbb{R}^{N-1}$. We define the set $\Lambda(n) \subset \mathbb{R}^N$ as the union of the sets

(a) $(\mu; n)$, where μ is a Young diagram with $n - |\mu| \in 2\mathbb{N}$,

(b1) $\Lambda_{N-2} + (\mu; n-4)$, where μ is a Young diagram with $n - 4 - |\mu| \in 2\mathbb{N}$,

(b2) $\Lambda_{N-1} + (\mu; n-3)$, where μ is a Young diagram with $n - 3 - |\mu| \in 2\mathbb{N}$,

(c) $2\Lambda_{N-1} + (\mu; n-6)$, where μ is a Young diagram with $n - 6$ boxes.

2. The set $\Lambda(n)$ can also be described as a set of pairs (μ, r), where μ is as in the cases (a)-(c) above, and $r = 0, 1, 2$ depending on whether we are in case (a), (b) or (c). In this notation, $\Lambda(n)$ does not depend on N if $N > n$, and is referred to as the generic labeling set.

Proposition 3.4. *There exists an algebra $I_n(r, q)$, defined over the field $\mathbb{Q}(r, q)$ of rational functions in the variables r and q with the following properties:*

(a) It is a quotient of $\mathbb{Q}(r, q)\mathcal{B}_n$, with the images of the generators satisfying the same cubic equation as the generators of the q-Brauer algebra $C_n(r, q)$,

(b) It is a direct sum of full matrix algebras, whose simple components are labeled by the elements of $\Lambda(n)$,

(c) For given $\lambda \in \Lambda(n)$, a simple I_n-module W_λ restricted to I_{n-1} decomposes as a direct sum $\oplus_{\lambda'}$, where the summation goes over those elements $\lambda' \in \Lambda(n-1)$ for which $\lambda - \lambda'$ is a straight weight for E_N with $N > n$; this restriction rule is independent of the choice of N, $N > n$.

Remark 3.5. 1. At this point we do not have a presentation of these algebras via generators and relations. We can write down some relations, however. Let us denote the images of the braid generators in $I_n(r, q)$ by g_i again. Then they satisfy relation $(R1)$ for the q-Brauer algebra. Moreover, they also satisfy the relation

$(RI2)$ $\quad e_i(g_{i-1} - r^{-1})e_i((q^2 + qr)g_i - r)(qg_2 + 1) = 0 \quad$ for $2 \leq i \leq n - 1$.

The relation above is most likely not in its simplest form; it is possible to write down a slightly weaker relation in shorter form. The braid relations together with $(R1)$ and $(RI2)$ do indeed determine the algebra I_3; however, already for $n = 4$ an additional relation is necessary. As the decomposition into full matrix rings is known, with the dimensions given by inductive formulas, one can compute the dimension of I_n fairly easily for small n. The numbers for $1 \leq n \leq 5$ are equal to 1, 3, 19, 205 = $5 \cdot 41$ and 2821 = $7 \cdot 13 \cdot 31$. The algebras I_n contain the q-Brauer algebras and Hecke algebras as quotients.

2. It should be possible to find an integral basis for the algebras I_n as it was done for q-Brauer algebras (see Remark 2.5) and Hecke algebras. One possible approach would be to adapt the approach sketched by Lusztig in [21], 27.3.10 (for type D) to our setting, which would give an integral basis for $r = q^{2N-3}$. It then remains to show that one can replace powers q^{2N-3} by r in the matrices of the generators such that the resulting expressions are independent of the choice of N (for N suitably large).

To establish the relationship between the just defined algebras and exceptional Lie algebras, recall that we obtain a representation of the braid group $\mathcal{B}_n$ on $V^{\otimes n}$ via $\check{R}$-matrices. In particular, for a given dominant weight λ, the intertwiner space $\mathrm{Hom}(V_\lambda, V^{\otimes n})$ becomes a $\mathcal{B}_n$-module with $\sigma_i(f) = R_i \circ f$ for all $f \in \mathrm{Hom}(V_\lambda, V^{\otimes n})$.

Proposition 3.6. *If N is sufficiently large (say $N > n$), we obtain a faithful representation of $I_n(q^{2N-3}, q)$ on $\oplus_\lambda \mathrm{Hom}(V_\lambda, V^{\otimes n})$, where the summation goes over all elements $\lambda \in \Lambda(n)$; here a generator g_i acts via left multiplication by $q^{1/(N-9)} \check{R}_i$. Moreover, each Hom-space is an irreducible $I_n(q^{2N-3}, q)$-module.*

3.3. **Restriction to finite E_N.** As mentioned earlier, we will be interested in when an irreducible highest weight module appears for the first time in $V^{\otimes n}$ for the finite E_N types.

Lemma 3.7. *Let notations be as in Lemma 3.2, and let $N \in \{6, 7, 8\}$.*

The irreducible highest weight module V_λ appears for the first time in the n-th tensor power of V, where n is as in the description of Lemma 3.2(a)-(b), and where $n = \nu_N + Nk_1 + 3k_2$ for case (č) with k_1 and k_2 determined by $k(\lambda) = 3k_1 + k_2$ and $0 \leq k_2 \leq 2$.

Observe that except for case (č) in Lemma 3.2 the dominant weights are of the same type as the objects described in Definition 3.3. For case (č) we define the map

(3.4) $\quad \lambda = k(\lambda)\Lambda_{N-1} + (\nu^{(D)}; \nu_N) \quad \mapsto \quad k_2\Lambda_{N-1} + k_1([1^N]; N) + (\nu^{(D)}; \nu_N),$

where $k(\lambda), k_1$ and k_2 are as in the previous lemma, and where $[1^N]$ is the Young diagram with N boxes in its only column. It follows immediately from the definition that this map is injective and maps the weights in case $(\tilde{c})$ into a subsect of the generic set $\Lambda(n)$ as defined in Definition 3.3. We denote the subset of objects in $\Lambda(n)$ coming from the highest weights in $V_{new}^{\otimes n}$ together with Young diagrams containing $\leq 2N - 2$ boxes in the first 2 columns for case (a) by $\Lambda(n)^{(N)}$. The inclusion of additional Young diagrams for case (a) is a consequence of the fact that the q-Brauer algebra $C_n(q^{2N-3}, q)$ is a quotient of $I_n(q^{2N-3}, q)$ and Remark 2.2.

Proposition 3.8. *Let* $N \in \{6, 7, 8\}$, *and let* $V_{new}^{\otimes n}$ *be the maximum direct summand of* $V^{\otimes n}$ *which only contains irreducible subrepresentations which have not already appeared in* $V^{\otimes k}$ *for* $k < n$. *Moreover, let* $I_n^{(N)}$ *be the algebra generated by the R-matrices* $\check{R}_i$, $i = 1, 2, \ldots n - 1$, *acting on* $V_{new}^{\otimes n}$. *Then*

(a) The algebra $I_n^{(N)}$ *is semisimple, except possibly if* q *is a root of unity, with its simple components labeled by the elements in* $\Lambda(n)^{(N)}$.

(b) The dimensions of the simple $I_n^{(N)}$*-modules can be determined inductively by a restriction rule similar to the one for the generic case, i.e. for* I_n.

Remarks 1. The precise combinatorics for the last proposition is currently being worked out. In particular, the interesting question is being studied how the $\mathcal{B}_n$-module $\mathrm{Hom}(V_\lambda, V^{\otimes n})$ decomposes into a direct sum of simple I_n-modules for $\lambda \in \Lambda(n)$. It appears that one can obtain the whole commutant of the quantum group action on $V_{new}^{\otimes n}$ from the $\check{R}$-matrices and additional elements occuring in the $N - 1$-st and N-th tensor power.

2. After having analyzed the decomposition of $V_{new}^{\otimes n}$, it is not hard to determine the decomposition of the full tensor product for type E_6 and E_7. In both cases the first representation which is not new is the trivial representation, which appears in $V^{\otimes(9-N)}$ for $N = 6, 7$, and no other 'old' representation appears at the same time. This suffices to show that the $\check{R}$-matrices together with the additional elements needed for the new part generate the commutant for the whole tensor product. As usual, the situation is more complicated for E_8, where both the trivial representation and V itself appear in the 2nd tensor power of V. It follows from this that the simple representations appearing in the n-th tensor power of V are labeled by $\bigcup_{i=1}^n \Lambda(i)$. As soon as the homorphism spaces $\mathrm{Hom}(V_\lambda, V^{\otimes n})$ for $\lambda \in \Lambda(n-1)$ are understood, one can again obtain the complete decomposition of $V^{\otimes n}$.

3.4. **Types F_4 and G_2.** We now briefly sketch how one can consider F_N and G_N-series quite analoguous to the E_N-series. We shall not go into too many details here, as it will turn out that the representation theory for F_4 and even more so for G_2 is considerably easier than the E_N cases.

Let (ϵ_i) be the standard basis for $\mathbb{R}^N$, with $N \neq 5$. We supply $\mathbb{R}^N$ with a symmetric bilinear form for which $\langle \epsilon_i, \epsilon_j \rangle = \delta_{ij}$ for $i \neq N$ and $\langle \epsilon_N, \epsilon_N \rangle = 1/(5 - N)$. Hence the form is positive definite for $N < 5$ and has signature

$(N-1, 1)$ for $N > 5$. The simple roots are given by

$$(3.5) \qquad \alpha_i = \begin{cases} \epsilon_i - \epsilon_{i+1} & \text{if } i \le N-2, \\ 2\epsilon_{N-1} & \text{if } i = N-1, \\ (5-N)\epsilon_N - \sum_{i=1}^{N-1} \epsilon_i & \text{if } i = N, \end{cases}$$

It is now easy to check that $\check{\alpha}_i = \alpha_i$ for $i \le N-2$ and $\check{\alpha}_i = \alpha_i/2$ for $i = N-1, N$. The fundamental weights are given as the basis dual to $(\check{\alpha}_i)_i$ with respect to our chosen bilinear form. One can write them down explicitly as $\Lambda_i = i\epsilon_N + \sum_{j=1}^{i} \epsilon_i$ for $i < N$, and $\Lambda_N = 2\epsilon_N$. As in the E_N-case, we take as fundamental module $V = V_{\Lambda_1}$. We also define the gradation of a $\mathfrak{g}(F_N)$-module with a highest weight with respect to the generator f_N as before in the E_N-case. In this case, the pieces of our gradation are sp_{2N-2}-modules.

Lemma 3.9. *Let $V = V_{\Lambda_1}$.*

(a) We have $V(0) \cong V_{[1]}$, the vector representation of sp_{2N-2}, $V(1) \cong V_{[1^{N-2}]}$, and $V(2)$ is isomorphic to the direct sum of simple sp_{2N-2}-modules V_λ labeled by Young diagrams with i rows with 2 boxes and $2k-1$ rows with 1 box, with $i + 2k \le N$, and $k \ge 1$.

(b) For $N = 4$, $V(2) \cong V_{[1]}$, and $V(i) = 0$ for $i > 2$, and all its Littelmann paths are given by straight lines, except for 2 paths of the form $0 \to -\alpha_i/2 \to 0$ for $i = 1, 2$. In particular, $\dim V = 6 + 14 + 6 = 26$.

Having the paths available, it is easy to compute the decomposition of tensor products of V, at least for a small number of factors. It turns out that the new part of $V^{\otimes n}$ for F_4 is already completely determined by the representation theory of sp_6. More precisely, we have

Proposition 3.10. *The decomposition of $V_{new}^{\otimes n}$ for $\mathfrak{g}(F_4)$ coincides with the decomposition of the vector representation of sp_6. In particular, in the quantum case, the braid representations given by $\check{R}$-matrices, acting on $V_{new}^{\otimes n}$ factor through the q-Brauer algebras for Lie type C_3, i.e they factor through the algebras $C_n(-q^7, q)$.*

For a G_N-series, one defines a gradation in terms of $\mathfrak{g}(A_{N-1})$-modules. So, in particular, the decomposition of the new part for $\mathfrak{g}(G_2)$-modules is given by sl_2-modules, and the corresponding braid representations factor through a well-known quotient of the Hecke algebras, the Temperley-Lieb algebras. However, in order to get the decomposition of the whole tensor product, one encounters similar (but less complicated) difficulties as for E_8. This opens the way to treat at least the new part of tensor powers of certain representations uniformly for all Lie types in the following sense:

Remark 3.11. Assume that the algebras $I_n(r, q)$ can be defined via canonical bases (see Remark 3.5). Then there exists for any Lie type X_N a module V generating the representation category of $U_q\mathfrak{g}(X_N)$ (except for $\mathfrak{g}$ of orthogonal type, where we get a subcategory) such that the algebra generated by the $\check{R}$ matrices acting on $V_{new}^{\otimes n}$ is a quotient of $I_n(q^{e(X_N)}, q)$, for some integer $e(X_N)$; for classical types, the statement even holds for the full tensor product $V^{\otimes n}$.

The intertwining algebras $\text{End}(V^{\otimes n})$ resp. $\text{End}(V_{new}^{\otimes n})$ can be described easily in terms of these braid quotients.

3.5. The classical limit. For the classical case $q = 1$ the $\check{R}$-matrices degenerate to permutation matrices. However, all its eigenprojections are still well-defined. Moreover, our surjectivity proofs depended on the non-vanishing of certain q-$6j$-symbols. These can be computed explicitly, and one can check that they do not vanish in the limit. Hence, by defining I_n as the algebra generated by the eigenprojections of the generators g_i, we also obtain a well-defined algebra for $q = 1$ with the same dimension. In particular, all the statements above also hold for the classical case.

3.6. Vogel-Deligne approach. Our work on exceptional Lie algebras was strongly influenced by Deligne's article [7] which in turn was based on work by Vogel [31]. There seems to be little point in reproducing Deligne's very readable article, so only a very short account of this approach is given. Here a number of remarkable properties are listed which are shared by small tensor powers of the adjoint representation of an exceptional Lie algebra, as well as a few classical types (namely A_1, A_2 and D_4). So, at least for the exceptional types, the second tensor power of the adjoint representation has exactly 5 simple mutually nonisomorphic summands. Uniform formulas can be given for their dimensions and the value of the Casimir acting on them, only depending on one parameter. Similar results have also been obtained for the 3rd tensor power.

It is possible to generalize the dimension formulas to include q-dimensions in the quantum setting (for the second tensor power). Moreover, in this case one can give a conceptual proof of these formulas using the rigidity of representations of $\mathcal{B}_3$ up to dimension 5 (see [28]); the latter means that any simple representation of $\mathcal{B}_3$ is already uniquely determined by the eigenvalues of their standard generators, with choices of a 2nd and 5th root of the product of the eigenvalues for dimensions 4 and 5 respectively.

The results in [7] [8] and [31] suggest that it might be possible to study the decomposition of tensor powers of the adjoint representation of exceptional Lie algebras in a uniform way. This would be interesting as it would deal with the whole tensor product and not just parts of it, as it is done in this paper. For E_8 the adjoint representation coincides with our module V. So it is conceivable that there exists a connection between these 2 approaches. The following remarks make this speculation a little bit more precise; the observations regarding the Vogel-Deligne approach actually provided some of the motivation for the current approach:

Remark 3.12. Let $\mathfrak{g}$ be the $U_q\mathfrak{g}$-module corresponding to the adjoint representation of an exceptional Lie group, and define $\mathfrak{g}_{new}^{\otimes n}$ as before $V_{new}^{\otimes n}$. Then the $\check{R}$-matrices, restricted to $\mathfrak{g}_{new}^{\otimes n}$ satisfy a cubic equation; the algebra generated by $\check{R}_1$ and $\check{R}_2$ can be considered as a specialization of our algebra I_3, except possibly for an additional 1-dimensional representation. So assuming a similar uniform behaviour exists for the braid representations on $\mathfrak{g}_{new}^{\otimes n}$ (for $\mathfrak{g}$ belonging to the exceptional series) as for $V_{new}^{\otimes n}$ in our approach, the corresponding algebra would be a generalization of our algebras I_n. '

References

[1]   H.H. Andersen, Tensor Products of Quantized Tilting Modules, Commun. Math. Phys. **149** (1992) 149-159.

[2]   H.H. Andersen, J. Paradowski, Fusion categories arising from semisimple Lie algebras, Comm. Math. Phys. **169** (1995) 563-588.

[3]   J. Birman; H. Wenzl, Braids, link polynomials and a new algebra, Trans. AMS **313** (1989) 249-273.

[4]   Ch. Blanchet, Hecke algebras, modular categories and 3-manifolds quantum invariants, preprint GT/9803114.

[5]   R. Brauer, On algebras which are connected with the semisimple continuous groups, Ann. of Math. **63** (1937), 854-872.

[6]   A. Bruguières, Reconstruction des certaines catègories monoidales d'aprés Kazhdan and Wenzl, notes on seminar talks, University of Paris.

[7]   P. Deligne, La série exceptionnelle de groupes de Lie. C. R. Acad. Sci. Paris Sér. I Math. **322** (1996), no. 4, 321–326.

[8]   P. Deligne; R. de Man, La série exceptionnelle de groupes de Lie. II. (French) C. R. Acad. Sci. Paris Sér. I Math. 323 (1996), no. 6, 577–582.

[9]   V. Drinfeld, On almost cocommutative Hopf algebras, Leningrad Math. J. **1** (1990) 321-343.

[10]   M. Finkelberg, An equivalence of fusion categories, GAFA, **6** (1996), 249-267.

[11]   R. Häring-Oldenburg, The reduced Birman-Wenzl algebra of Coxeter type B, J. Alg. **213** (1999) 437-466

[12]   M. Jimbo, A q-analogue of $U(gl(N+1))$, Hecke algebras and the Yang-Baxter equation, Lett. Math. Phys. **10**, (1985), 63-69.

[13]   V. Jones, Index for subfactors. Invent. Math. 72 (1983), no. 1, 1–25.

[14]   V. Jones, A polynomial invariant for knots via von Neumann algebras. Bull. Amer. Math. Soc. (N.S.) 12 (1985), no. 1, 103–111.

[15]   A. Joyal; R. Street, Braided tensor categories, Adv. Math. **102** (1993) 20-78.

[16]   D. Kazhdan; G. Lusztig, Tensor structures arising from affine Lie algebras. III, IV. J. Amer. Math. Soc. **7** (1994), no. 2.

[17]   D. Kazhdan; H. Wenzl, Reconstructing monoidal categories, Adv. in Soviet Math., **16** (1993) 111-136.

[18]   A. Kirillov Jr, On an inner product in modular categories, J of AMS, **9** (1996) 1135-1170.

[19]   P. Kulish, N. Reshetikhin, E. Sklyanin, Yang-Baxter equation and representation theory, Lett. Math. Phys, **5** (1981) 393

[20]   P. Littelmann, Paths and root operators in representation theory. Ann. of Math. (2) **142** (1995), no. 3, 499–525.

[21]   G. Lusztig, Introduction to quantum groups, Birkhäuser

[22]   J. Murakami, The Kauffman polynomial of links and representation theory, Osaka J. Math. **24** (1987) 745-758.

[23]   Moore, Gregory; Seiberg, Nathan Classical and quantum conformal field theory. Comm. Math. Phys. **123** (1989), no. 2, 177–254.

[24] H. Morton; A. Wassermann, A basis for the Birman-Wenzl algebra, preliminary preprint 1989/2000, University of Liverpool

[25] R. Orellana: H. Wenzl, q-Centralizer algebras for spin groups, in preparation.

[26] N. Reshetikhin, Quantized universal enveloping algebras, the Yang-Baxter equation and invariants of links, LOMI preprint, 1988

[27] J. Stembridge, Computational aspects of root systems, Coxeter groups and Weyl characters, www.math.lsa.umich.edu/ jrs/papers/carswc.ps.gz

[28] I. Tuba; H. Wenzl, Representations of the braid groups B_3 and of $SL(2, \mathbb{Z})$, Pacific J. (to appear).

[29] V. Turaev; H. Wenzl, Semisimple and modular categories from link invariants, Math. Ann. **309** (1997) 411-461.

[30] V. Turaev, Quantum invariants of knots and 3-manifolds, de Gruyter, 1994.

[31] P. Vogel, Algebraic structures on modules of diagrams, preprint.

[32] A. Wassermann, Operator algebras and conformal field theory. III. Fusion of positive energy representations of LSU(N) using bounded operators. Invent. Math. 133 (1998), no. 3, 467–538.

[33] H. Wenzl, Hecke algebras of type A_n and subfactors, Inv. math. **92** (1988) 349-383.

[34] H. Wenzl, Quantum groups and subfactors of Lie type B, C and D, Comm. Math. Phys. **133** (1990) 383-433

[35] H. Wenzl, C^* tensor categories from quantum groups, J. of AMS, **11** (1998) 261-282.

[36] H. Weyl, The classical groups, Princeton University Press.

[37] E. Witten, Quantum field theory and the Jones polynomial, Comm. Math. Phys. 121 (1989) 351-399.

Department of Mathematics
UC San Diego
La Jolla, CA 92093-0112
USA

Printed in the United States
by Baker & Taylor Publisher Services

Printed in the United States
by Baker & Taylor Publisher Services